T0094175

Amplifying Voices in UX

SUNY series, Studies in Technical Communication

———————

Miles A. Kimball, Charles H. Sides, Derek G. Ross,
and Hilary A. Sarat-St. Peter, editors

Amplifying Voices in UX

Balancing Design and User Needs in Technical Communication

Edited by

AMBER L. LANCASTER and
CARIE S. TUCKER KING

SUNY
PRESS

Published by State University of New York Press, Albany

© 2024 State University of New York

All rights reserved

Printed in the United States of America

No part of this book may be used or reproduced in any manner whatsoever without written permission. No part of this book may be stored in a retrieval system or transmitted in any form or by any means including electronic, electrostatic, magnetic tape, mechanical, photocopying, recording, or otherwise without the prior permission in writing of the publisher.

For information, contact State University of New York Press, Albany, NY
www.sunypress.edu

Library of Congress Cataloging-in-Publication Data

Names: Lancaster, Amber L., editor. | King, Carie S. Tucker, editor.
Title: Amplifying voices in UX : balancing design and user needs in
  technical communication / edited by Amber L. Lancaster and Carie S.
  Tucker King.
Description: Albany, NY : State University of New York Press, [2024] |
  Series: SUNY series, studies in technical communication | Includes
  bibliographical references and index.
Identifiers: LCCN 2023029845 | ISBN 9781438496740 (hardcover : alk. paper) |
  ISBN 9781438496757 (ebook)
Subjects: LCSH: User-centered system design—Case studies. | User interfaces
  (Computer systems)—Case studies. | Communication of technical
  information—Case studies.
Classification: LCC QA76.9.U83 A48 2024 | DDC 004.2/1—dc23/eng/20240116
LC record available at https://lccn.loc.gov/2023029845

10 9 8 7 6 5 4 3 2 1

Contents

PART ONE

PART TWO

PART THREE

Illustrations

Tables

Figures

Preface

The field of Technical and Professional Communication (TPC) is broad, and exploration and research related to the field, and to usability in particular, have expanded beyond the original scope of technical writing. Usability and user experience (UX) research continue to be primary emphases in Technical Communication (TC) because of the focus on audience, the diversity of users, and the creation of technical deliverables. In response to these emphases, this book highlights the need for TPC to extend UX design practices beyond translating and tailoring for local users to broader global users, while considering the diversity of user uniqueness, customization desires, all stakeholders, and social needs. The book's contributing authors address this expanding need in various areas related to social issues; to diversity, inclusion, and equity; to user empowerment and advocacy; to accessibility; to stakeholder agency; and to empathy, unity, and reconciliation.

Over the past 20-plus years, usability research has expanded in depth and breadth to include many topics:

- cultural, intercultural, and language considerations (e.g., Dragga, 1999; Gonzalez & Zantjer, 2015; Matheson, 2019; St.Amant, 2013; Yu, 2012);

- social action (e.g., Ding, 2009; Miller, 1984; Pickering, 2019);

- accessibility (e.g., Youngblood, 2013; Zdenek, 2011);

- participatory culture and user agency (e.g., Agboka, 2013; Bannon & Ehn, 2013);

- environmental awareness (e.g., Frost, 2013; Simmons, 2007); and

- advocacy (e.g., Durá, 2018; Martin et al., 2017; Rose, 2016; Walton et al., 2016).

More recently, usability research has embraced focused themes at conferences and in flagship journals and publications. For example, ACM Special Interest Group for Design of Communication (Association for Computing Machinery SIGDOC, 2020) concentrated on UX, inequities, and injustices in communication design as its 2020 conference theme, which carried into the 2021 theme of "advocacy, accountability, and coalitions" that asked our field "to consider how and who we advocate for, how we hold ourselves and each other accountable, and how we form and nurture coalitions" (p. 5). Similarly, at its 2020 conference, Council for Programs in Technical and Scientific Communication (CPTSC) examined how TPC "programs can increase inclusion and promote anti-racism" and saw continuation of TPC program challenges (as noted in the call for proposals) "intersect with ongoing issues related to equity, inclusion, and justice that have long concerned and will continue to impact TSC [technical/scientific] programs" (CPTSC, 2021). Association for Teachers of Technical Writing (ATTW) focused its 2021 conference theme on "Language, Access, and Power in Technical Communication," with much of the presented scholarship focused on inclusion and social justice areas of TPC (Haas, 2021). Additionally, our field is publishing works on topics of user marginalization in TPC and health and medical discourse in special issues (e.g., *Technical Communication Quarterly, 30*, 2021; *Technical Communication, 69.4,* 2022) and books dedicated to topics of socially just work in TPC (Walton & Agboka, 2021), and researchers launched a new journal dedicated to social justice (i.e., *Technical Communication and Social Justice*).

Even with so much new UX scholarship, the field seems imbalanced, and we must ask ourselves if we are hyper-focusing our research and practice in ways that could exclude secondary and tertiary audiences. Although we are excited to see the broadening scope and intensity of focus on UX, we see the need for more scholarship that highlights inclusive design that considers and empowers *all* voices. In this book, scholars seek answers to many questions about extending UX research and practice:

- How can we extend localization practices and principles to increase user agency, to support user advocacy, and to expand deliverables for global use?

- How can TPC programs and instructors integrate values practiced in the field—e.g., service, advocacy, participatory design—into the classroom to prepare students to think critically, engage globally, and focus on human experience in the work they do?

- In a research-driven field, how can TPC education prepare professionals-in-training to practice empathy for understanding diverse user experiences and to design for inclusion (for all voices) in their products?

- How can designers engage and empower user groups to reflect and respect diverse voices?

- How can technical and professional communicators work to establish balance in the voices they promote, thus building unifying experiences for diverse user groups?

The authors' responses suggest diverse answers and improvements to our field.

To address these questions, this edited collection provides readers with original thought and scholarship in TPC and UX as they relate to environmental communication, crisis communication, discerning design, curriculum development, and equity in higher education. Our Introduction provides a foundation for critically examining UX research and practice through a lens of balance, or equilibrium, in UX—what we term *equilibriUX*. With this new term, we offer a new perspective to situate balance in design and UX.

Following the Introduction, the text is divided into three parts.

Part One (Chapters 1–5) addresses issues of UX curriculum design, including disabilities and accessibility, empathy skill building and UX job readiness, design thinking and service-learning models, and design justice and co-creation. Chapter 1 presents a case of design thinking as it can be used to address wicked problems in TPC and more specifically in a service-learning and pedagogy framework. In this chapter, the authors share various measures to introduce this innovative course design, to encourage designers to consider empathy and take on iterative processes that allow them to synthesize and thus to influence future technical communicators in their application of UX. Chapter 2 resists "checking the boxes" of universal design and advocates for students to learn balance in design through their TEACH Framework: Teaching accessibility and usability, Emphasizing empathy, Amplifying student agency, Contextualizing design and rhetorical awareness, and Highlighting balance. Through the five-point pedagogical framework of TEACH, they describe curriculum and provide examples to illustrate how others can apply this framework in their own curricula. Chapter 3 examines theoretical connections between UX design, universal design, and disability justice to identify and counter ableist influences in online courses. In this way, curriculum designers consider all students' needs, create balance in course construction, and design accessible spaces that prioritize accessibility. Chapter 4 dignifies the presence of human emotion in design and in instruction, recognizing that instructors and students are emotional beings who rationally apply UX principles in their deliverables. The authors share a pedagogical approach to teach students to integrate accessibility to their designs in working with visual language. Chapter 5 addresses that UX evolves as people experience the product, and designers should apply empathy for users in the ongoing process. The authors address how instructors can teach future technical communicators how to develop empathy throughout the design process, going beyond the *what* of empathy in instruction to the *how* of teaching and designing with empathy. The chapters in Part One create balance between humanity, designers, and UX by encouraging empathy, acknowledging the influence that TPC instructors have in cultivating empathic design, and empowering curriculum design to allow for *equilibriUX*.

Part Two (Chapters 6–10) critically examines UX methods and approaches (narratives, storytelling, networked creation, etc.) for designing effective medical and health communication: mental health intervention programs, support community resources, and patient-centered discourse. Chapter 6 presents a case study, framed by interviews of the mHealth app Medscape, created by Global North designers, as a tool that Global South (GS) health professionals use, specifically in Nepal. Through this case study, the author addresses what constitutes perceived ease of use, usefulness, and credibility per the mHealth app designs to address GS practitioners needs and expectations. Chapter 7 presents a case study of advanced-directive forms and UX to establish advocacy for those who are facing potential end-of-life choices and care. Considering forms available from a majority of states in the US, the authors identify UX challenges, including different instruction types and inconsistent definitions and options, and they recommend making advanced-directive forms more accessible, readable, and consistent. Chapter 8 addresses mental health resources, focusing on the application of TPC in National Collegiate Athletic Association (NCAA) resources for coaches and athletes as they address eating disorders, a challenge that faces many collegiate athletes. Analyzing two NCAA resources and considering how these resources help coaches educate athletes, the study analyzes users who previously have not been addressed in TPC. Chapter 9 identifies key strategies to construct effective higher-quality audio descriptions to create higher levels of connectedness and feelings of social inclusion among people who are blind and Deafblind or who have low vision in a large-scale public intervention. The authors suggest high-quality audio descriptions may significantly reduce secondary health risks associated with related social exclusion such as anxiety, depression, and chronic pain. Addressing this public-health issue, audio description "presents an alternative way to build human environments, with equivalent public discourse simultaneously flowing through multiple, multimodal channels, not all eyesight-oriented." Chapter 10 moves the discussion of healthcare to China and the BabyTree app—a tool for new and expecting mothers, physicians, small-business owners, and the Chinese government. Serving a spectrum of users, the BabyTree app incorporates *equilibriUX* into its design, balancing perspectives, voices, and agency as stakeholders provide dynamic interactions and diverse contexts, including social, cultural, economic, reproductive, and medical exigencies. The author frames this "walkthrough method" with UX and usability studies to investigate the structure of the app and note how the tool is created for user engagement as well as China's social and medical exigencies. Part Three focuses on medical, health, and wellness tools and relationships, further addressing the need for empathic design, identifying new spaces and communities for UX evaluation, and creating *equilibriUX* by expanding discussions of agency.

Part Three (Chapters 11–14) addresses technology and UX in labor, social, and political movements in TPC, with consideration of equity issues, visual impairments, safety vulnerabilities, and civic engagement. Chapter 11 dignifies teaching faculty, sometimes called "contingent" or "non-tenure-track faculty," who are engrossed in teaching and thus may be less involved in the construct of UX research and development and of courses that teach design. These instructors are establishing the foundation of future practitioners and scholars; the field can empower teaching faculty by addressing the inequalities in higher education and in the field to improve curriculum design and establish *equilibriUX* in curriculum design. In Chapter 12, the authors analyze Stacey Abrams's Georgia gubernatorial campaign communication to illustrate how UX design can support community voter rights, particularly in diverse U.S. communities, as they apply rhetorical and UX design theories and consider Abrams's documentary and articles. Chapter 13 focuses on environmental TC and different stakeholders in communication to share the Beekeeper's Companion app. This is a climate-smart information communication technology for development (ICT4D) app to support women beekeepers in Lebanon and to help them improve hive management, boost hive and honey productivity and quality, and improve resilience. In this way, the beekeeper program also helps to address women's opportunities and literacies and creates balance through customization and localization in the app's design. This chapter demonstrates how the tool's construction and participatory design and communication through organizations such as HiveTracks, ICARDA, and icipe are serving the users. Chapter 14 presents the story of China's DiDi Hitch, a ridesharing mobile app that portrayed women with sexual rhetoric, and analyzes the app's context and content, which led to the rapes and murders of two female passengers who were presented as objects of desire rather than as consumers or users. This chapter proposes feminist UX design heuristics for designers to balance design purposes and to center user agency, autonomy, and safety. Part Three addresses projects that are seeking to create *equilibriUX* by establishing communication standards that focus on the safety and well-being of users while acknowledging that the outcomes can inclusively benefit society as a whole.

Our objective is to introduce new concepts and methods for usability testing and UX to encourage practitioners, researchers, and scholars to create balance in their designs and processes. For that purpose, we are delighted to offer the following chapters.

References

Agboka, G. Y. (2013). Participatory localization: A social justice approach to navigating unenfranchised/disenfranchised cultural sites. *Technical Communication Quarterly, 22*(1), 28–49. https://doi.org/10.1080/10572252.2013.730966

Association for Computing Machinery SIGDOC. (2020, October 5–9). *SIGDOC '20: Proceedings of the 38th ACM International Conference on the Design of Communication.* ACM. https://dl.acm.org/action/showFmPdf?doi=10.1145%2F3380851

Bannon, L. J., & Ehn, P. (2013). Design matters in participatory design. In J. Simonsen & T. Robertson (Eds.), *Routledge handbook of participatory design* (pp. 37–63). Routledge.

CPTSC. (2021). *CPTSC Conference 2021 Call for Proposals.* https://conference.cptsc. org/index.php/2021/04/10/cptsc-2021-envisioning-tech-comm programs-in-a-new-era

Ding, H. (2009). Rhetorics of alternative media in an emerging epidemic: SARS, censorship, and extra-institutional risk communication. *Technical Communication Quarterly, 18*(4), 327–350. https://doi.org/10.10.1080/1057225090 3149548

Dragga, S. (1999). Ethical intercultural technical communication: Looking through the lens of Confucian ethics. *Technical Communication Quarterly, 8*(4), 365–381.

Durá, L. (2018). Expanding inventional and solution spaces: How asset-based inquiry can support advocacy in technical communication. In G. Y. Agboka & N. Matveeva (Eds.), *Citizenship and advocacy in technical communication: Scholarly and pedagogical perspectives* (pp. 23–39). Taylor & Francis.

Gonzalez, L., & Zantjer, R. (2015). Translation as a user-localization practice. *Technical Communication, 62*(4), 271–284.

Haas, A. (2021, May 14). ATTW 2021 Conference Update. https://attw.org/ 2021_conference

Martin, S., Carrington, N., & Muncie, N. (2017). Promoting user advocacy to shift technical communication identity and value. *Technical Communication, 64*(4), 328–344.

Matheson, B. (2019). Beginning with Ganesha: The founding and early history of the Society for Technical Communication in India. *Journal of Technical Writing and Communication, 50*(3), 289–307. https://doi.org/10.1177/004 7281619873139

Miller, C. R. (1984). Genre as social action. *Quarterly Journal of Speech, 70*(2), 151–167.

Pickering, K. (2019). Emotion, social action, and agency: A case study of an intercultural, technical communication intern. *Technical Communication Quarterly, 28*(3), 238–253. https://doi.org/10.1080/10572252.2019.1571244

Rose, E. J. (2016). Design as advocacy: Using a human-centered approach to investigate the needs of vulnerable populations. *Journal of Technical Writing and Communication, 46*(4), 427–445. https://doi.org/10.1177/0047281616653494

St.Amant, K. (2013). What do technical communicators need to know about international environments? In J. Johnson-Eilola & S. A. Selber (Eds.), *Solving problems in technical communication* (pp. 479–499). University of Chicago Press.

Walton, R., & Agboka, G. Y. (2021). *Equipping technical communicators for social justice work: Theories, methodologies, and pedagogies.* Utah State University Press.

Walton, R., Mays, R. E., & Haselkorn, M. (2016). Enacting humanitarian culture: How technical communication facilitates successful humanitarian work. *Technical Communication, 63*(2), 85–100.

Youngblood, S. A. (2013). Communicating web accessibility to the novice developer from user experience to application. *Journal of Business and Technical Communication, 27*(2), 209–232.

Yu, H. (2012). Intercultural competence in technical communication: A working definition and review of assessment methods. *Technical Communication Quarterly, 21*(2), 168–186. https://doi.org/10.1080/10572252.2012.643443

Zdenek, S. (2011). Which sounds are significant? Towards a rhetoric of closed captioning. *Disability Studies Quarterly 31*(3). http://dsq-sds.org/issue/view/84

Acknowledgments

I am grateful to the contributing authors whose work provided the groundwork for this edited collection. Your research and scholarship have established foundational ideas for the field of Technical and Professional Communication and will inspire scholars for many years to come. I know I am inspired! I have learned so much from your research and from the process of collaborating with you on this edited collection. I am also thankful to have had an amazing coeditor, Dr. Carie S. Tucker King. Thank you for your brilliant ideas and the many hours of stimulating conversations about UX, design, social justice, user advocacy, and of course editing! Finally, I want to thank all those who have supported this project in a number of ways: reviewers for their thoughtful criticisms and feedback on the manuscript, Kimberlee Britton for her careful APA reviews and editing in early stages of development, and Oregon Tech colleagues for their encouragement and ongoing accountability checks. Your support has been invaluable to the success of this project.

Amber Lancaster

I am grateful to the authors and reviewers who submitted work for this coedited collection and trusted Amber and me with the integrity of their research. We are thrilled with the outcome—the quality exceeds our greatest expectations. I also appreciate my colleagues at The University of Texas at Dallas, who have allowed me to prioritize this project and have worked around our deadlines. Most of all, I want to thank my family members, who have listened to me rave about the research and the process, and Dr. Amber Lancaster, who truly is an incredible writing and editing partner.

Amber, thank you for authentic communication, flexibility, and creativity; it is a privilege to collaborate with such an astute and ethical professional, but even more, it is a privilege to call you "friend."

Carie S. Tucker King

Introduction

EquilibriUX—Designing for Balance and User Experience

Amber Lancaster and Carie S. Tucker King

In March 2020, the COVID-19 global pandemic shifted the way that the world communicates; circumstances required that most citizens of the world shift their communication to virtual environments. With new communication needs, usability became a greater necessity, as many users seeking to maintain relationships and to work remotely while maintaining social distance and respecting quarantine were required to conduct communication through digital means. We predict that, with this shift, usability testing and design principles must also shift; most workers have learned to use tools to accomplish their work, expecting tools to be user-friendly and usable. At the same time, the world experienced numerous conflicts and the global environment changed. The US experienced racial conflict (Chavez, 2020; Sugrue, 2020); Hong Kong saw political upheaval (Barron, 2020; Chor, 2021); Italy experienced unanticipated mortality (Chirico et al., 2021; Modi et al., 2021); nations around the world (many underserved and poorly prepared) experienced natural disasters and political unrest (Omer, 2020; Thompson, 2020); and Japan experienced a shortage of healthcare providers to care for COVID patients and to vaccinate (Du & Katanuma, 2021), all while preparing for and hosting the world at the 2020 Summer Olympics (Yoneoka et al., 2022). One U.S. presidential administration pushed vaccinations through testing and approval in record time (Mango, 2021; Vazquez & Carvajal, 2021), and another U.S. presidential admin-

1

istration surged forward to make vaccinations available to U.S. citizens and to less fortunate nations (Samuels, 2021; Stevens & Ahmed, 2021). In response to the shift to remote work, the global pandemic and need for information, and the world's conflicts as well as unprecedented natural disasters, researchers were and continue to be called to expand usability testing and application as well as perspective on effective usability principles. As Technical and Professional Communication (TPC) expands its research and scope of users, it also calls for expanded user-centric focus with efforts to establish equality in design for all users.

At the time of preparing this edited collection, our nation is experiencing a slowly decreasing number of COVID cases and thus considers the pandemic to be under control, although new strains of the coronavirus have been identified, and citizens around the world are being encouraged to receive vaccination boosters (U.S. Food and Drug Administration, 2023). Having experienced three years of a pandemic, we find ourselves emotionally exhausted and longing to return to normalcy, but we also seek to apply our experiences to create better ways of living, doing, and working. Perhaps we are nearing a pinnacle in some aspects, but we also see the timeliness in making real change.

In a collegial conversation one day in early 2021, we were discussing our concerns about TPC and usability research. We acknowledged the influence of the COVID pandemic and the world's many influential current events through the lens of TPC. As we conversed, we identified the importance of balance in design, which allows designers to respect others and gives designers the benefit of the doubt (Lancaster & King, 2022, p. 2). We also expressed concern over the growing dissension in the field, our country, and the world, and we talked about our desire for reconciliation, empathy, and harmony. As we continued to talk, we shared experiences that had influenced our concerns and our design philosophies.

In this conversation, the term *equilibriUX* was birthed; Amber began to advocate for balance, and Carie, then teaching a scientific publications and communication skills class that had just completed examining chemistry writing, brought up *equilibrium*. The conversation inspired us to discuss recent literature, and we both voiced concern that the field was becoming perhaps hyper-focused and was not considering new research on studies that could create unity, reconciliation, and balance. We reached out to other colleagues in TPC and in usability research to ask if they shared our concerns, and the consensus was, "Yes!" The result of our search for balance and new perspectives is this edited collection.

From 1999 to 2003, I (Amber) had a very personal experience with failed UX, which drove my passion to improve the design of communication—and ultimately to publish this book. As a first-generation college student and daughter of a blue-collar worker, I saw firsthand the effects of marginalized voices and the grave (and fatal) effects that omission of UX in communication design can have on the people relying on communication to do their work. I was a graduate student then, and my dad was a Ford Motor Company factory employee. In 1999, a boiler explosion at the Rouge plant (in Dearborn, MI) took the lives of several factory workers and injured many others. I remember the outrage and heartache that so many people in the Ford community felt. But it was not until 2003 when, as a doctoral student, I began researching the explosion that I truly understood and empathized with the Ford factory community. I spent days in the Michigan OSHA reading room sifting through and examining documents from the investigation. Learning that the explosion tragedy likely could have been avoided with better communication was heartbreaking: if better workplace communication practices were in place, if factory workers had participated in the design of their workplace communication, if technical information had been made more accessible, if employees' voices had been heard. These were harsh realities to face, and it took me nearly two decades to publish my research on this case (Lancaster, 2018). It was a project I had hoped to make my dissertation, but the emotional aspects, my ties to the Ford community, and my respect for my father's retirement status hushed my desires to pursue it more publicly. With the passing time came healing notions, but almost 20 years later, I still feel strongly that these lives matter, that including their voices in UX matter, and that social justices and information rights matter. I find myself wanting to advocate harder for those voices to be heard, but I struggle with being heard myself. With this edited collection, I hope we progress and continue pushing to achieve agency and balance in the design of communication and that we accomplish greater equality in UX practices.

Being raised in a military family, I (Carie) was blessed to move around the United States and the Far East, as my father served as a communications officer for the U.S. Armed Forces. Early on, I learned about the diversity of different areas of the world and the beauty of language, culture, history, and humanity. Our family experienced some jolts of cultural change as we moved from San Antonio, Texas; to Montgomery, Alabama; to Honolulu, Hawaii; and then to Tokyo, Japan. The military community is also diverse, and my friends in these various locations had different ethnic, religious, and cultural backgrounds—a kaleidoscope of humanity. I appreciated the joys of being culturally educated in vivo, but I also witnessed the ugliness of prejudice and the horror of hate, and I personally experienced the challenges of being considered an outsider—specifically a "haole" and then a "gaijin." When I returned to the United States for college, I struggled with culture shock and was more aware of challenges. (I had to learn to drive on the left side of the road, calculate mph, consider different clothing sizes, and learn regional language and accents.) Then, as a graduate student, I was exposed to the power of truly universal design. My passion for empathic design expanded as I studied usability and design, particularly as they relate to medical health; in my research, I continue to learn the power of virtual tools designed for global audiences (varying per age, education, ethnicity, geographic location, and perspective of disease and medicine) as I seek to dignify diverse patients' and users' voices (King, 2017). Now, serving at a university that is nationally ranked for its diversity and international student population, I am constantly expanding my perspective about the value of inclusive design. These experiences inspire me to advocate for my users and pursue designs with a global worldview. In that way, I advocate to create balance in the polyphony so that unheard users are a design priority, but designers are trusted, secondary users are considered, and UX is reconciled with designs created with kindness for humanity. The result is a balance in design: what we have termed *equilibriUX*. I hope this edited collection is a step to expand this conversation in TPC.

Practical and Academic Relevance:
Moving Beyond "Localization"

Localization has been defined as "creating or adapting an information product for use in a specific target country or specific target market" (Hoft, 1995, p. 11). In practice, localization increases the likelihood that interface, design, and communication messages will be received in intended and favorable ways. However, localization can also create barriers and challenges, particularly when secondary and tertiary audiences seek to use tools that do not meet their needs and expectations. Any product that is well designed and audience-centric will likely generate positive experiences for target users and likely show a positive return on investment. However, measuring what counts as a positive return becomes more complex when we consider more than the target user and see the diverse variables defining "user needs." If "the majority" is no longer "the metric" for making design decisions, whose voice gets heard, whose voice influences design decisions, and whose voice is manifested in the final product? Should we always design for a targeted country, market, or user?

Localization considers design choices related to the users' specific cultural expectations but also limits the globalization of the design (Alexander et al., 2017). Localization also overlooks research of underdeveloped nations whose user citizens have not been considered in user experience (Acharya, 2018). As scholars, we note that localization limits designers, designs, and users per geographic and cultural designations; therefore, when considering usability, we propose to expand universal design.

Effective design that meets the expectations of diverse users—e.g., per culture, age, gender, location, ethnicity, education—requires that we work with users as codesigners to design products that they can use (Acharya, 2018), but it also requires that we seek to create just-in-time design and interactive influences that connect diverse users with design. Scholars call for unique research and design principles, particularly because users differ across cultures (Acharya, 2018). But design cannot meet all the needs of users for international access. We also want to dignify the designers who have insight on the products they are creating: to give them the benefit of the doubt in design decisions.

In considering this challenge, we considered the metaphor of a chemical reaction—the blending of compounds (cultures and perspectives) in an experiment (a design) to create a reaction (a resulting product from the designer) that is safe and effective (that benefits the user).

In chemistry, when both elements are countered in a beneficial or neutral response, balance, or *equilibrium*, is achieved. *Equilibrium* is (1) "a state of intellectual or emotional balance" and "a state of adjustment between opposing or divergent influences or elements" and (2) "a state of balance between opposing forces or actions that is either statis (as in a body acted on by forces whose resultant is zero) or dynamic (as in a reversible chemical reaction when the rates of reaction in both directions are equal)" ("equilibria," n.d.). Equilibrium requires analysis and evaluation of interacting influences to achieve balance; that is, all variables are connected and dependent to maintain a state of balance.

"Le Chatelier's Principle" (also called "Le Chatelier-Braun Principle" [Smith, 2020]) notes that "if a stress is applied to a system at equilibrium, the equilibrium will shift to counteract the stress" (Treptow, 1980, p. 417). More specifically, "When a system in dynamic equilibrium is acted on by an external stress, it will adjust in such a way as to relieve the stress and establish a new equilibrium" (Norwich, 2010, para. 1). In considering responses, reactions, and influences, chemistry anticipates the different elements and their reactions when they are involved in an interaction. Each element has characteristics that direct it to respond to the situation. The concept then indicates a counter to imposed influences, rather than a negation of influences. In anticipating reactions in chemistry, the scientist must consider all influences as well as participate and counter reactors by instigating counter-reactors, rather than negating reactions. These reactors include temperature, mass, and pressure (Lower, 2021), and the model can be applied to physics, physiology, and linguistics (Norwich, 2010) as well as biology and economics (Smith, 2020) and nutrition (Henry & Camps, 2018). Per the principles of equilibrium, the balance of reactions is key.

We argue that this model can also be applied to localization and UX—a state of balance we call *equilibriUX*. Usability must exist with balance as the goal. With this balance—a product with beneficial engagement and interaction from a variety of users and satisfaction from the designer—is decreased "stress." It is in this state of balanced design we achieve equilibrium in UX. As we apply the metaphor, we encourage designers, researchers, and instructors to also consider primary audiences but also plan on and know secondary and tertiary audiences who engage through a product or design, so they can anticipate potential interactions, evaluate contexts, and analyze outcomes to create balance. That balance involves stakeholders but also product, design, and development.

EquilibriUX:
A New Model to Achieve Agency and Balance

We propose applying this model to TPC, and more specifically to UX, which has a history of adopting terms and principles from other fields (Sánchez, 2016, para. 3) and adapting practices to create its own. *EquilibriUX* describes usability design and testing, not to globalize a design by eliminating characteristics that relate to users to negate reaction but to integrate reactors of character and influence to establish balance and to respect diversity in design. That balance allows for the expertise and cultural competence of the designer and also respects diverse users' voices and perspectives to empower users and create balance in the influences of design. It also considers the agency and needs of diverse users, including students, instructors, women, underrepresented populations, and users with needs for accommodation. (This book addresses these users and more.)

EquilibriUX also considers subjective data from users (Sawyer & CDRH Work Group, n.d.) and, as users tell their stories, this balance "giv[es] power away" (Bacha, 2018, p. 222) to dignify the voices of users. In creating balance, we consider their needs and integrate additional characteristics—e.g., accessibility, plain language and expand our designs to allow for "local" needs but also anticipate that, in a global environment, our products (particularly those online) can be accessed and valuable to those we previously might have "othered." *EquilibriUX* results when designers gather different user stories and testimonies (as encouraged by Bacha) and apply those stories as reactants to the design process to create balance between the voices and needs of designer, client, and users (or potential users, as identified by personae). In seeking this balance, this equity, we also seek action to advocate for underrepresented communities and users (social justice, per Jones, 2016).

EquilibriUX focuses on use rather than content (aligning with Sun [2012], as cited in Acharya [2019]) and decreases power struggles to prioritize and include preferences and expectations from a diverse community of users, giving voice to *all* users. The principle also requires that designers know communities and aspire to cultural and intercultural competence "to understand localization practices, politics, inequalities, and social justice issues, especially in those countries where human rights are violated and privileged groups of people have access to information technology" (Acharya, 2019, p. 22).

The principle requires tolerance of conflicting opinions and experiences and design with a human element (Dragga & Voss, 2001), something that the world appears to be lacking. Some products allow for localization—when the user community is limited to a particular corporate or geographic setting (e.g., local news in Seoul, South Korea). However, other products, particularly those that are globally accessible, are best designed with a broad perspective of intended user communities. (This call responds to Sullivan's 1989 original call to move beyond a narrow definition and practice of usability as well as revisiting, per Johnson et al., 2007.) These products may also be used locally but by those who are not typically "local," such as tourists, immigrants, visitors, and new arrivals in a locality. In building a "bridge" between diverse users with user-friendly design, we require balance from diverse participant communities, and balance requires collaboration, relationship, empathy, and engagement.

In discussing *equilibriUX* outside the field of TPC, we have connected with a similar mindset. For example, in a conversation with a librarian (who studied usability as a graduate course), we noted the example of community and academic library websites in the United States. These sites can be used by the local community to identify scheduled events, to reserve and access library holdings, and to connect with experts who manage data and resources. However, these sites also serve as a cover, or face, for the local community and a portal for those in other communities and nations who are seeking resources. In this way, in a post-pandemic world, community and academic libraries anticipate that users beyond their previous user population access their site and depend on it for information and resources, and libraries must anticipate this user population and integrate design elements in to create balance between its local users, its librarians, and its secondary audiences.

Some design choices are more obvious than others: e.g., plain language (Plain Language Act of 2010; Plain Language Action and Information Network, 2011). In medical and health communication, for example, we see a call for plain language to meet the needs of a wide readership, and the World Health Organization (n.d.) has encouraged designers to simplify language to communicate with clarity and concision. Plain language allows a broader community to access technical documentation (Cheung, 2017) and thus is an important characteristic of design with *equilibriUX*.

Other design choices are complex because diverse users are influenced by culture, local standards, and preset notions or practices (e.g., individualistic versus collectivistic society; Hall et al., 2004). Balance requires that both influences are considered.

Usability must be emphasized in design because a well-designed and audience-centric product creates a return on investment for the designer and originating organization. However, usability also calls designers to consider the integrity of their work and serve as advocates for users. The idea of designer as advocate is not new but also is not universally accepted; the idea does consider that, if usability involves ongoing development of a design with ongoing analysis, users' rights and interests are an ethical responsibility of the designer (Human Factors & Ergonomics Society, 2020; IESBA, 2019; User Experience Professionals Association International, n.d.). Usability as advocacy considers digital transformation and the user's experience as well as the designer's observation and interpretation and thus requires diverse perspectives and expertise that influences the user's experiences and recognizes each user as a unique individual.

As technical and professional communicators have considered their users, they have considered the characteristics of their users, allowing those characteristics to influence human-centered design and to establish usability and the user's experience as centered on representatives of users, who are influenced and defined by culture (St.Amant, 2015). Culture and communication, the essence of being human, are intertwined and thus should be considered in the construction of words and visual designs of documents and tools to ensure that user experience is a positive and relevant means. In considering construction, TPC scholars consistently call for designers to consider the user experience—through narrative inquiry (e.g., asking users to respond to design; Jones, 2016) and participation in the design process (Agboka, 2013; Bannon & Ehn, 2013; Getto, 2014; Johnson et al., 2007; Moore & Elliott, 2015; Oswal, 2014; Spinuzzi, 2005): to know the user, to ensure that the user has input in design, and to create a more user-centric approach to the design process.

Users cannot always participate in the design process. However, the process should not exclude users in communities who may face participation challenges: for example, distance; cultural differences (Hall et al., 2004); migrant status (Rose et al., 2017); language barriers; or disabilities (Oswal, 2019). Every effort should be made to include voices from all user groups. How do we accomplish this, though, when historically these voices have been marginalized?

TPC scholars have called for social justice to be a focus of technical communicators, considering contexts that cross cultural, disciplinary, and organizational lines and expanding research to advocate and consider users who are underrepresented (Walton & Jones, 2013). This call has been focused even more to establish social justice as an objective of human-centered design, with feminist theory—one that "embraces concepts and considerations of equality and justice" (Jones, 2016, p. 477)—as one potential framework. However, a variety of theoretical approaches have been considered, with the goal of strongly encouraging technical communicators to be trained in these areas (Cleary & Flammia, 2012). Designers must respond to this call for social justice without focusing only on one subset of the user population. Instead, they need to consider the polyphony of user voices—to increase the dignity of the voices of the previously unheard without silencing other voices. In this way, design embraces balance and respects the value of *all* voices, *all* users, *all* stakeholders.

From Here to There: *EquilibriUX* in Practice

When the user community shares geographic location and culture, the designer can localize the design to ensure that the user's needs, expectations, and preferences are considered. Localization requires that technical communicators be culturally competent. They must pay "attention to the characteristics and needs of a particular culture, population, or even individual" (Breuch, 2015, p. 114) to build that competency. In that way, they are able to understand "material culture and members' practices" (Bannon & Ehn, 2013) and build relationships and collaborate with community strategists—to cultivate "a global network of people with diverse skills, identities, and experiences, covering a range of organizations, cultures, languages, and geographical locations . . ." (Shivers-McNair & San Diego, 2017, p. 100) and to create a culturally focused participatory design process. Even in localizing, they pursue balance in design by considering the diversity of users. In this process, technical and professional communicators are encouraged to involve members of a cultural community to gain insight into cultural priorities that may use specific design elements related to navigation, color, and text (Alexander et al., 2017, p. 78) to design with the culture's prominent standards and expectations such as design complexity (p. 81) or thought and browsing habits (p. 84). Culture does not always align with national, geographical, or religious alignment but can also consider organizational cultures (Eriksson & Eriksson, 2019).

Blending the global characteristics of online communication and the localized needs of users and their specific cultures, "glocalization" (Robertson, 1995) does acknowledge the broader, universal audience and also considers the particulars of localized design to create a "balance" (Breuch, 2015, p. 114). Scholars have called for adapting regional products to create usability that expands the usefulness of products across the globe (Acharya, 2019), expanding a product's usefulness to "resource-constrained settings" (p. 8).

In this tone, as TPC embraces the principle of *equilibriUX*, we consider concerns that research related to usability and localization has become controversial. At times, the call for social justice may integrate political and social value in design work and thus may reflect the polarization of the United States and the world. At other times, the TPC field narrative, rather than being unifying, appears to be battling internally when the field needs to be unified. However, we see research and scholarship as a tool to move the TPC field forward, strengthening what technical communicators do agree on: empowering users and considering the agency and voices of all users are ways to "level the playing field" and to dignify those who are underrepresented or oppressed without silencing those who have not experienced such submersion. Research and scholarship can unify us by demonstrating leadership from the field of TPC, integrating empathy, kindness, and inclusion for *all* in user-centered design.

If our research is left unpublished, if the knowledge is ignored because of the chaos it might create, we will never achieve *equilibriUX*. We will never know the countercultures—those who think, research, and perceive differently than we do (from which we all learn and grow)—to create balance in the voices involved. When we consider only the primary stakeholders in design, we deemphasize empathy and balance for the greater good. But when we silence those voices who have been prominent in the past, we defy balance and only shift the imbalance from one set of voices to another. Without *equilibriUX*, we will never be *truly* inclusive; for inclusion requires that *all* voices are heard and that *every* user is equal, even if they are different.

Extending Conversations: *EquilibriUX* in this Collection

We recognize the TPC field's conversations are moving us in the direction of achieving *equilibriUX*, and more research and scholarship will continue to advance our field. In this collection, we offer what we hope to be even more expansive action in TPC—to acknowledge and embrace inclusion and intentional user consideration and involvement in design and also to consider *all* users as valuable, including some populations that continue to be overlooked in usability and UX design. We have considered how to categorize the included projects, and, considering the authors' objectives, we have expanded the conversation to address *equilibriUX* in professional training and curriculum design, in medical and health tools and narratives, and in civic and social projects; the diversity of voices is inspiring.

Shared purpose is a powerful motivation. Shared purpose in designing TPC curriculum invests in the future of the field with a focus on the "localization" of curriculum for TPC students. Expanding curriculum to focus on participatory design and users' cultural representation but also considering humanity allows technical communicators to integrate empathy and heart into curriculum design and thus in our goals for social justice and user advocacy. The motivation then is not political or social but personal and relational, with intentional focus on the diversity of all users. (After all, *every* user is a unique individual.)

The consideration of all academic voices in TPC is vital to the dignity of our field. Our community-college faculty who teach first- and second-year (not "lower-level") students are able to create strong community-centered foundations for TPC's role in education. Our teaching-track and teaching faculty in higher education bring valuable expertise, pedagogical knowledge and experience, and passion to their classrooms and are an important part of TPC in higher education. (We choose not to use "contingent," as these instructors' skills are not accidental and should not depend on circumstances, and we choose not to use "non-tenure track" because these faculty are not "non" entities. Perhaps TPC can begin advocacy within its own ranks by establishing new terminology that eliminates the "non" and dignifies the work that these qualified instructors do in our field.) Without excluding, we can dignify these voices, which may be overlooked in publications, even though these instructors are well qualified and educated.

More and more, TPC is empowering patients and communities to improve health, to thrive, and to survive natural disasters and circumstances. We provide systems for communication between stakeholders, including those who engage in disasters from within and from outside the context. TPC can unite in investing in the health of communities around the globe and in learning from localized experiences to expand our abilities to care—for those who struggle with mental health issues, those who have rare conditions that qualify them with special needs, those who may be struggling to empower their families with instructions on how they want to live and die, and those who have served and who have received care during the recent pandemic and the continuing evolution of patient-physician communication. We also continue the call to expand UX research beyond U.S. borders and those of Global North nations as we consider research of users in various nations around the world. TPC and UX must embrace the global and post-pandemic emphasis on all audiences worldwide. In this way, we dignify other voices: in this collection, beekeepers in Lebanon, healthcare providers in Nepal, new and expecting mothers and physicians in China, rideshare passengers in China, the designers who create tools for these users, and the systems that ensure that the tools are available and functioning.

Technical and professional communicators can create *equilibriUX* in considering technical and international challenges that have not been addressed. We can seek to establish how TPC can invest in the management of mental disabilities and challenges that include visual impairment; the field can ask how technology can be used to better the lives and usability of tools for those who experience physical and mental challenges. We can consider how virtual communication, such as ride-share apps, must consider the safety of those who use the tools—in an effort to prevent harm with effective, strategic, and predicting usability testing. The field can consider how to design social and political campaigns to ensure that all participants in U.S. electoral systems have voices. And we can investigate ways to ensure that we are taking care of the earth, resources, and life to prolong life across the planet.

After inviting scholars to submit to this collection, we were delighted by the diversity, the freshness, the respect, the passion, and the compassion that the scholars relayed in presenting their practices, programs, and research. The submissions came from diverse scholars—diverse in age, gender, ethnicity, citizenship, level of education, geographic location, and notoriety in the field of TPC. The expertise of some authors is balanced by the newness of other voices—which gives us great hope for the future of the field.

Thus, with this edited collection, we seek to build balance: to celebrate TPC and the diversity of those who study, research, and practice in the field and to challenge TPC scholars, instructors, researchers, and practitioners to embrace and practice *equilibriUX*: that balance of sometimes controversial content to create harmony and to participate in engaged and diplomatic discourse. We build on the scholars who have come before us, expanding their call but uniting our field to dignify *all* users, *all* voices, *all* communities. This expansion of the call is idealistic, but it is realistic in that, by integrating an awareness of humanity as well as empathy, compassion, and inclusion for all humans (without bias, retribution, or division), TPC can improve to be and be known as a field for agency and balance through words, pictures, and intent.

References

Acharya, K. R. (2018). Usability for user empowerment: Promoting social justice and human rights through localized UX design. In *SIGDOC '18: Proceedings of the 36th ACM International Conference on the Design of Communication*, 6; 1–7. https://doi.org/10.1145/3233756.3233960

Acharya, K. R. (2019). Usability for social justice: Exploring the implementation of localization usability in Global North technology in the context of a

Global South's country. *Journal of Technical Writing and Communication,* 49(1), 6–32. https://doi.org/10.1177/0047281617735842

Agboka, G. Y. (2013). Participatory localization: A social justice approach to navigating unenfranchised/disenfranchised cultural sites. *Technical Communication Quarterly,* 22(1), 28–49. https://doi.org/10.1080/10572252.2013.730966

Alexander, R., Murray, D., & Thompson, N. (2017). Cross-cultural Web usability model. In Bouguettaya A. et al. (Eds.), *Web information systems engineering—WISE 2017.* WISE 2017. Lecture Notes in Computer Science, vol. 10570. Springer. https://doi.org/10.1007/978-3-319-68786-5_6

Bacha, J. A. (2018). Mapping use, storytelling, and experience design: User-network tracking as a component of usability and sustainability. *Journal of Business and Technical Communication,* 32(2), 190–228. https://doi.org/10.117/1050651917746708

Bannon, L. J., & Ehn, P. (2013). Design matters in participatory design. In J. Simonsen & T. Robertson (Eds.), *Routledge handbook of participatory design* (pp. 37–63). Routledge.

Barron, L. (2020, August 4). How Beijeng's national security crackdown transformed Hong Kong in a single month. *Time.* https://time.com/5874901/hong-kong-national-security-law-timeline

Breuch, L. M. K. (2015). Glocalization in website writing: The case of MNsure and imagined/actual audiences. *Computers and Composition,* 38, 113–125.

Chavez, N. (2020). The year America confronted racism. *CNN.* https://www.cnn.com/interactive/2020/12/us/america-racism-2020

Cheung, I. W. (2017). Plain language to minimize cognitive load: A social justice perspective. *IEEE Transactions on Professional Communication,* 60(4), 448–457. https://doi.org/10.1109/TPC.2017.2759639

Chirico, F., Nucera, G., & Szarpak, L. (2021). COVID-19 mortality in Italy: The first wave was more severe and deadly, but only in the Lombardy region. *Journal of Infection,* 83(1), e16. https://doi.org/10.1016/j.jinf.2021.05.006

Chor, L. (2021, June 16). From packed streets to silence: Documenting the fall of Hong Kong. *The Guardian.* https://www.theguardian.com/global-development/2021/jun/16/from-packed-streets-to-silence-documenting-hong-kong-authoritarianism

Cleary, Y., & Flammia, M. (2012). Preparing technical communication students to function as user advocates in a self-service society. *Journal of Technical Writing and Communication,* 42(3), 305–322. https://doi.org/10.2190/TW.42.3.g

Dragga, S., & Voss, D. (2001). Cruel pies: The inhumanity of technical illustrations. *Technical Communication,* 48(3), 265–274.

Du, L., & Katanuma, M. (2021, February 8). Why Japan's world-class health system buckled under COVID. *Bloomberg.* https://www.bloomberg.com/news/articles/2021-02-08/covid-exposes-weakness-in-japan-s-top-tier-health-care-system#xj4y7vzkg

"Equilibria." (n.d.). *Merriam-Webster Dictionary.* https://www.merriamwebster. com/dictionary/equilibria

Eriksson, P. E., & Eriksson, Y. (2019). Live-action communication design: A technical how-to video case study. *Technical Communication Quarterly, 28*(1), 69–91.

Frost, E. A. (2013). Transcultural risk communication on Dauphin Island: An analysis of ironically located responses to the deepwater horizon disaster. *Technical Communication Quarterly, 22*(1), 50–66. https://doi.org/10.1080/ 10572252.2013.726483

Getto, G. (2014). Designing for engagement: Intercultural communication and/ as participatory design. *Rhetoric, Professional Communication, and Globalization, 5*(1), 44–66.

Hall, M., De Jong, M., & Steehouder, M. (2004). Cultural differences and usability evaluation: Individualistic and collectivistic participants compared. *Technical Communication, 51*(4), 489–503.

Henry, C. J., & Camps, S. G. J. A. (2018). Chapter 3.2—Advanced nutrition and dietetics in nutrition support. In M. Hickson, S. Smith, & K. Whelan (Eds.), *Energy requirements in nutrition support.* https://doi. org/10.1002/9781118993880.ch3.2

Hoft, N. L. (1995). *International technical communication: How to export information about high technology.* Wiley.

Human Factors & Ergonomics Society. (2020). *Code of Ethics.* https://www.hfes. org/about-hfes/code-of-ethics

IESBA. (2019, June 26). *Global Ethics Board Launches eCode: Takes Usability and Accessibility of Code of Ethics to Next Level* [Press Release]. https://www.ethics board.org/news-events/2019-06/global-ethics-board-launches-ecode takes-usability-and-accessibilitycode?email_version=9eda1dc90810f216942fcc79 42d126fca8eb70c4

Johnson, R. R., Salvo, M. J., & Zoetewey, M. W. (2007). User-centered technology in participatory culture: Two decades "beyond a narrow conception of usability testing." *IEEE Transactions on Professional Communication, 50*(4), 320–332. https://doi.org/10.1109/TPC.2007.908730

Jones, N. N. (2016). Narrative inquiry in human-centered design: Examining silence and voice to promote social justice in design scenarios. *Journal of Technical Writing and Communication, 40*(7), 471–492. https://doi. org/10.1177/0047281616653489

Lancaster, A. (2018). Identifying risk communication deficiencies: Merging distributed usability, integrated scope, and ethics of care. *Technical Communication, 65*(3), 247–264.

Lancaster, A., & King, C. S. T. (2022). Introduction: Localized usability and agency in design: Whose voice are we advocating? [Special issue]. *Technical Communication, 69*(4), 1–6.

Lower, S. (2021, March 3). Le Chatelier's Principle [Online]. *Chemistry LibreTexts.* https://chem.libretexts.org/@go/page/3532

Mango, P. (2021, May 4). The truth about Trump's Operation Warp Speed. *Yahoo News!* https://news.yahoo.com/truth-trump-operation-warp-speed-103059422.html

Modi, C., Böhm, V., Ferraro, S., Stein, G., & Seljak, U. (2021). Estimating COVID-19 mortality in Italy early in the COVID-19 pandemic. *Nature Communications, 12,* 2729. https://doi.org/10.1038/s41467-021-22944-0

Moore, K. R., & Elliott, T. J. (2015). From participatory design to a listening infrastructure: A case of urban planning and participation. *Journal of Business and Technical Communication, 30*(1), 59–84. https://doi.org/10.1177/1050651915602294

Norwich, K. H. (2010). Le Chatelier's principle in sensation and perception: Fractal-like enfolding at different scales. *Frontiers in Physiology, 1,* 17. https://doi.org/10.3389/fphys.2010.00017

Omer, S. (2020, December 9). 8 of the worst disasters in 2020. *World Vision.* https://www.worldvision.org/disaster-relief-news-stories/worst-disasters-2020

Oswal, S. K. (2014). Participatory design: Barriers and possibilities. *Communication Design Quarterly 2*(3), 14–19. https://doi.org/10.1145/2644448.2644452

Oswal, S. K. (2019). Breaking the exclusionary boundary between user experience and access: Steps toward making UX inclusive of users with disabilities. In *Proceedings of the 37th ACM International Conference on the Design of Communication.* Portland, OR: ACM. https://doi.org/10.1145/3328020.3353957

Plain Language Act of 2010, Public Law 111-274, 5 U.S.C. 105. (2010). https://www.govinfo.gov/content/pkg/PLAW-111publ274/pdf/PLAW-111publ274.pdf

Plain Language Action and Information Network. (2011). *Federal plain language guidelines.* https://www.plainlanguage.gov/media/FederalPLGuidelines.pdf

Robertson, R. (1995). Glocalization: Time-space and homogeneity-heterogenity. In M. Featherstone, S. Lash, & R. Robertson (Eds.), *Global modernities* (pp. 25–42). Sage.

Rose, E. J., Racadio, R., Wong, K., Nguyen, S., Kim, J., & Zahler, A. (2017). Community-based user experience: Evaluating the usability of health insurance information with immigrant patients. *IEEE Transactions on Professional Communication, 60*(2), 214–231. https://doi.org/10.1109/TPC.2017.2656698

Samuels, M. (2021, January 13). Biden's COVID vaccine rollout plan. *Boston University School of Public Health.* https://www.bu.edu/sph/news/articles/2021/bidens-covid-vaccine-rollout-plan

Sánchez, F. (2016). The roles of technical communication researchers in design scholarship. *Journal of Technical Writing and Communication, 47*(3), 359–391.

Sawyer, D., & CDRH Work Group. (n.d.). *Do it by design: An introduction to human factors in medical devices.* Washington, DC: U.S. Department of Health and Human Services. https://elsmar.com/pdf_files/FDA_files/DOITPDF.PDF

Shivers-McNair, A., & San Diego, C. (2017). Localizing communities, goals, communication, and inclusion: A collaborative approach. *Technical Communication, 64*(2), 97–112.

Simmons, W. M. (2007). *Participation and power: Civic discourse in environmental policy decisions.* State University of New York Press.

Smith, W. R. (2020). A precise, simple and general Basic Le Chatelier Principle based on elementary calculus: What Le Chatelier had in mind. *Journal of Mathematical Chemistry, 58,* 1548–1570. https://doi.org/10.1007/s10910-020-01140-3

Spinuzzi, C. (2005). The methodology of participatory design. *Technical Communication, 52*(2), 163–174.

St.Amant, K. (2015). Introduction to the special issue: Cultural considerations for communication design: Integrating ideas of culture, communication, and context into user experience design. *Communication Design Quarterly, 4*(1), 6–22. https://doi.org/10.1145/2875501.2875502

Stevens, H., & Ahmed, N. (2021, April 28). Timeline: How the vaccine rollout progressed during Biden's first 100 days. *The Washington Post.* https://www.washingtonpost.com/politics/2021/04/23/biden-vaccine-timeline

Sugrue, T. J. (2020, June 11). 2020 is not 1968: To understand today's protests, you must look further back. *National Geographic.* https://www.nationalgeographic.com/history/article/2020-not-1968

Sullivan, P. (1989). Beyond a narrow conception of usability testing. *IEEE Transactions on Professional Communication, 32*(4), 256–264.

Thompson, A. (2020, December 22). A running list of record-breaking natural disasters in 2020. *Scientific American.* https://www.scientificamerican.com/article/a-running-list-of-record-breaking-natural-disasters-in-2020

Treptow, R. S. (1980). Le Chatelier's Principle: A reexamination and method of graphic illustration. *Journal of Chemical Education, 57*(6), 417–421.

U.S. Food and Drug Administration. (2023, March 14). COVID-19 bivalent vaccine boosters. https://www.fda.gov/emergency-preparedness-and-response/coronavirus-disease-2019-covid-19/covid-19-bivalent-vaccine-boosters

User Experience Professionals Association International. (n.d.). *UXPA professional code of conduct.* https://uxpa.org/uxpa-code-of-professional-conduct

Vazquez, M., & Carvajal, N. (2021, December 22). Biden offers rare praise of Trump during COVID speech. *CNN News.* https://www.cnn.com/2021/12/21/politics/biden-trump-covid-vaccine-booster/index.html

Walton, R., & Jones, N. N. (2013). Navigating increasingly cross-cultural, cross-disciplinary, and cross-organizational contexts to support social justice. *Communication Design Quarterly, 1*(4), 31–35. https://doi.org/10.1145/2524248.2524257

World Health Organization. (n.d.). Principle: Understandable. *Use plain language.* https://www.who.int/about/communications/understandable/plain-language

Yoneoka, D., Eguchi, A., Fukumoto, K., Kawashima, T., Tanoue, Y., Tabuchi, T., Miyata, H., Ghaznavi, C., Shibuya, K., & Nomura, S. (2022). Effect of the Tokyo 2020 Summer Olympic Games on COVID-19 incidence in Japan: A synthetic control approach. *British Medical Journal Open*, *12*(9), e061444. https://doi.org/10.1136/bmjopen-2022-061444

Part One

Pedagogical Topics

When we consider how to create balance and unity in the field of TPC but also in the processes we use and the designs we create, we must step back and consider how we are preparing the next generation of technical communicators. How are we teaching soft skills of empathy, of accessibility, and of value for all voices to ensure that we address the discord in the field? The best way is to start with instruction, designing curriculum that emphasizes balance and empathy and integrating these values into our classrooms. With that, we present five chapters that focus on this objective with innovative approaches to pedagogical issues.

Part One

Pedagogical Topics

Chapter 1

Design Thinking in Localized Service-Learning

Innovating Solutions through Empathy, Research, and Advocacy

Jason Tham and Rob Grace

Technical Communication (TC) scholars as well as UX design practitioners have steadily embraced "design thinking" as a solution-driven, human-centered approach to solving complex problems in sociotechnical (people- as well as technology-oriented) situations. Both a *mindset* and a *method*, design thinking offers a framework for understanding human needs and creating innovations following an empathy-first philosophy. Design thinking's empathic foundation is characterized by the Hasso Plattner Institute of Design at Stanford University (also known as the d.school [2018]) in terms of seven main attributes:

- Focus on human values: Empathy for the people you are designing for and feedback from these users is fundamental to good design.

- Be mindful of process: Know where you are in the design process, what methods to use in that stage, and what your goals are.

- Radical collaboration: Bring together innovators with varied backgrounds and viewpoints. Enable breakthrough insights and solutions to emerge from the diversity.

- Bias toward action: Design thinking is a misnomer; it is more about doing than thinking. Bias toward doing and making over thinking and meeting.

- Embrace experimentation: Prototyping is not simply a way to validate your idea; it is an integral part of your innovation process. We build to think and learn.

- Show don't tell: Communicate your vision in an impactful and meaningful way by creating experiences, using illustrative visuals, and telling good stories.

- Craft clarity: Produce a coherent vision out of messy problems. Frame it in a way to inspire others and to fuel ideation. (d.school, 2018)

These seven attributes form a mindset for problem solving that encourages designers to center human concerns when applying the widely adopted design-thinking workflow to innovate solutions. This workflow, as Figure 1.1 shows, is meant to be an iterative (recursive/cyclic) process that embodies the attributes outlined by the d.school (2018). Designers gather insights about user needs and their contextual concerns to define a working problem statement. Based on a focused direction, designers ideate creative yet user-centered solutions to be made tangible via prototyping exercises. These solutions are then tested and validated by real-world users who also give feedback for improvement. After reaching a saturation point, the polished solution exits the design-thinking workflow and moves into production and implementation.

Figure 1.1. The Popular Design Thinking Process. *Source*: Created by the author.

Although design thinking is only one of many innovative approaches in product development and service design, it is celebrated by popular tech companies such as IBM (IBM Design, 2022); Apple (Thomke & Feinberg, 2009); and Google (Pferdt, 2019). Businesses have considered design thinking to be a catchall model for growth and change, including customer relations, human resource management, leadership and stakeholder relations, project management, and corporate social responsibility (Brown, 2009). Also, design thinking has gained traction in cultivating positive transformation in nonbusiness sectors, such as education (Koh et al., 2015; Wise, 2016) and healthcare design (Altman et al., 2018; Roberts et al., 2016; Valentine et al., 2017). For TC, design thinking affords a structured yet flexible framework to practice socially responsive and localized strategies across multiple industries (Tham, 2021a). To prepare future practitioners to engage confidently with design thinking, we posit that this practice should be implemented in TC courses.

Recently, instructors in TC courses have integrated design thinking with service-learning[1] pedagogy to explore the potentials of this design-centric method in helping students learn and practice empathy toward those who are affected by *wicked* issues (e.g., Bay et al., 2018; Tham, 2018, 2021b; Wible, 2020). The term "wicked problems" was coined by design and urban-planning scholars Rittel and Webber (1973) and popularized in communication scholarship by design theorist Buchanan (1992). Although the wicked-problem framework provides a heuristic for characterizing complex issues as design challenges that embody incomplete, contradictory, and evolving requirements, it tends to generalize social issues as ill-defined global problems. As technical communicators know, to understand and devise practical solutions, more than a blanket problem description is needed: also, the designer must have a situated awareness of local needs and requirements.

To prepare future technical communicators and designers to extend their evaluation of problems to local contexts, we have integrated design thinking instructions into TC courses such as professional report writing and UX research via a service-learning model. This case study of pedagogies includes core phases of design thinking—empathy, definition, ideation, prototyping, testing, and iteration—to let students examine critically the struggles facing their local clients and to tinker meaningfully with available means for creating solutions that address those needs.

In this chapter, we (1) provide an overview of the landscape of design thinking as a normalizing *methodology* (both a mindset and a method) for user-centered problem-solving, (2) situate TC problems as wicked problems that require localized thinking and strategies, (3) describe our pedagogical motivations and approaches, (4) detail our students' journeys and results, and (5) discuss the implications of design thinking as an innovative yet locally relevant tactic to generate solutions for TC purposes.

In showcasing the examples, we aim to complicate the existing relationships between strategic problem solving and user advocacy. Given the emergent concerns about the role of technical communicators in advocating for socially just and responsible actions, we hope to shed light on the ways in which design thinking may provide a viable framework for achieving equilibriUX through empathy, research, and passionate advocacy.

Empathy in Design Thinking

Whereas design thinking stresses *empathy* as a unique value proposition that differentiates it from other macro solution-driven models like Agile and Sprint, Lean Startup, ADDIE, and Six Sigma, empathy is not exclusive to design thinking. Designer scholars have traced the concept of empathy in business and economics literature where, for more than 20 years, empathy has been of interest to management and innovation studies (Brown, 2008; Fontaine, 2001; Köppen & Meinel, 2015; Pavlovich & Krahnke, 2012). Specifically, research rooted in philosophy and psychology has identified empathy as the capacity "to share, to experience the feelings of another person" (Greenson, 1960) and to comprehend the situations of others imaginatively and emotionally (Rogers, 1975). For this reason, we address the long history of empathy as part of design research and literature but also note more recent trends or lenses that this collection highlights, such as equilibriUX.

One of the most common distinctions made by those who applaud empathy as a social competency is how it differs from sympathy. Conventional understanding has it that sympathy comes from one's acknowledgment of another's experience without the necessity for action. An example would be this: Diya is aware of the financial difficulties that his friend Meghan is experiencing; however, Diya has no means to help Meghan nor offer any consolation. Diya, in this case, is sympathetic toward Meghan. By contrast, empathy by definition should invoke *action*. To empathize is to participate in a shared or mirrored experience so that an undesirable situation can be mended. Therefore, to empathize with Meghan, Diya would need to advocate for solutions that help Meghan overcome her financial challenges—like helping her with scholarship applications, finding a side gig, or gathering donations.

Celebrity author Brené Brown (2011) goes as far in her TED Talk as to say that "empathy fuels connection; sympathy drives disconnection" for the reason that empathy requires *feeling together with*, not just *feeling for*, someone, whereas sympathy is considered a one-sided reaction that often only serves the morale of the sympathizer rather than the needs of the sympathized.

In addition to instinctive, affectionate reactions to other people's experiences, empathy in design thinking is cognitive, "where one *understands* how others may experience the world from their point of view" (Gasparini, 2015, p. 50, emphasis added). This social cognition motivates practitioners to take stock of their own positionality and understand someone else's perspective. By adopting the perspectives of others, designers attempt to reconstruct specific experiences that are more desirable *by others*, not themselves.

This guiding principle for design thinking manifests itself in forms of UX research. Empathy techniques in UX include active listening methods such as ethnography or field studies, contextual inquiry, user interviews, focus groups, surveys, and participatory design workshops. These methods allow UX researchers to discover real-world experiences of the users for whom they are designing. The data gathered from these processes can be translated into actionable insights that guide design directions (Alrubail, 2015).

Empathy Is Always Local(ized)

If empathy is a means to acknowledge and experience people other than oneself, then *intersubjectivity* should be taken into serious account: "The concern of design culture with social issues has come to comprehend the very basis of society, i.e., the nature of social relationships" (Devecchi & Guerrini, 2017, p. S4360). This relational emphasis calls attention to the multiplicity of roles and agents in a design scenario, as well as the sociality of all parties involved: "Empathy is the condition of a connection rather than a fusion self-other" (p. S4361). This *connection*, akin to Brown's (2011) interpersonal association, signals the need to account for multiple stakeholders in a given design challenge and to consider contrasting or conflicting interests within the same context. Take a healthcare quality-improvement effort for example: engagement, uptake, and implementation can be understood differently by various stakeholders like patients, physicians, nurses, investors, and system administrators as particular roles desire distinctive needs (Norris et al., 2017). A patient's need is certainly much different from an investor's mission.

Empathy in design thinking reveals multiple stakeholders with contradictory needs and, as a result, reveals the impossibility of a single solution. Following this concern, designers and communicators adopt the *wicked problem* frame to guide their invention. For TC instructors who integrate design exercises via service-learning, this multiplicity of wickedness presents new layers of challenge to teaching and learning.

Next, we explore the layered challenges and ripple effects these challenges present to TC and design pedagogy, with particular attention to service-learning course design and community partnership setup.

Wicked Problems in Design and Service-Learning Pedagogy

Design thinking practitioners—whether in UX or other design industries—subscribe to the notion of wicked problems, as defined by Rittel and Webber (1973), which includes the following characteristics:

1. Wicked problems have *no definite formulation* because they operate in a continuous feedback loop with the surrounding environment.

2. Wicked problems have *no stopping rule* because they cannot be reduced to their base variables.

3. Wicked problems contain *no criteria for correct solutions* in that no objective ways exist for judging whether solutions are right or wrong, good or bad. New realities simply emerge from the actions taken as a result of the solution concept.

4. Wicked problems provide *no way to test the quality of a proposed solution* since an experiment can never be completely reset to the original starting conditions.

5. Wicked problems assume *no ultimate test of a solution*, only implementation.

6. Once committed to a solution and a plan of action, the *consequences* to addressing wicked problems *are permanent* and cannot be undone.

7. Wicked problems do not have a *completely fixed list of the permissible activities*, only the ones constrained by prescription or proscription.

8. Wicked problems *do not depend on well-defined solutions*, just options that result in new emergent trends.

9. Every wicked problem is unique, making SOPs [standard operational procedures] and doctrine *conditional instead of authoritative.*

10. The wicked-problem solver has no right to be wrong since their *actions have real consequences* to the lives of others.

Designers who find themselves in the position of innovating with design thinking need to abandon any idealistic assumptions about the problem at hand and engage empathy as a strategy for inventing user-centered solutions that have ripple effects on the people they serve as well as the extended community. Buchanan (1992) famously argued that design thinking for wicked problems "is not thinking directed toward a technological 'quick fix' in hardware but toward new integrations of signs, things, actions, and environments that address the concrete needs and values of human beings in diverse circumstances" (p. 21).

In other words, design thinking urges practitioners to consider implications beyond the most immediate audience and context. Because wicked problems cannot be reduced to singular or definitive variables (see number 2 above; Rittel & Webber, 1973), designers need to continuously experiment and iterate based on results with multiple audiences. Diversity of circumstances is uncertain, so an integrative approach is necessary—a synthesis of "the ideas of designers and manufacturers about their products; the internal operational logic of products; and the desire and ability of human beings to use products in everyday life in ways that reflect personal and social values" (Buchanan, 1992, p. 20).

From Client-centered to
Stakeholder-centered Service-Learning

Given this multiplicity of stakeholders, designers should not center their efforts merely around the requirements prescribed by the client, who represents only a part of the design context described by Buchanan (1992). A client-centered design approach, although pleasing and desirable by organizations, seldom recognizes the true experiences of local users. Client-centered design may address user issues observed through the client's perspectives, but the observations are often profit driven. Adopting the design thinking model for solving UX problems means a departure from client centeredness toward a stakeholder-oriented approach in which the multiplicity of audience, context, and community must be foregrounded. Using Norris et al.'s (2017) healthcare context, a stakeholder-centered design approach would consider the stakes of users—patients, physicians, nurses, investors, and system administrators—across multiple-use contexts and their communal objectives and challenges.

TC educators who integrate service-learning with their pedagogy face a similar concern. Even 20 years ago, scholar-teachers already had identified that service-learning projects need to develop reciprocal relationships between students and the community partners (Bowden & Scott, 2003; Dubinsky, 2002). Rather than meeting the needs of the client, the project should address the needs in a larger social context. Attending to the ambiguity of social problems can also cultivate in students what Pope-Ruark (2014) called *metic intelligence*—the rhetorical flexibility to face unprecedented challenges. And in the cultivation of such intelligence, instructors need to strike a balance between client and student-learning goals. Strong community partnership requires meeting both the interests of the client as well as the students (Kimme Hea & Shah, 2016). The client's goal, as made apparent, is only a piece of the larger service-learning picture.

Yet, when partnering with local clients who have very specific goals, instructors may find it difficult to prioritize social goals over the client's requirements—for the reason that community partnership is hard to forge (Yusof et al., 2020). For projects with multiple stakeholders, a client-centered approach would appear to be less complicated and thus more favorable by instructors with limited time and resources to facilitate service-learning pedagogy. This reality is itself a wicked problem for educators and their respective institutions. A balancing act in designing fruitful service-learning experiences for multiple interests is seemingly impossible—but should it be so? As teachers who have experimented with ways to connect students with real-world design challenges, we believe there are strategies to achieve the balance of users' agency (equilibriUX).

We include the two approaches to integrating service-learning with TC courses—client-centered design and stakeholder-centered design—and discuss the affordances and constraints in both kinds of pedagogy. In distinguishing stakeholder-centered design from client-centered design, we argue that adopting a stakeholder-centered approach to service-learning requires a localized, empathic design model such as design thinking. This approach, powered by empathy-driven research and stakeholder advocacy, can serve as an applied pedagogy across TC programs. In the next section, we describe our case study and qualitative methods in collecting and analyzing student experiences.

Finding Ways to Localize: Two Pedagogical Approaches

In this section, we begin by introducing four service-learning projects that attempted to address wicked problems by balancing students' needs to empathize with multiple, local stakeholders and design communication deliverables for individual service-learning clients. We then describe the focus groups (Institutional Review Board approved study) and thematic analysis that we employed to understand students' experiences using design-thinking strategies during these projects. Our analysis was guided by the following research questions:

1. To what extent did design thinking afford students the opportunity to empathize with their service-learning clients and/or other stakeholders involved with their projects?

2. How can we promote stakeholder-centered design in addition to client centeredness in service-learning pedagogies?

During fall 2021, we supervised the following service-learning projects while teaching courses on professional writing and human-centered design courses at Texas Tech University:

- Project 1: Launching a mental health social media campaign. Students created an Instagram account for the university counseling center and designed multimedia content for a

4-month campaign on mindfulness, stress management, and counseling services that were available to students. The service-learning client for the project was a psychologist who was serving as the director of the counseling center.

- Project 2: Conducting a survey on sustainability literacy. Students designed and piloted a survey on sustainability literacy—knowledge about sustainability topics and challenges such as climate change, biodiversity loss, and environmental resource depletion (STARS, 2021). The survey instruments and results from a pilot administered to students from multiple departments were submitted to the service-learning client, an assistant director of the university's office of sustainability.

- Project 3: Communicating the opening of a recycling center. Students designed and administered a short survey to gauge students' knowledge of on-campus recycling opportunities. Results of the survey were used to create advertisements for a new on-campus recycling center. The survey results and advertisements were submitted to the service-learning client, an assistant director of the university's office of sustainability.

- Project 4: Evaluating the usability of a nonprofit organization's website. Students critiqued and conducted user tests on the design of a local nonprofit organization's website. Based on users' feedback, students provided redesign recommendations to the organization's director in both written and recorded oral presentations.

At the conclusion of each project, we conducted focus groups with each project group. Our questions focused on students' use of design-thinking methodology in their projects and the extent to which design thinking created opportunities to empathize with multiple stakeholders while also designing communication deliverables for their service-learning client. The focus groups were conducted via Zoom video conference, recorded, and later transcribed for analysis. In addition, we took notes during these sessions to note unexpected topics and guide close reading of the transcripts.

We analyzed themes that emerged during the focus groups through an iterative process of inductive coding. This process involved open coding to identify emergent themes in the data and subsequent coding to group and identify relationships between these themes (Glaser & Strauss, 2012). This process was guided by the following sensitizing concepts (Bowen, 2006):

- Empathy: A cognitive competency that motivates action-based advocacy for stakeholders.

- Wicked problems: Issues or challenges that do not have direct solutions and that have ripple effects to multiple aspects of experience involving the stakeholders.

- Client centeredness: Activities to understand and design solutions that meet the needs of service-learning clients.

- Stakeholder centeredness: Activities to understand and address the needs of diverse, local stakeholders affected by wicked problems and/or solutions students design to meet the needs of service-learning clients.

In the section below, we present the results of our analysis.

Student Responses

In this section we reflect on our experience teaching service-learning projects during fall of 2021. We present themes observed in student feedback to answer the question: "To what extent did design thinking help students empathize with their service-learning clients and/or other stakeholders involved with their projects?" We describe how each course involved a balance between client- and stakeholder-centered service-learning design and the localized design thinking required during each service-learning project. Lastly, we highlight students' experiences, including the challenges they experienced striking this balance.

EMPATHY

If empathy involves understanding and advocating for stakeholders implicated in a design context, students demonstrated empathy in two distinct ways in their service-learning projects. Naturally, most students began the project by attempting to empathize with their service-learning client. Adopting a standard UX design approach, each student group sought to understand their client's experience of the problem motivating the design project and advocate for solutions that would meet their client's needs.

However, some *client-centered* students empathized only with their client whereas other, *stakeholder-centered* students reached out to a broader group of people implicated in the design context. The latter identified stakeholders who, like their client, also encountered the motivating problem but, unlike their client, experienced the problems in different ways. These students also identified stakeholders who might be impacted by the solutions proposed to meet their clients' needs. By contrast, client-centered students limited the scope of empathy to a single individual or small group of professionals.

Importantly, we observe that the narrow and broad scope of empathy demonstrated, respectively, by client-centered students and stakeholder-centered students, correspond with distinct activities of empathizing undertaken by students that led to different design outcomes. Students who adopted a client-centered approach worked reactively in response to their client's expressed needs and preferences. The groups produced a limited set of requirements and, from the client's perspective, often satisfactory design solutions. By contrast, students who adopted a stakeholder-centered approach worked more proactively than other groups by reaching out to multiple people in different ways. These research activities resulted in the identification of multiple, sometimes contradictory (i.e., "wicked"), project requirements that helped students appreciate the role of their client as a service provider to diverse audiences and, in turn, the need to design services for these audiences, rather than merely client-approved products. We further detail these two approaches toward empathy in the following sections.

WICKED PROBLEMS

In their responses, we observed the following:

1. Some students described their service-learning projects as wicked problems while other students did not.

2. To the extent that students perceived their service-learning projects as wicked problems, they were more likely to empathize with other stakeholders involved with their projects.

3. Students who did not describe their projects as wicked problems were less likely to empathize with other stake-

holders and more likely to focus only on their clients' needs and requirements.

Although we did not expect students to understand their service-learning projects as confronting "wicked problems" per se, we observed that some students discussed their projects as resembling wicked problems while other students did not. When reviewing students' feedback, the concept of the wicked problem offered us a frame for understanding the sensitivity of students to the complexity of the issues they were attempting to solve, the difficulty of defining the problem or its solution, and the meaningful consequences they have or might have on people's lives.

Some students struggled to pin down the definition of the problem or possible solutions. At the outset of the project on sustainability literacy, one student admitted that his group "generally just had no idea what was going on." Other students noted the difficulty of pinning down a problem when working with a client. Two students talked about the difficulty they experienced working with a moving target: "It's frustrating at times. During our project, the client had worked on the redesign of the website as we were testing the old design. They would change things that we're like, wait, that was an issue that we were planning on testing"; "Today we noticed that they completely changed the 'Donate' page. Just different from the screenshots that we used in our presentation." Students expressed a general feeling of unfamiliarity with the project topic: "My group members and I did not understand the topic of sustainability ourselves." This feeling was compounded by the open-endedness of the project and the nature of service-learning pedagogy: "I think at the beginning we were all very confused with what we were actually doing for the project. I didn't realize that in the end we were actually creating something for our client that they needed. I thought it was a fictional idea to allow us to complete the project." Similarly, among groups evaluating the usability of nonprofit websites, "Our group initially struggled to understand how we could contribute to the [deliverable] because of our lack of knowledge within the website creation space." For most students in our classes, service-learning was a new experience. "I've never done anything like this before," concluded another student.

Some students did not know how to judge whether the solution they were developing would be successful. Three students shared this question:

> Our client gave us a very rough idea of what she wanted us to do but was never specific enough for us to determine a good product to give back to her. . . .

We were not exactly sure where to start with the project and felt like we did not have a ton of guidance. We asked for help and got help, but I believe the project overall had very ambiguous instructions, which left us all not knowing what to do.

There were several attempts made to try and get the project done and make something that we thought we were supposed to. However, every time we thought we knew what we were supposed to do we were told we had the wrong idea.

These comments speak to the need for clear instructions and scaffolding that exist in any learning environment. When students "asked for and got help," they received instruction on problem definition, requirements gathering, and design-thinking approaches. However, as students need to design a product for a real client in service-learning projects, part of the work in such service-learning projects is to define the problem and solution for the client who may not know exactly what they need or how to measure its success.

These challenges were echoed across other project groups, such as the group tasked with surveying university students to effectively communicate on-campus recycling policies: "We found it challenging not knowing the baseline knowledge of students. It made the project very general as we had to start from scratch. . . . This made it kind of challenging to brainstorm what advertisements we wanted to create. Also, none of us had made ads before, and they had to be based on the various knowledge points." This feeling was shared by students developing a social media campaign for the university counseling center: "The main purpose of our assignment was to create a proposal to our client, the SCC [Student Counseling Center], that would promote their services via social media. However, as none of us were familiar with the SCC and the services that it offered, it made things difficult as there were never any sure-fire answers."

Lacking background knowledge on the topic, having never taken a service-learning course before, and faced with defining a project deliverable that would be used by actual people, students began the project feeling bewildered. These comments reflect the common experience of students facing an unfamiliar problem they cannot solve alone. In classroom situations, students often look for guidance from instructors and peers. In our service-learning projects, we observed two orientations to design thinking that shaped student practices of empathy: client centeredness and stakeholder centeredness.

CLIENT CENTEREDNESS

Client centeredness describes different ways that students used design thinking to empathize with a single stakeholder: the service-learning client to whom they were assigned for their project. For example, the group working on the social media campaign described how they approached the scope of empathy required for their project: "When we were first introduced with our Instagram project and what the SCC [Student Counseling Center] was looking for in a new Instagram presence, my group and I were somewhat confused of how they wanted for this to be achieved in the grand scheme of content to post and how to represent the SCC to their liking." Many students described a similar approach to design thinking that extended practices of empathy to their client but to no one else. Consequently, these students often understood the "grand scheme" of design requirements based on explicit feedback from the service-learning client. These students, in turn, tended to be those who discussed the problem motivating their project in terms of easily defined objectives and solutions without recognizing tradeoffs and challenges that emerged when understanding the points of view of multiple, local stakeholders. That is, these students were less likely to identify any "wicked" characteristics of the service-learning project.

Overall, we observed two themes of client centeredness across students:

1. Narrow scope of requirements gathering

2. Reactive project work

As we emphasize in the discussion, client centeredness is a critical aspect of service-learning and design thinking. Students need to learn to define and solve problems for people (e.g., clients) in contexts in and out of the classroom. However, as we emphasize in the discussion, client centeredness is necessary but insufficient for design thinking. Mindsets and methods to empathize with multiple, local stakeholders improves design outcomes, especially for the immediate client of a design team.

First, many students narrowly focused on understanding their client's requirements for the communications deliverable they were tasked to design:

> The first problem we faced was at the very beginning when we really were not sure what the client wanted from us. Our

biggest question for her was what did she want to know and discover from our survey to create advertisements for the recycling center?

My original plan for the project was completely different from what I had envisioned earlier. The information we needed was nowhere to be found, and there wasn't any way to acquire that information other than earlier in the semester when we interviewed our client.

Although problem definition is an important part of service-learning projects, students rarely attempted to look beyond the explicit requirements and ideas provided by their clients. This is important because the success of clients' communication products depended on awareness of and empathy with multiple stakeholders related to the service-learning project.

Second, students worked reactively for the client. Students followed clients' schedules when planning and working to design their communication deliverables, often against clients' expectations for proactive, independent work groups. On the one hand, they effectively worked around their clients' busy schedules to assist them. On the other hand, they often waited until they received a "green light" from their client before moving on to new tasks:

Another challenge we faced was waiting for comments and revisions from our sustainability office contact. We understood that she was busy, but we found ourselves at a standstill waiting for the green light or notes so we could continue.

In the beginning it was hard to set up a time to meet with [client], but from multiple emails and scheduling we were able to meet with her, hear her thoughts, and get the ball rolling.

One challenge we faced was trying to get in contact with [client] from The Office of Sustainability. Often there were times where we would email her asking questions about the survey or asking for feedback. With only two people working in the office, and part-time at that, there were a few occasions where it was difficult getting a quick response.

Students reacted to client feedback but also neglected opportunities to engage other stakeholders and resources that would allow them to overcome some project delays and improve their ability to empathize with and evaluate feedback received from their clients.

Stakeholder Centeredness

By contrast, students also demonstrated what we refer to as "stakeholder centeredness": practices of design thinking to empathize with multiple stakeholders, including their service-learning clients. We observed three ways students adopted a stakeholder-centered mindset:

1. Recognizing the client as a service provider.

2. Reaching out to stakeholders.

3. Designing services, not products.

Recognizing the client as a service provider. Client centeredness helps students focus on clients' needs. However, some students recognized that their clients were serving others (external stakeholders), and so the students focused their requirements gathering to address the needs of these stakeholders. Often this involved questioning the feedback or current approach adopted by their clients. For instance, when asked about challenges, a student expressed concern about the client's actual ability to serve the stakeholders they were aiming to serve. She was concerned about the accessibility of the service mentioned on the client website and how the information actually gets relayed to the people who needed it. This student's concern is an instance of empathy that directs the focus on stakeholders rather than the client: "Yea, and I was wondering, do they offer taxi vouchers to those who can't drive [to the food shelf]? Do they stop seeing people at 7 p.m.? In Lubbock, you pretty much can't walk anywhere. I am just curious how the community is really served by our partner." Another student said that the current content on the website doesn't directly speak to the needs of the population of community members who needed the client service. The student noted that it's important, for testing purposes, to find representative test participants to provide feedback about the content. That way, the design team can learn directly from the point-of-view of a representative audience: "It was actually hard to see how much of the service is available to people because of how it's organized on the website. It's also hard to get people who have been in that position [of receiving service] to speak to the actual experience."

Reaching out to stakeholders. Students contacted stakeholders they learned about through background research on the problem. This is, of course, the priority of empathizing in design thinking approaches. A student pointed out the importance of centering external feedback rather than internal guesses that helped gather realistic reactions to a product design (in this case, a website): "I think that because of all the different opinions and ideas we have taken for the same website, it really shows the affordance of taking others' directions into account, because, honestly, all the different perspectives that came up throughout the project really made me appreciate it (other people's views) . . . without them I would have missed small things that otherwise wouldn't have been brought up." Another student confessed that, as a website manager, he was learning how to care more about stakeholders' needs rather than the designer's: "So now when I'm working on a site [. . .] I pretend like I am the customer, unlike before I'm like, oh, this would be cool, this would look good, but now I'm like, is this going to be easy for them to find . . . is that going to throw them off . . . is it easy to navigate?" Other than the pilot survey we conducted in class, we also tried reaching out to the TechNews and other organizations that my group was involved with on campus to produce more results.

Taking initiative and not relying solely on their client, students mobilized their personal networks to reach out to stakeholders to define the problem and accomplish their project goals. These networks included professors and students in their classes and members of their extracurricular clubs:

> We decided it would be best if we could advertise the survey through our own organizations and clubs, as we were all involved in at least one. This would draw out more majors from different backgrounds, which we were successful in when it came time to administer the survey.

> Our group and I were also unsure how to spread the word around campus for students to take the survey. We were considering going to our business classes and asking there, but

then that would create a survey bias by only having business students as participants . . .

Ongoing background research was also critical to students' ability to identify and reach out to multiple stakeholders. Lacking experience in the area of their projects, most groups conducted background on the problem as a first step: "To address the problem of the unfamiliarity regarding the [Student Counseling Center] we each did comprehensive research. By reading the SCC's website, talking with other students who had used the SCC's services in the past, and meeting with our client via Zoom we were able to become more familiar with the SCC."

Designing services, not products. When students recognized that their client is a service provider, they recognized that the deliverables they were tasked to create would be judged on how much they improved the quality of services clients offered to stakeholders including university students, faculty, and staff and the people using nonprofit websites. One student noted the invisible layers in design that needed to be considerate of front-end users' experience. She talked about how content design is not just what one sees on the screen, signaling an awareness for stakeholder needs: "I am not second-guessing myself, but I'm giving it a lot more thought now [when using an information interface] . . . there's a lot more to this than what I see on the surface. When I open up something, like this website, . . . and start picking it apart, then I think—wait, there's a lot more to this, there's more behind all these. And I had to stay focused on the [UX] issues that we are tackling."

What We Learned: Balancing Act for Service-Learning

Given our observations of the students' mindsets and ways of engaging with their client projects, we posit that a stakeholder-centered approach to addressing social problems can enrich students' experience with service-learning rather than just client centeredness. Accordingly, and in response to RQ2, we discuss here some pedagogical strategies that promote stakeholder-centered design thinking that service-learning instructors may consider for their pedagogies. Letting students experience how to encounter, define, and address wicked problems in a service-learning context is, indeed, a balancing act for instructors who need to coordinate logistics between community partners and the classroom, as well as other institutional units pertaining to the service-learning effort. Resounding the theme of equilibrium in user-experience design in this collection, we encourage instructors to consider the following approaches to find balance in facilitating student learning and client relations.

First, clarify to students and service-learning partners that their projects are wicked in nature. Because we found in this case study that students who perceive their client projects to be wicked problems are more likely to empathize with the stakeholders, such clarification can be beneficial to all parties of the project if the instructor foregrounded the concept of wicked problems and how to engage with them productively. Spending time to learn about the characteristics of wicked problems may also help students think about local problems critically, as complex issues tend to be tied to local contexts or situations. It also can prompt students to dig deeper into issues beyond surface assumptions, which may lead them into an empathic mode more naturally.

That said, we do understand that empathizing is not an easy practice and is difficult to sustain throughout a project. We are cognizant about the need to create more systematic processes to help students practice empathy. So, what do students need to empathize with service-learning clients and other stakeholders involved with their projects? From our interviews and observations, we noted three important facilitators of empathy:

1. Projects should encourage an in-person or intimate experience.

2. Projects should provide a methodical structure to user experience research.

3. Projects should promote reflective discussions about service-learning experiences.

Being able to engage with clients and stakeholders directly is perhaps the most crucial factor for empathic involvement in service-learning projects. So, the first facilitator of empathy is to *encourage in-person or intimate experiences*. Many students have reported that the distant nature of their projects made them confusing and hard to really relate. Additionally, our circumstances were complicated by the recent health pandemic. While not all service-learning projects need to be done in person, an intimate experience between the students and the stakeholders of their projects is necessary to promote empathy. This may seem like a simple setup, but it can in fact take time and effort to carefully arrange opportunities for students and stakeholders to interact. Some interactions may be formal (like a client meeting and site visitation), while others can be impromptu or less orchestrated (such as conversations or discussions that take place during ideation and testing sessions).

For courses that are delivered in hybrid or fully online formats, instructors may use various meeting technologies and social communication tools to increase interactions between students and stakeholders. Clients may "drop in" during synchronous class meetings and maintain connections with students using messaging apps or online forums in course management systems. The volume and frequency of these interactions may need to be greater than the in-person version of the course to create a better sense of presence digitally. To ensure personal security and privacy outside the class meetings, instructors need to put in place communication policies that keep client interactions visible and accountable.

In comparing our experience teaching the two TC courses featured in this chapter, we noted the second important facilitator of empathy that is to *provide methodical structure to students in UX research* so they can actively practice empathy. Teaching the design-thinking workflow both as a philosophy and methodology gave students the language to describe their process. Moreover, what design thinking promotes as desirable attributes of empathic problem solving can give clear guidelines to students as they navigate through the wicked problems. UX methods like contextual inquiry, qualitative interviews, affinity mapping, user personas, journey mapping, and other ways to gather user insights can give students directions to understanding their respective stakeholders. This way, students can maintain compassion and responsiveness to the people they are serving through the service-learning projects.

Lastly, as with any critical pedagogies, we consider the third facilitator of empathy to be the *promotion of reflections on service-learning experiences* as they are imperative to students' growth. Stakeholder-centered mindsets can be cultivated ongoingly through intentional conversations or discussions about students' current experience with service-learning and what they hope to achieve next. Instructors play a key role here in moderating reflections and sharing of vulnerable feelings toward the students' projects. Depending on the stage of the projects, clients may participate in these conversations to listen for questions and feedback. To encourage open sharing, students should not be evaluated for the content of their reflections. They may, however, be guided toward reflections that demonstrate empathy and actionable suggestions for continuous improvement, which emulates design thinking's iterative problem-solving mentality.

As we have posited earlier in the chapter, design thinking is an adaptable model for service-learning pedagogies. Having learned about the ways to localize wicked problems and strive for equilibrium in teaching and learning, we offer the following heuristics for using design thinking to achieve stakeholder-centered design (Figure 1.2).

This set of heuristics can be modified to suit the needs of different service-learning courses. It can also be adapted for the client's requirements for deliverables. Here we provide some recommendations for TC courses like grant writing, usability studies, and internships.

Design thinking in grant writing. Students learning to work with stakeholders in a grant-writing context can apply the design-thinking heuristics to produce stakeholder-centered grant proposals. Considering the proposal as a deliverable that affects stakeholder success, instructors can guide students in the process of assessing client needs and understanding the contexts of the particular wicked problem(s). Parallel to the empathy phase of design thinking, students work to identify the balance between actual user needs and the desirable niche the client may seek to occupy via the grant proposal process. Especially for projects that aim to propose new products or platforms to solve existing problems, the design-thinking heuristics can be used to build tangible prototypes of the proposed solution as a way to demonstrate to funders the market need and "de-risk product development" (Sharp, 2019).

Design thinking in usability studies. As part of our case study here, we have attempted to show the viability of design thinking in user research and usability studies courses. Because these courses usually require students to adopt a user-centered design mindset, design thinking seems like an obvious complement to performing user research. In addition to

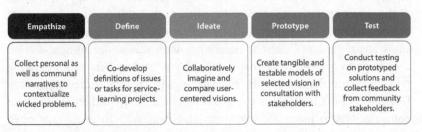

Empathize	Define	Ideate	Prototype	Test
Collect personal as well as communal narratives to contextualize wicked problems.	Co-develop definitions of issues or tasks for service-learning projects.	Collaboratively imagine and compare user-centered visions.	Create tangible and testable models of selected vision in consultation with stakeholders.	Conduct testing on prototyped solutions and collect feedback from community stakeholders.

Figure 1.2. A Stakeholder-centered Service-learning Model. *Source*: Created by the author.

teaching students the methods for researching user behaviors and attitudes, the design-thinking heuristics can support instructors in guiding students to situate usability studies within the larger ecology of technology use. Often, usability studies and user research focus narrowly on the specific settings in which a tool is applied by a user and how effectively and satisfactorily the user could use the tool to solve particular problems. Our design-thinking heuristics seek to expand the scope of user studies beyond the immediate use context—into the greater community. Students can learn to consider the user as multifaceted individuals who have different stakes in the product use given the tasks they perform in different roles. In the example of our case study, students identified a few roles the same individual could play when interacting with a website—as a volunteer, a donor, a facilitator, or all the above—that can influence their experience with the website differently in different roles.

Design thinking in technical communication internships. Our design-thinking heuristics can also support capstone courses including internship or externships where students participate in real-world experiential learning. In applying the general design-thinking model in studying student internship experience, McConnell (2020) found that design thinking provided a mindset that can help students stay motivated during the internship search process, identify their options, adjust their methods based on options that became known to them, and effectively self-promote to achieve an internship position. Design thinking can empower students to empathize with themselves and their own strengths (and weaknesses). And during the internship, our heuristics can be deployed as a basis for understanding workplace culture, human relations, stakeholder needs, and performing assigned roles. Students can use an empathic lens to view and get a sense of the local dynamic in the workplace—who does what and why, what matters most to whom, etc. Through definition strategies, students can practice articulating the issues within wicked problems that they are attempting to solve. Using ideation, prototyping, and testing methods, students can create tangible, testable, and even shippable products that directly affect the outcomes of the projects they engage with.

Conclusion

This chapter considered design thinking as a viable, flexible model for integration with service-learning TC courses that can support critical and meaningful engagement with wicked problems. Through an explication of our pedagogical approaches in getting students to evaluate and respond to local issues, we have documented how students dealt with their respective clients and projects, and the ways a design-thinking-powered approach can promote empathic and critical connections between students and their projects, resulting in a stakeholder-centered mindset. This mindset is desirable for TC and UX courses where students learn to advocate for users and stakeholders in socially just and responsible ways. In the end, we offered an adaptable design-thinking framework that may serve as a viable set of heuristics for achieving equilibrium in teaching, service-learning, and community partnership.

Our case study was affected partially by the COVID-19 pandemic and our local community conditions. Also, our data comprised only the student narratives from our respective classes; they are not meant to be generalized across the field. Future studies may explore the impact of design thinking on service-learning pedagogies by honing specific phases of design thinking—empathy, definition, ideation, prototyping, and testing—or by collecting narratives from community partners. Another prospect for this kind of study is the role of technologies in facilitating student learning and their engagement with the projects. TC and design scholars may examine how technologies amplify or limit empathy efforts when addressing wicked problems.

Note

1. We follow Dubinsky's (2002) and Garza's (2013) intentional use of "service-learning" (hyphenated) to place equal emphasis on the service component of the experience and the learning outcomes for the student.

References

Alrubail, R. (2015, June 2). Teaching empathy through design thinking. *Edutopia*. https://www.edutopia.org/blog/teaching-empathy-through-design-thinking-rusul-alrubail

Altman, M., Huang, T. T. K., & Breland, J. Y. (2018). Design thinking in health care. *Preventing Chronic Disease, 15,* E117. https://doi.org/10.5888/pcd15.180128

Bay, J., Johnson-Sheehan, R., & Cook, D. (2018). Design thinking via experiential learning: Thinking like an entrepreneur in technical communication courses. *Programmatic Perspectives 10*(1), 171–200.

Bowden, M., & Scott, J. B. (2003). *Service-learning in technical and professional communication.* Longman.

Bowen, G. A. (2006). Grounded theory and sensitizing concepts. *International Journal of Qualitative Methods, 5*(3), 12–23.

Brown, B. (2011). *The power of vulnerability* [Video]. TED Talks. YouTube. https://www.youtube.com/watch?v=iCvmsMzlF7o

Brown, T. (2008). Design thinking. *Harvard Business Review 6*, 84–92.

Brown, T. (2009). *Change by design: How design thinking transforms organizations and inspires innovation.* HarperCollins.

Buchanan, R. (1992). Wicked problems in design thinking. *Design Issues, 8*(2), 5–21.

Devecchi, A., & Guerrini, L. (2017). Empathy and design: A new perspective. *The Design Journal, 20*(1), S4357-S4364. https://doi.org/10.1080/14606925.2017.1352932

Dubinsky, J. (2002). Service-learning as a path to virtue: The ideal orator in professional communication. *Michigan Journal of Community Service Learning, 8*, 61–74. http://hdl.handle.net/2027/spo.3239521.0008.206

Fontaine, P. (2001). The changing place of empathy in welfare economics. *History of Political Economy, 33*(3), 387–409.

Garza, S. (2013). *Adding to the conversation on service-learning in composition: Taking a closer look.* Fountainhead Press.

Gasparini, A. A. (2015). Perspective and use of empathy in design thinking. *ACHI 2015: The Eighth International Conference on Advances in Computer-Human Interactions* (pp. 49–54). IARIA.

Glaser, B., & Strauss, A. (2012). *The discovery of grounded theory: Strategies for qualitative research.* Aldine Transaction.

Greenson, R. R. (1960). Empathy and its vicissitudes. *International Journal of Psychoanalysis, 41*, 418–424.

Hasso Plattner Institute of Design at Stanford University. (2018). *Design thinking bootleg*, d.mindsets (archived). https://dschool.stanford.edu/resources/design-thinking-bootleg

IBM Design. (2022). *Enterprise design thinking.* https://www.ibm.com/design/approach/design-thinking

Kimme Hea, A. & Shah, R. W. (2016). Silent partners: Developing a critical understanding of community partners in technical communication service-learning pedagogies. *Technical Communication Quarterly, 25*(1), 48–66. https://doi.org/10.1080/10572252.2016.1113727

Koh, J. H. L., Chai, C. S., Wong, B., & Hong, H-Y. (Eds.) (2015). *Design thinking for education: Conceptions and applications in teaching and learning.* Springer.

Köppen, E., & Meinel, C. (2015). Empathy via design thinking: Creation of sense and knowledge. In H. Plattner, C. Meinel, & L. Leifer (Eds.), *Design thinking research: Building innovators* (pp. 15–28). Springer.

McConnell, K. (2020). A design thinking approach to increasing student efficacy in the internship search process. In *Proceedings of ASEE: At Home with*

Engineering Education (pp. 1–15). June 22–26, 2020. Springer. https://peer. asee.org/a-design-thinking-approach-to-increasing-student-efficacy-in-the-internship-search-process.pdf

Norris, J. M., White, D. E., Nowell, L., Mrklas, K., & Stelfox, H. T. (2017). How do stakeholders from multiple hierarchical levels of a large provincial health system define engagement? A qualitative study. *Implementation Science, 12*(98), 1–13. https://doi.org/10.1186/s13012-017-0625-5

Pavlovich, K., & Krahnke, K. (2012). Empathy, connectedness and organisation. *Journal of Business Ethics, 105*(1), 131–137.

Pferdt, F. G. (2019). *Design thinking in 3 steps: How to build a culture of innovation.* Think with Google–Future of Marketing. https://www.thinkwithgoogle.com/future-of-marketing/creativity/design-thinking-principles

Pope-Ruark, R. (2014). A case for metic intelligence in technical and professional communication programs. *Technical Communication Quarterly, 23*(4), 323–340. https://doi.org/10.1080/10572252.2014.942469

Rittel, H., & Webber, M. (1973). Dilemmas in a general theory of planning. *Policy Sciences, 4*(2), 155–169. https://link.springer.com/article/10.1007/BF01405730

Roberts, J. P., Fisher, T. R., Trowbridge, M. J., & Bent, C. (2016). A design thinking framework for healthcare management and innovation. *Healthcare, 4*(1), 11–14. https://doi.org/10.1016/j.hjdsi.2015.12.002

Rogers, C. R. (1975). Empathic: An unappreciated way of being. *Psychologist, 5,* 2–5.

Sharp, C. (2019). How using a design thinking framework can strengthen grant applications. *Medium.* https://medium.com/@c.sharp/how-using-a-design-thinking-framework-can-strengthen-grant-applications-db56b7fc6060

STARS. (2021). *The sustainability tracking, assessment & rating system.* https://stars.aashe.org/resources-support/help-center/academics/sustainability-literacy-assessment/#what-kind-of-questions-address-sustainability-literacy

Tham, J. (2018). Learning from making: A design challenge in technical writing and communication. In L. Angeli & T. K. Fountain (Eds.), *Proceedings of ACM Special Interest Group on Design of Communication (SIGDOC) Conference.* https://dl.acm.org/citation.cfm?id=3233935

Tham, J. (2021a). *Design thinking in technical communication: Solving problems through making and collaboration.* Routledge.

Tham, J. (2021b). Design thinking in first-year composition: Writing social innovation into service-learning pedagogy. *Currents in Teaching and Learning, 13*(1), 11–24.

Thomke, S. H., & Feinberg, B. (2009). *Design thinking and innovation at Apple.* Harvard Business School Case 909-066. https://www.hbs.edu/faculty/Pages/item.aspx?num=36789

Valentine, L., Kroll, T., Bruce, F., Lim, C., & Mountain, R. (2017). Design thinking for social innovation in health care. *The Design Journal, 20*(6), 755–774. https://doi.org/10.1080/14606925.2017.1372926

Wible, S. (2020). Using design thinking to teach creative problem solving in writing courses. *College Composition and Communication, 71*(3), 399–425. https://eric.ed.gov/?id=EJ1310114

Wise, S. (2016, February 8). *Design thinking in education: Empathy, challenge, discovery, and sharing.* Edutopia. https://www.edutopia.org/blog/design-thinking-empathy-challenge-discovery-sharing-susie-wise

Yusof, N., Tengku Ariffin, T. F., Awang-Hashim, R., Nordin, H., & Kaur, A. (2020). Challenges of service learning practices: Student and faculty perspectives from Malaysia. *Malaysian Journal of Learning and Instruction, 17*(2), 279–309.

Chapter 2

TEACH—Standardizing Accessibility with Usability Pedagogy in an Introductory Technical and Professional Communication Course

JESSICA NALANI LEE AND SUSHIL K. OSWAL

Accessibility often is an afterthought in user experience (UX) design and is treated separately from usability. Accessibility testing occurs late in the design and development process and takes the form of check marking boxes or "compliance testing" for web accessibility standards (benchmarking)— Quality Matters (QM), Web Content Accessibility Guidelines (WCAG), or the U.S. Government's Section 508 Standards (Oswal & Melonçon, 2017; Power et al., 2012; Sandhu, 2011; W3C.org, 2018). Accessibility testing for benchmarking demonstrates accessible aspects but overlooks root causes of accessibility problems. Late-stage accessibility testing leaves little time to uncover design flaws due to high professional production costs. Designers and developers often end up retrofitting accessibility features that rarely meet the needs of disabled users (Wentz et al., 2011).

Similarly, curricula in design fields—content development, page layout, and web design—have largely failed to integrate accessibility, with few exceptions (Huntsman, 2021; Sonka et al., 2021). Such curricula in design have taken place with individual initiatives rather than through programmatic change. As reported in Technical and Professional Communication (TPC) pedagogical literature, only a small number of TPC courses have integrated accessibility as core curriculum: e.g., case studies (Browning & Cagle, 2017; Mahaffey & Walden, 2019); closed captions in presentations (Clegg, 2018); disability and accessibility discussions (Zdenek, 2020); integration of accessible designs in web and built environments (Palmer et al., 2021); and introducing accessibility to novice TPC students (Young-blood, 2013). Additionally, researchers have integrated some disability and accessibility discussions in their scholarship: e.g., closed captions (Butler, 2016, 2018); disability and accessibility (St.Amant, 2018); and making disability visible (Wilson, 2000).

Lancaster and King (2024) set up the goal of honoring the "diversity of user uniqueness, customization desires, and social needs" and thus of building balance—equilibriUX—for this collection (p. xiii). In this chapter, we argue for a vision of accessibility as an integral aspect of usability in design practice and pedagogy. We thus advocate for teaching accessibility bound to usability so that we honor disabled and nondisabled users' needs, and both sets of needs are reconciled at the early design stage. If we teach our students to design content, page layout, and UX architecture from this inclusive standpoint, they develop a more holistic and empathic understanding of users' needs. We teach usability as a multifaceted concept that necessarily involves accessibility in a five-point pedagogical framework: TEACH—(1) Teach accessibility and usability as a bound unit, (2) Emphasize empathy for diverse users, (3) Amplify student agency in product/analysis decisions, (4) Contextualize website evaluations with audience and rhetorical awareness, and (5) Highlight need for balancing design decisions by stressing the critical social model of disability. We expand on this five-point pedagogical framework, providing examples from the courses we have taught to illustrate how instructors might apply our framework.

The Intro Service Course and Resisting the "Checklist" Trap

Significant to our curriculum is situating usability and accessibility testing within an introductory or service TPC course. Including accessibility in students' initial impressions of their understanding of technical communication (TC) establishes foundations for students to develop habits of mind that resist the "checklist" trap for achieving accessibility.

Though pedagogical scholarship has emphasized the relevance of disability and accessibility to TPC curriculum (Wilson, 2000), existing literature fails to teach this concept of accessibility as an integral part of usability where our pedagogical framework, TEACH, comes in. For example, Putnam et al. (2016) published a study of teaching accessibility in university-level design programs in the United States. Browning and Cagle (2017) advocated for teaching a case study method in TPC to avoid essentializing and/or othering accessibility best practices. Miller (2019) argued for communicative access as a process of negotiation across individuals and modes than just as a process of reducing barriers. Oswal and Palmer (2022) suggested that we teach students accessible design as a component of our curriculum and reinforced the importance of involving disabled users in this work, thus introducing the relevance of participatory design to integrating accessibility in usability.

Our Five-Point "TEACH" Framework

Our approach to teaching usability and accessibility is distinct from practices noted in the TPC literature review above. One of the defining characteristics of our curriculum is problematizing existing distinctions between usability and accessibility testing methodology. This pedagogical understanding is based on our belief that accessibility cannot be separated from usability in design because disabled users are an integral part of the overall user population. In the subsections that follow, we discuss each of the five points, or goals, of our TEACH framework: (a) defining each of them, (b) explaining why we advocate for these, (c) describing how these goals avoid the "checking the box" trap for accessibility, (d) detailing one curriculum component from our own course to showcase application of the goal, and (e) demonstrating through our course narrative how instructors can adopt these goals—either fully or in piecemeal—and the curriculum they can teach.

ONE: TEACH ACCESSIBILITY AND USABILITY AS A BOUND UNIT

a. Challenging standard practices that consider the accessibility of a product as separate from its usability (i.e., it is possible for a product to be usable but not accessible) is important because there is always a risk of forgetting the disabled users in the design process.

b. Just as in design practice, we advocate accessibility as an integral part of usability pedagogy.

c. Recognizing accessibility as inextricable from usability defies any inclinations to view accessibility as yet another "box to check." Hutter and Lawrence (2018) demonstrated how accessibility is always already bound to usability as they emphasize the principles of accessible design for user-testing protocols when working with deaf users of American Sign Language (ASL) while refraining from descending into audism and technological paternalism.

d. Whereas Hutter and Lawrence (2018) focused on modifying the testing protocol for deaf users, our pedagogical framework goes further in asking students to integrate accessible design and disabled users in the very conceptualization of a website as well as later, during its testing.

e. TPC instructors can highlight the interconnectedness of accessibility and usability by bringing to students' consciousness how they define what is usable versus how they define what is accessible, explicitly asking them: Can a text truly be considered usable if it is not accessible to all users?

Two: Emphasize Empathy for Diverse Users

a. Empathy in the context of disability means being able to see the humanity of this population. Recognizing accessibility as an integral part of usability necessitates empathy for the humanity of diverse users.

b. We believe an attention to the array of forms interaction with a text can take is necessary for motivating accessibility work, as it personifies what can otherwise be seen as yet another "hoop" to jump through.

c. Students need an explanation about why popular checklist-based approaches to accessibility—QM, WCAG, and Universal Design (UD)—do not work. They address the machine side of accessibility but lack the crucial empathy for the users of the accessible products.

d. One way to discuss the identity of disabled users is to bring into class a discussion of adaptive user interfaces, such as screen readers, voice input systems, keyboard-only data input, and other assistive devices. This can assist students to see disabled people as active users of technology and help them develop empathy for the users of such technologies.

e. Reflecting on all audiences' differing user characteristics and the overall ethics of inclusion should also be a tenet of teaching students about accessibility. Instilling empathy for diverse users can begin with self-reflection, teaching students to understand themselves and what impacts them as users, to challenge the assumption of a generic, all-encompassing universal experience.

THREE: AMPLIFY STUDENT AGENCY IN PRODUCT/ANALYSIS DECISIONS

a. Our assessments of accessible usability are not confined to the classroom but rather taught to be communicated to audiences that can affect change. In this way, students experience the agency that comes with communicating what they have learned not just to demonstrate their own knowledge but to influence development in the world around them.

b. Experiencing this agency further guarantees application of inclusive design principles beyond our own course, as students are taught how to express their knowledge to make a difference.

c. Thus, students are motivated to go beyond what is expected by the instructor and promote an idea they believe in.

d. To achieve this goal, students could be asked to identify a specific purpose of the website they are analyzing and then make a judgment as an intermediary between the website designer and a particular subset of the intended audience.

e. Students should have agency in making rhetorical decisions in usability and accessibility work.

FOUR: CONTEXTUALIZE ACCESSIBLE USABILITY EVALUATIONS WITH AN AWARENESS OF THE RHETORICAL SITUATION

a. Fully understanding the context in which users interact with texts is important because nondisabled people have very limited knowledge of disabled users.

b. Students need to learn that the context of use for disabled people can be quite different because of the bodily differences, adaptive technologies, and their physical location.

    c. The goal should be to establish a framework that highlights the importance of context, problematizing a one-size-fits all approach to evaluating function. This specific emphasis on audience's context of use prevents students from over-generalizing users; it also keeps them from check-marking accessibility.

    d. We stress the context of use by asking students first to analyze a website for usability from their own perspective and then from the perspective of disabled users.

    e. Instructors can use any of the details about disabled users in the narrative below to teach context as it is tied to use.

FIVE: HIGHLIGHT NEED FOR BALANCING DESIGN DECISIONS BY STRESSING THE CRITICAL SOCIAL MODEL OF DISABILITY

    a. Understanding disability as rooted in societal structures and constructed environments than in the individual's physical or mental impairment alone is a fundamental premise of a critical social model of disability.

    b. We ground our understanding of accessible usability in the critical social model of disability in direct dissension to utilitarian approaches to accessibility because they do not center the disabled user. Such utilitarian approaches also overlook the exclusionary design of social and physical environments always already infused with a stigma about disability, disabled means of access, and adaptive technology.

    c. Utilitarian approaches are about figuring out the right content to populate the checklist for ensuring accessibility and not about user needs.

    d. One way the critical social model of disability can be included in the curriculum is by teaching Finkelstein's (1975) parable of the disabled village that we discuss later in this chapter.

    e. Grounding accessible usability in the critical social model of disability involves including disability and accessibility narratives in the day-to-day class conversation instead of using it as an additive to the curriculum.

Having described our five-point TEACH framework, we now discuss the design of our course around these concepts that emphasize the need for inclusivity, regard disability as a normal aspect of human life, and show how we give it a place in our curriculum.

Designing an Inclusive TPC Course: An Overview

We designed this course curriculum in a 2-year college setting as a 10-week introductory TPC course. Students enrolling in this course already have completed a required writing course. The accessible usability content appeals to students because it challenges their traditional view of a TPC course. The course described here was taught both face-to-face in-person and fully online asynchronous, depending on student demand, in pre-COVID times. The main differences between an in-person and online course in designing this accessibility-focused pedagogy are roughly the same that instructors deal with in any introductory TPC course—creating a social presence, explaining the syllabus and how this course differs from other courses, giving students a good understanding of the course outcomes, explaining how students will be learning traditional TPC genres—reports, proposals, etc.—while conducting usability and accessibility testing, and most importantly, keeping students on track from week to week.

A Course Showcase Operationalizing our Five-Point TEACH Framework

Note: The details for the narrative below are based on instructor notes and retrospective reflection by both authors on their teaching of this curriculum. The text from a student reflection has been quoted with permission. The same has been approved by the college's Institutional Review Board authority.

The course begins by introducing students to the course theme of accessibility in the syllabus. We spend the first 4 weeks of the course teaching general TPC concepts such as designing a writing plan and writing three professional emails to three different audiences, including learning to handle difficult situations. The overall goal of curricular instruction during the first four weeks is to provide students with a basic grounding in technical writing skills.

Table 2.1 provides a visual of the timeline of the usability and accessibility modules of our introductory TPC course. We will discuss the execution of this content in the next section.

Table 2.1. Timeline and Content for Usability and Accessibility Curriculum

Week	Pedagogical Framework Goal(s) Met	Module	Lectures	Reading and Writing Assignments
Week 5	Four: Contextualizing the Rhetorical Situation	Usability: Analyzing Websites	*Introduction to Usability *Introduction to Positionality	*Reading: "Usability and User Experience Research" (Getto, n.d.) *Discussion Post: Rhetorically Analyzing Websites
Week 6	Three: Amplifying Student Agency; Four; Five: Highlight Critical Social Model of Disability	Usability Testing	*Creating and Conducting a Usability Test	*Reading: "Designing Websites" (Johnson-Sheehan, 2012) *Discussion Post: Creating and Conducting a Usability Test
Week 7	Four	Usability Report	*Writing Your Usability Report	*Major Assignment: Usability Report
Week 8	One: Teach Accessibility and Usability as a Bound Unit; Two: Empathy for Diverse Users Three; Four; Five	Introduction to Accessibility	*From Usability to Accessibility *Experiencing Screen Reader Software *Creating Accessible Documents	*Reading: "The Disabled Village" (Finkelstein, 1975) *Worksheet: Inaccessible Document
Week 9	One; Two; Three; Four	Accessibility Testing	*Introduction to WAVE	*Reading: "Five Stages of Accessibility" (Featherstone, 2011) *Reading: "Guidelines for Accessible and Usable Web Sites: Observing Users Who Work with Screen Readers" (Theofanos & Redish, 2003) *Discussion Post: Accessibility Testing
Week 10	One; Two; Three; Four	Accessibility Report	*Writing Your Accessibility Report	*Major Assignment: Accessibility Report

Source: Created by the author.

INTRODUCING THE CONCEPT OF USABILITY AND USER-EXPERIENCE FOR TESTING

We begin with a working definition of usability as the process of "making sure that any kind of communication deliverable (e.g., a website, a handbook, a user guide) is intuitive, easy-to-use, [sic] and helps users achieve their goals" (Getto, n.d., para. 1). In this week and the next, we work on Goal 4: Awareness of the rhetorical situation to help students understand the general concept of usability and to practice evaluating the usability of a sample website. We have adapted Kynard's (n.d.) online rhetorical tool for analyzing websites for this activity, which helps students understand that websites are multifaceted environments with interactive content, page structures, and web designs. At this point, students also read "Usability and User Experience Research" from Moxley's open-access project (Getto, n.d.). To aid in comprehending this information, we present the material interactively by having students answer questions about the reading. Students also read a brief introduction on positionality by Lee (2016).

Learning to Conduct Usability Testing

During the sixth week, students first review a prerecorded lecture and a real student example scripted in that lecture to understand what it looks like to conduct a usability test. At this time, students also read a chapter on web design (Johnson-Sheehan, 2012). Students learn that, having rhetorically analyzed a website the previous week and listened to the lecture about the process of usability testing, they now have the agency (Goal 3) to create and conduct a usability test. The practical task of usability testing then involves these three steps:

1. **Deciding the type of website they want to test, which includes describing the general purpose their usability test will assess, as well as naming a specific audience.** We use Michael Albers's (2003) heuristic for "multidimensional audience analysis": consideration of the audience's "knowledge dimension, detail dimension, cognitive ability, and social or cultural aspects" (p. 268). Albers's dimensions help them contextualize the rhetorical situation (Goal 4), that is, analyze a website's intended audience to figure out what should be assessed to in turn determine whether the website works for that audience.

2. **Compiling a list of four to seven general traits that make sense given the type of website they want to evaluate.** Students are encouraged to go back to the previous week's discussions that rhetorically analyze websites. While reviewing these posts, students take note of common design and interactivity features for effective as well as ineffective sites. We encourage students to balance their own evaluation of design decisions with the viewpoints of others' who might not share their views (Goal 5). Students are also invited to apply Nielsen's (2012) five factors of judging whether a given communication deliverable is usable or not (Goal 3).

3. **Determining a numeric scale to use for tracing how well a website rates for each of the traits of the specific audience and purpose identified in step 1.** Students can use whatever scale they choose but need to be wary of an odd-numbered scale to avoid defaulting to evaluating criteria at the midpoint on the scale (Goal 3).

The key points students remember when creating a usability test is to make their test specifically relevant to the website's purpose and context of use (Goal 4). For example, assessing the learnability of a site that recruits volunteers for a project might look different than assessing the learnability of a site whose purpose is primarily to inform about a set of issues.

Reporting Usability Results

During the seventh week, students write a report based on the results of this testing. Through different iterations of this course, we have learned that teaching how to contextualize the rhetorical situation (Goal 4) provides a crucial foundation for the accessible usability curriculum moving forward. As we tell our students, you cannot tell if a website is effective at meeting the audience's needs unless the website speaks to a particular audience with a clear purpose.

Introducing and Interpreting Accessibility Concepts for Testing to Meet the Disabled Users' Needs

The eighth week shifts the focus to the main emphasis of the course, which is Goal 5: Highlight the need for balancing design decisions by stressing the critical social model of disability. To help students transition from usability to accessibility, we have students watch a multipart prerecorded lecture with transcript, "From Usability to Accessibility." Students are reminded that this introductory TPC course centers on the concept of accessibility because analyzing accessibility is a concrete way to grasp the fundamental concepts of user-centered webpage design and content development, also stressed in Johnson-Sheehan's website design chapter that they read recently. The lecture also problematizes the checklist approach to accessibility because such checking of accessibility by web designers and developers comes too late, is out of context, and does not particularly show an empathic treatment of disabled users in web development (Goals 2 and 4). Moreover, tackling the accessibility problems at this late stage ignores the fundamental axiom of starting with inclusive design.

Retrofitting accessibility also does not solve all problems and is costly. We stress that accessible content benefits all users, not just users with disabilities, with text messaging cited as just one example. We also make a quick business case for accessibility while emphasizing the ethical design imperative. We introduce students to some examples of disabilities—motor, visual, cognitive, and hearing—that can affect a person from experiencing the Internet the way people without disabilities do. We also stress the importance of hearing from those with disabilities themselves, rather than only relying on usability and accessibility experts. Students have to be primed to adopt the perspective of another by first reflecting on how their own backgrounds/past experiences affect how they view a site. Students are subsequently challenged to reflect on whether they are members of the intended audience and what aspects of their identity constitute that membership. Making students aware of their own identity and its influence on their experiences is to recognize and appreciate other users' identities that might differ from their own.

The lecture then addresses the questions of definitions of accessibility and Mace's concept of Universal Design (Story et al., 1998). Since accessibility is both a legal and technical term, different entities define it according to their location and purpose. Even when accessibility is being defined technically, academics and practitioners widely disagree on what should be included in accessibility. To add to the complexities of this term, disability studies scholars (see Kleege, 2014; Oswal, 2013)—many who are disabled and have an embodied understanding of accessibility barriers—have taken stances that expand the concept of accessibility beyond the legal and technical accessibility to include the social contexts of access barriers. We point out that legal or technical definitions of accessibility fall short of exhibiting a sense of social obligation or empathy toward disabled users (Goal 2).

To amplify the perceptions of accessibility by different entities, we rely on several definitions of accessibility from United States regulation agencies—the U.S. Department of Justice and U.S. Access Board—as well as definitions from other entities. We also discuss the legal positions taken by the U.S. courts. The Americans with Disabilities Act National Network (2022) has defined "accessible information technology" as "technology that can be used by people with a wide range of abilities and disabilities. It allows each user to interact with the technology in ways that work best for them." We share with students that the overly generalized language in this definition waters down accessibility through this vague description and does not serve any of the disability groups adequately (Goal 3). We then explain to students that the Section 508 definition is more direct in emphasizing the specific needs of users with sensory disabilities who are often overlooked in web design:

> An accessible information technology system is one that can be operated in a variety of ways and does not rely on a single sense or ability of the user. For example, a system that provides output only in visual format may not be accessible to people with visual impairments and a system that provides output only in audio format may not be accessible to people who are deaf or hard of hearing. Some individuals with disabilities may need accessibility-related software or peripheral devices in order to use systems that comply with Section 508. (ADA.gov, 2020)

We further deepen this discussion by sharing that in the South Carolina Technical College System (2013) decision, the court stated that

> "Accessible" means a person with a disability is afforded the opportunity to acquire the same information, engage in the same interactions, and enjoy the same services as a person without a disability in an equally effective and equally integrated manner, with substantially equivalent ease of use. The person with a disability must be able to obtain the information as fully, equally, and independently as a person without a disability. Although this might not result in identical ease

of use compared to that of persons without disabilities, it still must ensure equal opportunity to the educational benefits and opportunities afforded by the technology and equal treatment in the use of such technology.

After highlighting the education focus of this definition, we then explicate the Royal National Institute of Blind People's (RNIB) definition which stresses the concept of an accessible, multimodal website. RNIB has described "design for all" in relation to websites as "a single version of the Web site which is accessible to everyone" and that "well designed graphics and multimedia are a positive aid to using and understanding websites, and do not need to be sacrificed for accessibility" (Royal National Institute of Blind People, 2005). This detailed explication of the implementation is particularly relevant in this discussion to make the connection that, unlike RNIB, the U.S. Congress did not define web accessibility in the original statute in 1990 because the web was not yet born, but it also did not include web or web accessibility in the 2009 revision of the act (Americans with Disabilities Act, 1990; ADA.gov, 2020). Students learn that the outcome of this exclusion of web accessibility from the U.S. statutes is that the advocates can argue for accessibility only with favorable court decisions and ethical reasoning. We note that not all accessibility design experts understand that "accessibility is an experiential measure of quality; it is less a property of the Web content but rather a result of the interplay between the Web content, the browser, and potentially the assistive technology that some people with disabilities may be using to access the content" (Abou-Zahra, 2008, p. 103). This discussion helps us center the concept of UX in terms of accessible usability (Goal 1).

In addition to complicating what it means for a web page to be accessible, we also challenge students to critically question the concept of universal design when implemented in different contexts (Goal 4). We explain that universal design appears attractive in conceptual terms but has different implications for different settings. For example, on the one hand, in the context of built environments, it can often be the only accessibility solution where the infrastructure has to be designed for decades. On the other hand, websites are transient. Unfortunately, when universal design is implemented in web environments, it often ends up watering down accessibility and does not necessarily meet the needs of users with sensory disabilities. We explain that blind users are the ones most affected by inaccessible websites and that the six most common accessibility errors found on the top one million websites create barriers for this user group (WebAIM, 2022). One of these errors was present on 83% to 85% of these one million websites during the past 4 years. Because the HTML5 coding in the World Wide Web 2.0 environment is overwhelmingly visual and employs many sensory tools—audio, video, animation, and so on—it creates many accessibility barriers for users with visual, auditory, and cognitive disabilities. Using the example of websites coded in HTML5, we show how these attractive-looking pages do not necessarily match Berners-Lee's vision: "The power of the Web is in its universality. Access by everyone regardless of disability is an essential aspect" (Berners-Lee, 1997). We communicate that accessibility is not a monolithic concept and is also not a static concept. It changes with new technologies and the marketplace. Both legal and industry environments also affect how accessibility is implemented by government entities, educational institutions, and business and can create barriers to building accessible environments for everyone (Goal 5).

Enacting Accessibility from the Perspective of Disabled Users

At this point, we introduce our students to the work of the influential South African disability activist and writer Finkelstein (1975). We do this to help students understand how accessibility implementation by our governments, educational institutions, and industry might be well-meaning but not necessarily in line with what disabled users expect in accessibility (Goal 2). Finkelstein's short fictional narrative depicts a community of wheelchair users that has systematically organized a social and physical environment to function fully as productive community members. Finkelstein, however, deliberately makes this community an exclusive one, which does not consider the needs of nondisabled walking people. Therefore, the entire environment has been designed for wheelchair riders. Finkelstein stages his message about human society being so exclusionary to disabled people by bringing a nondisabled person into the village who bumps his head into everything because the wheelchair-height doors do not consider walking people's needs. To address the height problems of the nondisabled, Finkelstein's villagers design solutions from the perspective of wheelchair riders with an utter disregard for the needs of this nondisabled visitor.

Students find this narrative particularly thought provoking, as it poignantly highlights the absurdity of institutionally imposed accommodations on disabled users. They also realize the ethical imperative of including disabled users as active participants in design activity in conceptualizing and constructing social and physical environments (Goal 5).

Moving from Accessibility Theory to Practice

To engage students in the application of this theory, we teach students how to make an accessible Microsoft Word document for those who use screen readers and keyboard-only users (Goal 3). To accomplish this task, they receive the instructions for creating accessible Word documents through a prerecorded lecture with transcript. The second lecture of this eighth week gives students a chance to experience screen-reader software by viewing two separate before-and-after video clips. The first video demonstrates the screen reader software, JAWS, reading a syllabus that has not been formatted for accessibility. Then, to show how the accessibility features work in Word through special formatting, the second video recording demonstrates JAWS reading the formatted version of the syllabus.

Students practice formatting for accessibility with an inaccessible Microsoft Word document, and they revise it to be more accessible and usable for blind readers by using Microsoft's 2020 accessible Word documents guidelines (Goal 1). For this assignment, students achieve these five learning outcomes: (1) formatting headings so that the documents are easy to navigate with a screen reader <H> command, (2) creating alternative text descriptions of images for screen readers, (3) formatting a table that reads column headings with the row data so that the screen-reader user can visualize the table structure just by listening, (4) formatting lists so that a screen reader could read it accurately and effectively, and (5) creating links with meaningful labels that describe the content of the target page. This is a challenging assignment for many students because doing accessible design is new to them and this formatting exercise requires some understanding of how web pages are coded for making the content accessible to users employing user interfaces, such as screen readers, braille displays, keyboard-only interactions, and voice input.

If students do not own Microsoft Word, they are directed to use Office 365, which is free to students when they use their college email to sign up. Students are also invited to download the NonVisual Desktop Access (NVDA), a free screen reader from Australia (NVDA, 2022). Students conclude this exercise with a feedback form that invites them to request additional information on the topics discussed so that we could make those resources available (Goal 3).

This segment of the course is somewhat theory heavy, but it also shows students how complicated the notion of accessibility is. Students learn how different entities involved in accessibility matters—legislators, judges, executive branches of the government, educational institutions, and industry—might interpret the concept differently. Students also learn that not all interpretations of accessibility serve the needs of disabled users. Additionally, our accessibility exercise engages them in a serious practicum with the Microsoft Word document, which helps students understand that designing for accessibility does not always require a degree in web development or computer science, but it does require an understanding of how and why different elements in Word or a web page need to be coded for accessibility (Goal 4).

Learning to Conduct Accessibility Testing

Students begin the ninth week by interactively reading Featherstone's (2011) "Five stages of accessibility," which wittily compares the five phases many web developers go through before they are willing to incorporate accessibility into their work with Kubler-Ross's five stages of grief. We have students read Featherstone because his spoof cogently underlines that accessibility is something that most people desire in a website and is not an unachievable goal for designers and developers. At this stage, students also read Theofanos and Redish's (2003) study of screen-reader users, which highlights some key problems found in the manual testing of a website. They also offer a handy numbered list of guidelines for making websites accessible, which students find easy to remember.

To teach students how to begin assessing the interplay between assistive technology, web content, and the browser for usability, this week's curriculum also introduces students to Web Accessibility Evaluation Tool (WAVE), which is an automated web tool for analyzing web pages and identifying various accessibility issues (WAVE, 2022). WAVE identifies both website problems and website features. The problems are divided into two major categories: errors (egregious accessibility barriers) and alerts (problems affecting usability and accessibility). WAVE also highlights the design features of a webpage—revealing its structure—including what accessibility markup has been employed. WAVE's strength is in its ability to categorize these errors and alerts according to the three levels of WCAG 2.1. WAVE also provides pointers for remediating these problems. Students run their own test to learn what these problems mean (Goal 3).

In this automated testing experience with WAVE, students draw on Theofanos and Redish's (2003) reading for their discussion post this week, where they conduct a WAVE test on the same website that they previously analyzed for usability. To keep their analysis manageable, students focus on five errors from the automatically generated WAVE accessibility report. In their primary discussion posts, students must:

1. **Describe the errors, making sure to include detail on the context of the error in relation to the purpose of the website overall.** Asking students to pay particular attention to where the error appears on the website is meant to highlight accessibility as a facet of usability because these observations are meant to lay the foundation for understanding how accessibility errors can impede the overall purpose of the website (Goal 1).

2. **Name a matching guideline from the Theofanos and Redish (2003) reading for each of the five errors they list and explain how the guideline can be used to fix what WAVE has detected.** If students cannot find an applicable guideline, they explain how the error can be fixed in their own words, or ask the instructor about it in their post (Goal 4).

3. **Reflect on the website's overall accessibility.** Students explain their reaction to the number of errors WAVE found on the site and compare this evaluation to how they scored the same website for their usability test (Goal 2).

For their secondary posts, students respond to the primary post of a classmate by either agreeing or disagreeing with the guidelines their classmate recommended to fix what WAVE has detected, explaining their reasoning. They also inform their peer what guidelines they would recommend if their judgment is different from that of their peers. The week ends with students completing a feedback form inviting them to request more information on the week's topics of discussion.

In their final reflections on the course, students often cite this week as particularly exciting for them because many have never used a software program like WAVE before. Some confess to never having considered how the web would be navigated with ears (as opposed to eyes), or that the webpage itself needs to be designed to smoothly interact with an adaptive user interface like NVDA. This week's additional aim is to help students inhabit another person's perspective; thus, highlighting a key outcome of the course—showing empathy for the audience and helping them extract desired information quickly and easily (Goal 2).

Reporting Accessibility Results

During the tenth week, students finish the accessible usability curriculum by writing an accessibility report to the webmaster of the nonprofit website they had chosen for this testing. This report is either about the same website they analyzed for their usability report or a completely new nonprofit website. This option is provided to promote student agency (Goal 3). Students start with viewing a prerecorded lecture with transcript that deconstructs a previous student's accessibility report as an example to help them write their own report. A primary requirement for this accessibility report is that they present a compelling case to the webmaster that the website needs to be accessible and usable to accomplish a specific purpose because even 33 years after the passage of the Americans with Disabilities Act (1990) and thousands of web accessibility suits, most web developers and webmasters do not consider disabled users and their accessibility needs while designing and maintaining web pages and content (Goal 1).

We ask students to make the case to their audience for improving both usability and accessibility in relation to the most important purpose of the site; that is, providing the best user experience to all visitors to the site. After completing this accessibility report, students compose a concluding reflection on how working on the usability and accessibility projects has related to these core course outcomes: (1) produce writing that analyzes and evaluates how meaning is made in a particular context (Goal 4); (2) design and produce communications tailored to a specific audience, purpose, and the context-of-use (Goal 4); and (3) use the report genre features that show empathy for the audience and help them extract desired information quickly and easily (Goal 2).

TAKING STOCK OF WHAT WE LEARNED ABOUT TEACHING USABILITY AND ACCESSIBILITY

As our pedagogical narrative shows, by teaching accessibility as a bound concept with usability, we not only pedagogically center accessibility, but we also center disabled users as audiences of our students' work. This centering also helps us articulate all the five goals of our TEACH framework because disability, accessibility, and accessible usability all dovetail quite well with the underlying design value of accessible usability enshrined in this framework.

One of the questions this collection starts with asks, "In a research-driven field, how can TPC education prepare professionals-in-training to practice empathy for understanding diverse user experiences and to design for inclusion (for all voices) in their products?" (Lancaster & King, 2024, p. xv). This concept of empathy is implicit in different aspects of our accessible usability curriculum. Teaching an introduction to TPC curriculum that focuses on accessibility prepares TPC professionals-in-training to practice empathy and to establish balance. We teach students to take their disabled users seriously by helping them develop effective skills and strategies for addressing this specific, yet diverse, audience. Further, we stress the notion of balancing design (equilibriUX) with two design foci—accessibility and usability—that are so often separated in curriculum. Our goal in keeping accessibility and usability together is to create web development environments that in practice speak to the critical social model of disability, which demands that disabled people be integrated into society, including its web and built environments. Not too long ago, "usable" focused almost exclusively on design for the so-called normal bodies. Some highly visible cases for violations of disability laws against commercial entities such as Target and Domino's Pizza, as well as educational institutions like Harvard, MIT, and Pennsylvania State universities, have forced business and educational organizations to consider accessibility. Creating accessible and usable websites is as much a responsibility of business and industry as it is of those who educate the future web content developers and designers for these businesses. We have found that an accessibility-focused usability curriculum does help students develop a consciousness about the reality that users are diverse, and therefore accessibility has to be an integral part of usability. When we add accessibility early to our web page design, content, or structure, we also help students practice rhetorical analysis to write inclusively.

ENACTING A NEW TECHNÊ

Learning about usability in the context of disability also uncovers the interdisciplinary nature of TPC for our students. Hawk (2004) described this interdisciplinary interrelatedness as the "set of techniques that allows teachers to open up or reveal elements of a situation in a way that allows students to interact with and live in that distributed, ambient environment" (p. 385). Hawk asserted that TPC instructors should attempt at making "workplace contexts show up as sites of invention" (p. 385), and this post-technê can be manifested by teaching students to be sensitive to the ways in which various technologies must interact in the context of accessibility. Students need to learn that adaptive technologies—screen readers and voice-input systems—are add-on technologies necessitated by the inadequacies of the web user interfaces, such as Google Chrome and Microsoft Edge. Had the designers of these user interfaces integrated the needs of disabled users in their conceptualization of these browsers—several of the Apple products presently do so—we wouldn't have the need for screen readers and voice input systems. Complicating an understanding of usability through a consideration of the problematic nature of intertwining technologies for accessibility and calling out how developers often exclude disabled users from their designs can in turn incite a critical consciousness among our students about the ubiquitous presence of inaccessible technologies today.

As we recorded in the accessibility unit, students read a screen-reader user study by Theofanos and Redish (2003), in which the authors pointed out that, "to truly meet the needs of all users, it is not enough to have guidelines that are based on technology. It is also necessary to understand the users and how they work with their tools" (pp. 22–23). We stress to the students that the distinction Theofanos and Redish made is crucial, explaining that it is the difference between what can legally pass for accessibility online—checking compliance boxes—and how such so-called accessible online tools must actually function to be usable for those with disabilities. Our curriculum also reflects a pedagogy that already embodies the situatedness students must consider to be effective communicators and designers. This focus on web design allows for an overarching awareness of the many moving parts that must work together seamlessly for a web text to be accessible and usable.

As a correlative to this design idea, ending the course with an action-oriented proposal to the webmaster provides a nice way to highlight the "real-life" context in which genres are practiced. Students can use their newly learned knowledge about TPC genres and usability and accessibility for an activist cause beyond this course. As Oswal and Palmer (2022) argued, we believe that advocacy alone is not enough; we need to teach our students that disabled users have the same rights as all other users and consumers do and that students can become their allies for these rights through accessible usability.

Problematizing an Ideology of Normalcy

As we stated in our introduction and demonstrated in our pedagogy of usability and accessibility testing, accessibility is introduced only after we teach usability testing. This initial separation of accessibility from usability is deliberate and is meant to highlight a way of knowing that is obscure to some students. That is, we point out that most likely nondisabled students have been assessing the usability of texts by virtue of being a person without disabilities who interacts in the world. However, a key component of usability, accessibility, is often concealed to those who are not reliant on the adaptive technologies needed by others to experience a text. This revelation of their previous ignorance has proven to be particularly impactful for students. One of the students in his course reflection wrote:

> One component that was especially mind-blowing to me was the accessibility aspect of the class. The idea that someone was using the internet without being able to see was something that never occurred to me because I've never met anyone that was using screen readers to "see" the internet. Frankly, I was disappointed in my own ignorance since I like to think that I'm relatively aware of my surroundings, but I took for granted my own eyesight and the way I use the internet. The very idea of the internet is it's supposed to be a resource for anyone, and anyone means everybody, but I never considered those that could not see so this was an eye-opening revelation to me.

This unconscious assumption that one's own experience will be comparable to the majority's experience can easily go unchecked when products are designed to match one's own ways of interacting with the world. Students discover that what they initially assessed to be a usable site does not always consider the needs of users who employ adaptive technology to interact with digital environments. When students learn these two concepts in tandem—not concurrently—the shock of realizing their exclusion of disabled users during their usability testing of websites becomes a memorable lesson in inclusive design.

Perhaps one of the most frequent student comments on our accessible usability curriculum is some variation of the sentiment—"I enjoyed this class; it was completely different than what I expected it would be." It is an ambiguous statement, in some ways gratifying to know the content is mostly well received but in other ways curious for what is left unsaid. Is the enjoyment the result of a curriculum that defies stereotypical assumptions about TPC? Or is the enjoyment simply a consequence of encountering the unexpected? Whatever the case may be, we think it is vital to enact this pedagogy of accessibility we teach because it permits us to practice socially responsible TPC.

Overall Takeaways from this Course Design

1. In design, accessibility and usability need to be conceptualized side-by-side for inclusive treatment of all users.

2. Socially aware pedagogy also needs to teach accessibility and usability as a bound and contextualized concept.

3. This curriculum can be adapted for any introductory level TPC course irrespective of the location of the instructor in a high school or 2-/4-year institution because it bestows agency on students to explore, analyze, and evaluate users' needs and their contexts-of-use.

4. By stretching the boundaries of UX designs to include disabled populations, we add another 15% world-wide market segment that has remained underserved since the emergence of net-based commerce (Biddle, 2013).

5. If we highlight accessibility in our curriculum consistently, TPC can fashion itself as an accessible UX field that brings together the elements of human-computer interactions, content readability, and inclusively structured digital spaces where all users can come together.

6. Juxtaposing the definition of TPC with accessible usability at a nascent stage of learning technical communication principles can shape students' understanding of the field as a socially invested activity which also needs to be practiced in the workplace after they graduate.

7. To honor the multifaceted nature of individuals and the variety of their needs, and thus create balance, we must allot sufficient space in our lectures to define accessibility and all the contextual complexities that accompany it and also to give students agency to explore these differences through strategically designed assessment.

8. To embody the accessibility and empathy for disabled users we teach, we must be available to learn alongside our students and be open to confronting unforeseen circumstances, whether they be technical troubleshooting or institutional barriers.

Conclusion

By integrating accessibility into the usability and UX discourse, we exhibit how TPC instructors can work to center the marginalized voices our field wants to promote (Moeller, 2018). Our accessible usability TEACH framework for teaching usability and accessibility testing builds unifying experiences for a variety of users who access worldwide web employing diverse and adaptive user interfaces.

Lancaster and King (2024) sought to honor diverse users in this collection. Our chapter presents an ambitious curriculum for an introductory TPC course with the vision of a digital environment that everyone in our society could access, use, and enjoy. We are confident that if our colleagues in TPC adopt the proposed purpose and audience focused framework to accessible usability testing, they can introduce the concepts of inclusive design to their students, drill the concept that accessible usability and user experience is central to UX, and provide students with a foundation for user-centered design orientation. Once students have a basic understanding of why accessible UX is crucial for designers and have firsthand experience of accessibility barriers through usability and accessibility testing, they might become accessibility evangelists who will push for resources and designs in their workplaces to serve all users. As far as accessible UX designs are concerned, what is most important to users is that our students do not perceive disability as an anomaly but as a normal fact of life and living. If these designers of the future create a variety of designs conceptualized on a continuum of ability and these designs are implemented by developers for diverse bodies in their peculiar contexts of use employing the appropriate affordances of the product, the resulting user experiences will nevertheless be more satisfying than uniform factory designs.

References

Abou-Zahra, S. (2008). Web accessibility and guidelines. In S. Harper & Y. Yesilada (Eds.), *Web accessibility: A foundation for research* (pp. 79–106). Springer.

ADA.gov. (2020, February). *Rehabilitation Act. A guide to disability rights laws.* https://www.ada.gov/cguide.htm#anchor65610

Albers, M. J. (2003). Multidimensional audience analysis for dynamic information. *Journal of Technical Writing & Communication, 33*(3), 263–279. https://doi.org/10.2190/6KJN-95QV-JMD3-E5EE

Americans with Disabilities Act of 1990, 42 U.S.C. §12101 et seq. (1990).

Americans with Disabilities Act National Network. (2022). *Accessible information technology*. In Glossary of ADA Terms. ADA National Network. https://adata.org/glossary-terms#A

Berners-Lee, T. (1997). *World Wide Web consortium launches international program office for web accessibility initiative.* https://www.w3.org/Press/IPO-announce

Biddle, T. (2013). *User testing for web accessibility.* http://rss2.com/feeds/Six-Revisions/72

Browning, E. R. & Cagle, L. E. (2017). Teaching a "critical accessibility case study": Developing disability studies curricula for the technical communication classroom. *Journal of Technical Writing & Communication, 47*(4), 440–463. https://doi.org/10.1177/0047281616646750

Butler, J. (2016). Where access meets multimodality: The case of ASL music videos. *Kairos: A Journal of Rhetoric, Technology, and Pedagogy, 21*(1).

Butler, J. (2018). Embodied captions in multimodal pedagogies. *Composition Forum, 39.* https://compositionforum.com/issue/39/captions.php

Clegg, G. M. (2018). Unheard complaints: Integrating captioning into business and professional communication presentations. *Business and Professional Communication Quarterly, 81*(1), 100–122. https://doi.org/10.1177/2329490617748710

Featherstone, D. (2011, November 11). *Five stages of accessibility.* http://simply-accessible.com/article/five-stages-of-accessibility

Finkelstein, V. (1975). *To deny or not to deny disability—What is disability?* Independent Living Institute. https://www.independentliving.org/docs1/finkelstein.html

Getto, G. (n.d.) *Usability and user experience research.* https://writingcommons.org/section/research/research-methods/empirical-research-primary-research-scientific-research/usability-and-user-experience-research

Hawk, B. (2004). Toward a post-technê—or, inventing pedagogies for professional writing. *Technical Communication Quarterly, 13*(4), 371–392. https://doi.org/10.1207/s15427625tcq1304_2

Huntsman, S. (2021). Addressing workplace accessibility practices through technical communication research methods: One size does not fit all. *IEEE Transactions on Professional Communication, 64*(3), 221–234. https://doi.org/10.1109/TPC.2021.3094036

Hutter, L., & Lawrence, H. M. (2018). Promoting inclusive and accessible design in usability testing: A teaching case with users who are deaf. *Communication Design Quarterly, 6*(2), 21–30. https://doi.org/10.1145/3282665.3282668

Johnson-Sheehan, R. (2012). *Technical communication today* (4th ed.). Longman.

Kleege, G. (2014). What does dance do, and who says so? Some thoughts on blind access to dance performance. *British Journal of Visual Impairment, 32*(1), 7–13. https://doi.org/10.1177/0264619613512568

Kynard, C. (n.d.). *Rhetorical analysis of favorite websites*. Digi Rhetorics. http://www.digirhetorics.org/rhetorical-analysis-of-favorite-websites.html

Lancaster, A., & King, C. S. T. (Eds.). (2024). *Amplifying voices in UX: Balancing design and user needs in technical communication*. State University of New York Press.

Lee, J. (2016). *What is positionality?* Jessicanalanilee.wordpress.com

Mahaffey, C., & Walden, A. C. (2019). #teachingbydesign: Complicating accessibility in the tech-mediated classroom. In K. Becnel (Ed.), *Emerging technologies in virtual learning environments* (pp. 38–66). IGI Global.

Microsoft. (2020). Make your Word documents accessible to people with disabilities. https://support.microsoft.com/en-us/office/make-your-word-documents-accessible-to-people-with-disabilities-d9bf3683-87ac-47ea-b91a-78dcacb3c66d

Miller, E. L. (2019). Negotiating communicative access in practice: A study of a memoir group for people with aphasia. *Written Communication, 36*(2), 197–230.

Moeller, M. E. (2018). Advocacy engagement, medical rhetoric, and expediency: Teaching technical communication in the age of altruism. In A. M. Haas & M. F. Eble (Eds.), *Key theoretical frameworks: Teaching technical communication in the twenty-first century* (pp. 212–240). Utah State University Press.

Nielsen, J. (2012). *Usability 101: Introduction to usability*. NN/g Nielsen Norman Group. https://www.nngroup.com/articles/usability-101-introduction-to-usability

NVDA. (2022). NonVisual desktop access. NV Access. https://www.nvaccess.org/download

Oswal, S. K. (2013). Ableism: Multimodality in motion. *Kairos: A Journal of Rhetoric, Technology, and Pedagogy 18*(1). https://kairos.technorhetoric.net/18.1/coverweb/yergeau-et-al

Oswal, S. K., & Melonçon, L. (2017). Saying no to the checklist: Shifting from an ideology of normalcy to an ideology of inclusion in online writing instruction. *WPA: Writing Program Administration-Journal of the Council of Writing Program Administrators, 40*(3), 61–77.

Oswal, S. K., & Palmer, Z. (2022). A critique of disability and accessibility research in technical communication through the models of emancipatory disability research paradigm and participatory scholarship. In Schreiber & L. Melonçon (Eds.), *Assembling critical components: A framework for sustaining technical and professional communication* (pp. 243–267). The WAC Clearinghouse; University Press of Colorado.

Palmer, Z. B., Oswal, S. K., & Koris, R. (2021). Reimagining business planning, accessibility, and web design instruction: A stacked interdisciplinary collaboration across national boundaries. *Journal of Technical Writing and Communication, 51*(4), 429–467. https://doi.org/10.1177/0047281620966990

Power, C., Freire, A., Petrie, H., & Swallow, D. (2012). Guidelines are only half of the story: Accessibility problems encountered by blind users on the web. In *Proceedings of the SIGCHI conference on human factors in computing systems* (pp. 433–442). https://doi.org/10.1145/2207676.2207736

Putnam, C., Dahman, M., Rose E., Cheng J., & Bradford, G. (2016). Best practices for teaching accessibility in university classrooms: Cultivating awareness, understanding, and appreciation for diverse users. *ACM Transactions on Accessible Computing, 8*(4), 1–26. https://doi.org/10.1145/2831424

Royal National Institute of Blind People. (2005). *Background on web accessibility—Web access centre.* https://www.rnib.org.uk/advice/home-leisure-personal-care

Sandhu, J. (2011). The rhinoceros syndrome: A contrarian view of universal design. In W. Preiser & K. Smith (Eds.), *Universal design handbook* (2nd ed., pp. 44.43–44.11), McGraw-Hill.

Sonka, K., McArdle, C., & Potts, L. (2021). Finding a teaching A11y: Designing an accessibility-centered pedagogy. *IEEE Transactions on Professional Communication, 64*(3), 254–274.

Story, M. F., Mueller, J. L., & Mace, R. L. (1998). *The universal design file: Designing for people of all ages and abilities.* NC State University.

South Carolina Technical College System. (2013). *Resolution agreement: South Carolina Technical College System, OCR compliance review No. 11-11-6002.* https://www2.ed.gov/about/offices/list/ocr/docs/investigations/11116002-b.pdf

St.Amant, K. (2018). Reflexes, reactions, and usability: Examining how prototypes of place can enhance UXD practices. *Communication Design Quarterly Review, 6*(1), 45–53. https://doi.org/10.1145/3230970.3230976

Theofanos, M. F., & Redish, J. (2003). Guidelines for accessible and usable web sites: Observing users who work with screen readers. *Interactions, (X)*6, 38–51. https://redish.net/wp-content/uploads/Theorfanos_Redish_InteractionsPaperAuthorsVer.pdf

U.S. Access Board. (n.d.) https://www.access-board.gov

U.S. Department of Justice. (2020, February). *A guide to disability rights laws.* ADA.gov. https://www.ada.gov/cguide.htm

Web Accessibility Evaluation Tool. (2022). *Web accessibility in mind.* Utah State University. https://wave.webaim.org

W3C.org. (2018). *Web content accessibility guidelines (WCAG) 2.1.* http://www.w3.org/TR/WCAG

WebAIM. (2022). *The WebAIM million: The 2022 report on the accessibility of the top 1,000,000 home pages.* https://webaim.org/projects/million/#maincontent

Wentz, B., Jaeger, P. T., & Lazar, J. (2011). Retrofitting accessibility: The legal inequality of after-the-fact online access for persons with disabilities in the United States. *First Monday, 16*(11). https://doi.org/10.5210/fm.v16i11.3666

Wilson, J. C. (2000). Making disability visible: How disability studies might transform the medical and science writing classroom. *Technical Communication Quarterly, 9*(2), 149–161. https://doi.org/10.1080/10572250009364691

Youngblood, S. A. (2013). Communicating web accessibility to the novice developer from user experience to application. *Journal of Business and Technical Communication, 27*(2), 209–232. https://doi.org/10.1177/1050651912458924

Zdenek, S. (2020). Transforming access and inclusion in composition studies and technical communication. *College English, 82*(5), 536–544.

Chapter 3

Centering Disability in Digital Classes by Applying Disability Justice to User-Experience and Universal Design Practices

KRISTIN BENNETT

Recognizing the potential capacity for digital class design to reinforce ableism, Technical and Professional Communication (TPC) scholars have called for the application of disability studies[1] in the (re)design of more equitable digital courses (Oswal & Melonçon, 2014, 2017). Disability studies rejects notions of disability as lack and understands disability as an embodied experience and a relational phenomenon fostered by a bodymind's[2] (in)capacity to engage with certain spaces, technologies, and practices as they are designed. For example, previous TPC work has recognized the value of applying universal design (UD), a disability studies concept that prioritizes disabled users within design, to TPC user-experience (UX) practices to center disabled experiences (Oswal, 2019).

Although TPC's application of disability studies has proven beneficial to fostering accessible pedagogical design practices that prioritize disability (Hitt, 2018; Walters, 2010), it has not fully addressed how ableism may systemically influence digital course design. Disability justice thus complements disability studies by examining ableism as a systemic problem requiring social transformation led by disabled individuals (Berne et al., 2018). TPC scholars have applied disability justice in decisions related to curriculum design (Bennett, 2022; Wheeler, 2018). Yet, TPC has not applied the combination of UX and UD to specifically trace and challenge ableism's systemic influence in digital course design. This chapter thus illuminates the theoretical connections between UX, UD, and disability justice to demonstrate how instructors might design more equitable and socially just digital courses.

The following sections briefly situate the study in relevant literature, offer an overview of the chapter's methods and findings, and ultimately provide data-driven recommendations for instructors and designers of digital courses to prioritize disability justice through attention to UX and UD. Throughout this chapter, I use disability-first language (i.e., "disabled individuals") rather than person-first language (i.e., "individuals with disabilities") to prioritize disability as a desirable aspect of personhood. Person-first language may reinforce ableist notions that one is a person despite one's disability (Cherney, 2019). I thus use disability-first language to promote disability justice's framing of disability as innately human.

Literature Review

Historically, online classes have been credited for removing normative barriers through accessible tools like screen-reading technologies, asynchronous conversations, and flexible attendance (Kent, 2015). Broadly, accessibility reflects how well individuals "from a population with the widest range of characteristics and capabilities" can attain "specified goal[s]" using systems and products in certain contexts (ISO, 2014). Universities specifically have a "legal obligation" to facilitate access for disabled students by adjusting their "practices, facilities, or services" through individualized accommodations that allow these students to participate equally (Jung, 2003, p. 92). However, rather than transforming normative structures to include disabled folks, such practices often strengthen ableist systems by encouraging disabled individuals to align with able styles of engagement in spaces like digital class settings (Dolmage, 2017).

Previous TPC scholarship has used a combination of disability studies, UD, and UX to evaluate the capacity for digital classroom interfaces to facilitate access for disabled student users. Rather than something that is achieved, access functions as a complex "information ecology" that is continuously negotiated by individuals across diverse networks that "mediate . . . user experience" (Spinuzzi, 2007, p. 198). To better understand users' negotiations in digital courses, many TPC scholars have applied UX (Greer & Harris, 2018; Oswal, 2019; Petrie & Bevan, 2009). Although usability examines individuals' satisfaction with a product or environment's capacity to help them reach "specified goals with effectiveness, efficiency, and satisfaction" (ISO, 1998), UX analyzes both "task-oriented aspects and other non-task-oriented aspects . . . of eSystem use and possession, such as beauty, challenge, stimulation and self-expression" (Petrie & Bevan, 2009, p. 4). That is, usability testing assesses a user's perception of a digital interface's "effectiveness" in task completion, but UX evaluates the user's "subjective reactions," "perceptions," and "interaction[s]" (p. 4) when engaging with digital interfaces.

Privileging users as practitioners, producers, and citizens in technological spaces, UX helps us to consider how individuals engage with technology in unique ways (Johnson, 1998, p. 64). UX can thus aid designers of digital spaces in tracing "the complexities of human-technology relations" and their impact on individual agency (Clinkenbeard, 2020, p. 118). Specifically, UX has been applied in higher educational contexts through attention to a "student-experience mindset," which evaluates and redesigns technology in "ongoing" ways based on students' diversely embodied needs and experiences (Greer & Harris, 2018, p. 17). Despite UX's equitable potential, the field has been criticized for disregarding disabled individuals as passive users of technologies that are designed by able-bodied individuals (Melonçon, 2018; Oswal, 2019; Palmer et al., 2019). To counter these ableist assumptions, TPC has begun to integrate UX with UD. A design strategy advocated for by many disability studies scholars, UD anticipates as many bodyminds as possible at the forefront of design to promote access for all (Hitt, 2018). Specifically, UD challenges the concept of neutral design by investigating how normative design practices may privilege able bodyminds.

As does UD for general design, universal design for learning (UDL) rejects standardized, ableist pedagogies by prioritizing flexible teaching practices (Hitt, 2018) that "reduce barriers in the learning environment" (Wilson, 2017). Specifically, UDL urges instructors to provide course content across diverse media, to offer students multiple means for class participation, and to provide students with various ways to demonstrate knowledge (Dolmage, 2017, p. 145). To further UDL efforts, Dolmage (2017) specifically advocates for attention to the "five levels of deep accessibility." A concept theorized by Star Ford (in Dolmage, 2017), the five levels of deep accessibility promote design practices beyond wheelchair accessibility that also consider "sensory" and "cognitive" differences. Specifically, this concept encourages designers to evaluate how users move, feel, navigate, communicate, and actively participate in a specific space. Such considerations frame the access inspired by UDL as "transformative," or as an active, ongoing "process" in which designers and users of space collectively participate (Dolmage, 2017, p. 119).

Consequently, by coupling a student-experience mindset with UDL, designers of digital interfaces can gain further insight into the accessibility of digital courses and better understand students' embodied experiences in them. As users of digital interfaces, students possess integral design knowledge based on the unique ways they navigate digital courses. In turn, by applying a UDL-informed student-experience mindset, designers of digital interfaces can engage more socially just design methods that understand digital access as a collective effort that should be led by those most impacted by design.

Recognizing disability access as a matter of citizenship and, consequently, social justice, disability justice urges us to move away from framings of disability access as an individualized process and to instead understand it as a complexly systemic issue that requires collective efforts led by disabled individuals. Disability justice is a movement founded in the efforts of disabled people of color to resist "disability rights organizing's white-dominated, single-issue focus" (Lakshmi Piepzna-Samarasinha, 2018, p. 15). At its core, disability justice centers disabled ways of being and knowing; specifically, it understands disability as a dynamic and intersectional phenomenon impacted by other aspects of identity, such as race, gender, class, or sexuality (Berne et al., 2018). In evaluating digital course design through a UDL-informed student-experience mindset based in the five levels of deep accessibility, course designers can collaboratively work with students to interrogate systemic ableism present across digital courses and to pursue design tactics that align with disability justice.

Methods

This chapter thus applied a pedagogical case-study method (Hancock & Algozzine, 2017) grounded in UDL and a UX student-experience mindset. Specifically, graduate students were recruited (per protocol approved by the Institutional Review Board) to complete a survey and to participate in a series of three semi-structured interviews designed to build knowledge regarding digital class design based on their past experiences with online learning. Students were enrolled in an online degree-granting program offered by a large southwestern U.S. university. Because the program's courses are standardized, 7.5-week courses, all students experienced roughly the same digital curriculum. An initial inquiry survey was sent to students enrolled in a specific course during fall 2018. Survey questions have been excluded as they were used solely to identify interview participants.

Irving Seidman's (2013) three-interview series was used to conduct semi-structured interviews with participants identified by the survey; this facilitated the co-construction of knowledge with participants regarding their past and present experiences in digital classrooms to inform future digital class design. Each participant completed three 30- to 60-minute interviews from December 2018 to May 2019; participants were compensated $15 for their time. The first round of interviews asked participants to reflect on their previous university experiences, and the second set of interviews asked them to reflect on their online learning experiences at their current university, with the specific course they had been recruited from acting as a comparative baseline (refer to Appendix 3.A). During their third interviews, students were asked to reflect on their previous interviews to build knowledge regarding their online learning experiences. Questions for third interviews were constructed uniquely for each participant based on their first two interviews to promote this reflection; for example, one student was asked, "You mentioned in your second interview that the online program was like 'one big accommodation' for you, particularly because of the distractions you have experienced in the past with people speaking simultaneously in the classroom. Can you discuss the different ways that the online program has served you as an accommodation?" Because third-interview questions reflect personal information shared in the first two interviews, these questions are not included in this chapter.

Interview analysis consisted of four phases:

1. building a coding scheme for primary codes and subcodes;

2. analyzing and counting primary codes;

3. analyzing and counting subcodes; and

4. identifying prominent themes across student discussions.

In phase one of analysis, thematic coding was used to draw broader theoretical conclusions by examining relationships between concepts (Saldaña, 2016). Using thematic coding, I identified themes across participant discussions, addressing whether, and how, digital classes consider UDL by applying the five levels of deep accessibility, which include "movement," "sense," "architecture," "communication," and "agency" (Dolmage, 2017, pp. 118–119). Codes then were expanded to account for students' discussions of time and productivity since these were prevalent concerns. In phase two, each primary code was analyzed further through a series of secondary subcodes identified using in vivo coding (Saldaña, 2016), which involves prioritizing participant perspectives by building codes out of overlying discussion trends. Table 3.1 reflects these primary codes and secondary subcodes.

Table 3.1. List of Digital Space Codes and Subcodes

Primary Codes & Definitions	Subcodes & Definitions
Movement: Participants' physical navigation across digital space	• **Access:** Access to classroom movement • **No access:** Lack of access to classroom movement • **Linear movement:** Movement as linear • **Sideways movement:** Movement as non-linear, circulatory, and/or ongoing
Sense: Participants' emotional engagement in digital space	• **Satisfaction:** Satisfaction in digital space • **Dissatisfaction:** Dissatisfaction in digital space • **Stress:** Stress/tiredness in digital space • **Isolation:** Loneliness/isolation in digital space • **Belonging:** Belonging in digital space
Architecture: Participant discussions of built digital space	• **Class Design:** Digital class design • **Program Design:** Online MA program • **Institution design:** Institutional design • **LMS design:** LMS design
Communication: Participants' communication in digital spaces	• **Connection:** Connection, community, or understanding in digital space • **Disconnection:** Lack of connection, community, or understanding in digital space • **Monitored:** Communication in digital spaces as censored/monitored • **Supportive:** Communication in digital space as supportive
Agency: Participants' agency in shaping and participating in digital environments	• **Independence:** Actions completed without assistance • **Interdependence:** Actions completed with support of other people or resources • **No Agency:** Lack of agency and/or capacity to participate in digital space • **Autonomy:** Self-actualization from participation
Time: Participant references to time	• **Without time:** Lack of time • **Deadlines:** Deadlines • **Flexibility:** Flexible time • **Standard:** Standard timelines
Productivity: Participant discussions of productivity, money, or high grades	• **Capital value:** Productive value of grades and/or money • **Progress:** Class/program progress

Source: Created by the author.

Results and Discussion

Eight of 22 students responded to the survey; three respondents who discussed a range of positive and negative experiences with online classes were chosen for subsequent interviews. Although only two students in the initial survey identified as having disabilities, all students discussed conditions that they either classified as a disability or as disabling in subsequent interviews. The results of the three participants' interviews are discussed in this section, with their names being anonymized to Sally, Harry, and Lucy to protect the identities of each participant.

An analysis of codes and subcodes revealed a series of thematic negotiations that participants made in navigating digital classes. Identified themes and correlating codes included

1. independent autonomy: *architecture* and *agency*

2. assimilative participation: *communication, movement,* and *sense*

3. productive value: *productivity* and *time.*

This section offers an overview of these thematic negotiations based on coded data. Table 3.2 reflects an overview of code and subcode frequency per each of the three participants.

NEGOTIATIONS OF INDEPENDENT AUTONOMY

Participant interviews revealed negotiations between digital space design and feelings of autonomy. These negotiations were specifically apparent in my coding of participant conversations related to *architecture* and *agency*.

Architecture. Most frequent across participant discussions, the coded concern of *architecture* (348 instances) revealed a connection between the design of digital spaces and student agency in navigating them. Interview conversations regarding architecture were most focused on issues related to *class design* (277 instances). Student responses suggested tension between the capacity for digital spaces to foster student autonomy and a tendency for digital classes to compromise autonomy through demands for independence.

Conversations with all participants conveyed this tension, but it was most prevalent for Sally, who explained that ADD and sensory integrative

Table 3.2. Frequency of Codes and Subcodes

Codes & Subcodes	Sally	Harry	Lucy	Total
Code: Architecture	92	110	146	348
Subcodes:				
Class Design	76	101	100	277
Program Design	3	7	31	41
LMS Design	13	2	5	20
Institution Design	0	0	10	10
Code: Communication	38	52	93	183
Subcodes:				
Connection	19	24	35	78
Disconnection	13	17	30	60
Supported	1	10	14	25
Monitored	5	1	14	20
Code: Agency	35	33	112	180
Subcodes:				
Independence	9	12	41	62
No Agency	8	12	32	52
Interdependence	9	1	25	35
Autonomy	9	7	14	30
Code: Productivity	42	12	48	102
Subcodes:				
Capital Value	40	11	38	89
Progress	2	1	19	13
Code: Sense	30	17	52	99
Subcodes:				
Satisfaction	11	9	19	39
Stress	13	7	16	36
Dissatisfaction	6	1	9	16
Isolation	0	0	4	4
Belonging	0	0	4	4
Code: Time	16	49	18	82
Subcodes:				
Without Time	9	17	5	31
Flexibility	2	17	5	24
Deadlines	3	14	4	21
Standard	2	1	4	7

Codes & Subcodes	Sally	Harry	Lucy	Total
Code: Movement	26	5	40	71
Subcodes:				
Sideways Movement	20	4	22	46
Access	1	0	12	13
Linear Movement	5	1	3	9
No Access	0	0	3	3

Source: Created by the author.

dysfunction impacted her conversational engagement within in-person courses. Referencing it as an "accommodation in itself," Sally explained that digital learning allowed her to "create that environment that I need in my own home . . . so that I'm able to . . . give my attention where I want it to be." She explained that features like the discussion board provided her time to "focus my own thinking" before contributing to discussions. Unlike a physical classroom environment that demands more rapid participation, online learning allowed Sally to engage with the course material at her own pace. This indicated that the time and space fostered by digital courses may challenge ableism and privilege the class contributions of a range of bodyminds often silenced by in-person discussions.

Sally also discussed frustration with being unable to see her peers' posts prior to submitting her own due to the digital class design; she noted that it would be "reassuring" to review these in advance to ensure that her contributions aligned with those of her peers. This suggested that although discussion boards allowed Sally to autonomously shape her class experiences, these independent assignments reinforced peer-to-peer comparisons that limited her feelings of agency. Sally's experiences demonstrated that incorporating discussion boards in digital classes can facilitate interdependent knowledge construction; however, the independent nature of the posting process may encourage students to identify response differences as learning gaps.

As *architecture*'s secondary trends reflect, participant concerns related to individual class design were linked to larger concerns regarding *program design* (41 instances), learning management system (*LMS*) *design* (20 instances), and *institution design* (10 instances). For example, in discussing one application, Sally noted anxiety with the assignment submission process based on how that application had been integrated into the LMS. This suggested that an instructor's course design decisions are often constrained by available technologies.

Agency. My coding of participant discussions of *agency* (180 instances) also revealed their efforts in negotiating autonomy. Students frequently discussed digital classroom spaces as fostering *independence* (62 instances) along with *no agency* (52 instances). Although this implied a point of tension, since *independence* and a *lack* of agency seem contradictory, further analysis revealed a connection between the two. For example, Lucy described online learning as an autonomous experience because it provided her the time and space to make thoughtful insights. However, she expressed frustration with meeting instructor expectations. She referenced ongoing attempts to clarify one instructor's directions, relaying, "I had no faith in myself or the work." This response suggested that the student's experiences of agency directly correlated with her capacity to independently meet instructor expectations. Many understand independence as integral to autonomy; yet the former requires a self-sufficiency that is difficult for many, including many disabled students. Demands for self-reliance may thus marginalize disabled students who rely on external resources and disregard the integral nature of support in the achievement of autonomy. In promoting goals of independent autonomy, digital courses may thus contribute to institutional ableism. However, resisting these ableist impulses were *interdependence* (35 instances) and *autonomy* (30 instances). Although infrequent, these concerns embraced understandings of agency beyond productive independence.

NEGOTIATIONS OF ASSIMILATIVE PARTICIPATION

In addition, interviews relayed how student users of digital interfaces negotiate assimilative participation in digital classes. I identified this negotiation specifically through participant discussions of *communication, movement,* and *sense.*

Communication. *Communication* (183 instances) reflected participants' second most frequent concern. Specifically, participants discussed tension between experiences with communication as offering both *connection* (78 instances) and *disconnection* (60 instances). For example, Harry explained that he was able to connect with many of his instructors. He noted that one instructor told him, "I love that you communicate with me and tell me . . . what you think . . . because that really helps me." Although Harry's instructor validated his concerns, his disability accommodation for flexible time isolated him from his peers; he did not receive feedback from them, as they had "already made their feedback to everybody else" by the time he participated. He relayed that his peers did not "really see me" and "weren't really communicating with me." Harry's conversations revealed how digital spaces may unknowingly reinforce independent forms of engagement that exclude disabled bodyminds.

Tension in relation to communication also was indicated through students' frequent discussions of digital communication as both *supportive* (25 instances) of student needs, yet also as *monitored* (20 instances), or constrained by instructor expectations. For example, Lucy explained that she enjoyed discussion boards because they allowed her to connect with her peers based on their academic interests. However, she disclosed that one instructor monitored class conversation and encouraged students to heavily critique each other's work. She noted that in one instance the instructor replied to her peer response by publicly stating, "You and I both know this doesn't make sense. You don't need to validate that." Such instructor engagement may privilege certain perspectives as valuably productive and consequently silence others. As Lucy explained, "I noticed people didn't really do a lot of extra interacting, at least online, because . . . every time we'd respond, I were [sic] afraid . . . that [it] would count against our grade." In this way, this instructor's engagement stifled student voices who articulated nondominant perspectives.

Movement. Although the least frequent participant concern was *movement* (71 instances), this concern revealed an interesting tension in relation to student agency. Predominantly, students discussed their movement in digital courses as *sideways* (46 instances), though some noted linear (9 instances) movement as well. This suggested that aspects of digital course design may challenge ableist demands for linear progress. For example, Sally described her digital classes as "spiraling through [her] daily life," which demonstrated her ongoing and recursive relationship with digital coursework. However, she also preferred when instructors engaged in these spaces to redirect unproductive conversation so that the class could "move forward." Though this student celebrated *sideways* aspects of digital learning, she preferred "curated," professor-directed conversations.

Tensions regarding agency in online classes were likewise present in participant discussions of digital spaces as promoting *access* (13 instances) and *no access* (3 instances). Collectively, student participants highlighted the supportive nature of online learning. For example, Lucy indicated satisfaction with digital courses, having enrolled in this online program due to her living situation and work schedule. Her desire for online learning was further augmented due to personal circumstances that left her emotionally distressed. In discussing her capacity to do her coursework, she noted, "I was having a [difficult] time sleeping and couldn't even do my work . . . I would never have been able to get through anything without just the fact that I could go home and be depressed and be sad [while taking courses]." In this way, digital courses enabled Lucy to participate in class in what she described as a "vulnerable" state and to experience "sadness" and "depression" on her own terms. This implied that online digital interfaces may facilitate learning experiences for a range of dynamically vulnerable bodyminds.

Sense. Student discussions of their emotional experiences in digital courses, articulated by the *sense* code (99 instances), also indicated tensions related to agency. Participants predominantly noted feelings of *satisfaction* in their interviews (39 instances). Specifically, Harry referenced a positive experience with one instructor who communicated with him regularly. He explained, "She made me feel like I mattered in the class, so I felt like I was in a good space/place." This suggested that instructor behavior can directly impact the support students feel in online courses and the consequential agency they experience. However, feelings of *stress* (36 instances) were also common. This is unsurprising due to the 7.5-week length of the online courses and participants' emphases on productive independence. For example, Harry noted ongoing feelings of burnout from his courses, which suggested that standard deadlines may not support the agency of disabled students. Though present, less frequently discussed were feelings of *dissatisfaction* (16 instances), *isolation* (4 instances) and *belonging* (4 instances). However, when discussed, *dissatisfaction* was often associated with a *lack of agency*. This indicated a connection between students' satisfaction and autonomy in digital courses.

NEGOTIATIONS OF PRODUCTIVE VALUE

My findings also indicated participant concerns with negotiating their productive value in digital courses. These negotiations were demonstrated through the *productivity* and *time* codes.

Productivity. *Productivity* (102 instances) was common across participant discussions, specifically in relation to concerns for *capital value* (89 instances), or the monetary value of academic programs, as well as the value of certain grades. Although present, discussions of *progress* were less common (13 instances). For example, Sally noted the LMS's capacity to compare students' grades to the class average. She likewise discussed one application intended to increase student interaction through methods like polling and hashtags. As she explained, "You had to earn a certain number of points . . . per module" and certain activities qualif[ied] you for . . . so many points." While using this application, Sally disclosed that her class contributions became motivated by "the points, instead of . . . what I should be writing." She expounded how the tool also facilitated class comparisons, stating, "You can see the rankings of everybody in the class." Sally illustrated how LMS design features can promote peer-to-peer comparisons. Participant discussions thus revealed the complex connections between digital class design, LMS design, and students' understandings of their productive value.

Time. Although less frequently discussed, *time* (82 instances) was a common topic in participant interviews. Participants frequently communicated that they were *without time* (31 instances) and indicated concerns related to *deadlines* (21 instances). For example, Harry expressed that he always felt like he was "running out of time" since courses lasted 7.5 weeks. He explained that his accommodation for flexible deadlines "relieve[d] some of that stress and anxiety." Yet, he referenced attempts to achieve an unreachable sense of balance, even with his accommodation, as he felt perpetually "rushed." Harry likewise disclosed that his work schedule did not allow him to complete quizzes assigned by some instructors exclusively on the weekends. Because of this restricted schedule, he noted that he would occasionally have to miss work to complete assignments. Harry's interviews indicated that by holding students accountable to linear, time-based standards, digital courses may disregard students' embodied needs.

GUIDELINES FOR REVISING DIGITAL SPACES

Based on findings and theoretical insights from TPC, UDL, and student-experience design, this section offers guidelines for accessible course design. Because these guidelines arose from three students' broad experiences with digital courses, they are generalizable across a range of courses. Specifically, they encourage class designers to reject practices that may foster ableism by centering disability justice. Bennett and Hannah (2022) recommended applying disability justice to facilitate more socially just design. I argue here that disability justice values should be extended to the design of digital courses. My guidelines thus apply disability justice principles to address the insights reflected in the previous discussion. These include (1) embracing class design strategies that facilitate interdependence, (2) understanding accessible design as frictionally collective, (3) exchanging capitalist productivity in digital courses for an anticapitalist politic, and (4) positioning students as leaders in course design.

Embracing class design strategies that cultivate interdependence. As participant interviews revealed, pushes for independent productivity may hold students accountable to ableist standards that exclude disability. I therefore recommend that instructors promote interdependent knowledge construction between students. Although disability justice understands individual experience as influenced by larger sociopolitical contexts (Berne et al., 2018), my interviews revealed that digital classes may frame knowledge production as an independent phenomenon. Further, all participants noted that when instructors highlighted their class contributions, they felt validated; however, Sally's comments suggested that such moves may prioritize certain contributions over others. To support disability justice's understanding of knowledge as collectively constructed across a range of embodied experiences, I recommend that instructors and designers of digital classes expand considerations for what counts as knowledge-making. I thus recommend the following methods for fostering interdependent learning.

Designing digital courses to facilitate interdependent knowledge construction. As Sally discussed in her interviews, many instructors limit students' access to viewing the discussion board before posting to encourage original contributions. To help students to recognize that their posts may take a range of forms, instructors might include examples of student responses from their previous courses. Likewise, instructors might encourage student-to-student communication to facilitate interdependent learning by enabling student messaging in the LMS, creating a crowd-sourcing discussion board where students can share resources, or using an external application that supports student communications. In addition, instructors and/or class designers might design certain discussion posts that function as collaborative learning spaces where students collectively work through ideas discussed in readings and/or gain writing feedback. For example, instructors might use discussion boards for peer review to enable students to view each other's work and expand on it by offering feedback from their unique perspectives.

Designing discussion prompts that ask students to draw from subjective experiences. By designing discussion questions that ask students to connect classroom learning to their personal experiences, instructors and/or class designers can facilitate the co-construction of knowledge across different perspectives. Further, in their instructions and responses, instructors might draw from their own embodied experiences and situated knowledges. Likewise, when responding to students, they might validate a range of student discussions as equitably valid. Such actions legitimize a range of knowledges and help students understand knowledge construction as a uniquely embodied process.

Creating a discussion thread for sharing resources. I recommend that instructors and/or class designers offer students a space for collectively sharing information regarding campus and community support services. By crowdsourcing with one another, students can gain additional insight into available tools and recognize resources as integral to autonomy. Such actions likewise facilitate access to resources that can support a range of intersectional student identities.

In these ways, designers of digital courses can push back against independent standardizing goals that may disenfranchise disabled students and promote an interdependent sense of community across digital spaces. Such interdependent styles of learning are grounded in UDL, as they facilitate community building and collective agency for a range of students (Dolmage, 2017, p. 118). By positioning difference as a tool for expansive possibility rather than as a problem to be reconciled, instructors and students collectively can further the accessible potential of digital learning. Such methods can be adjusted based on an instructor's course loads and should be collaboratively pursued by both instructors and designers of digital classes.

Understanding accessible design as frictionally collective. My findings similarly indicated ableism's potential influence in restricting student participation in digital courses. I thus urge instructors to shift from understandings of access as "assimilation based" to understandings of access as "frictional noncompliance" (Hamraie & Fritsch, 2019, p. 10). Often, designers of digital interfaces offer access through accommodations that help students engage productively in ableist systems. For example, Harry expressed that the flexible time accommodation helped him manage standard course deadlines. However, such accommodations do not encourage class designers to critique how and why existing practices may not serve the needs of all students. Instructors and course designers might, therefore, understand access lapses like Harry's as frictional opportunities to critique the pedagogical status quo. I thus recommend that instructors or digital course designers apply UDL's flexible learning strategies, such as the following:

- **Offering multiple means of representation.** Instructors or course designers might incorporate pedagogical tactics that draw from "multiple modalities—vision, hearing, and touch" (Oswal & Melonçon, 2014, p. 288) across course materials. By providing information to students through diverse media, including primary sources and narrative content like opinion editorials; visual content like comics, video clips, or infographics; and audio content like podcasts, instructors and class designers can better ensure that course concepts are comprehensive for a range of students.

- **Facilitating diverse ways for students to demonstrate knowledge.** UDL encourages instructors to "provide multiple means of action and expression" and to make "assignments in different formats" to help students express their ideas in diverse ways (Oswal & Melonçon, 2014, p. 288). The range of assignments that instructors or course designers offer might thus consider not only students' learning differences and interests, but likewise their unique access to time or resources. For example, students might have a choice to complete a final reflection assignment as a typed narrative essay, a video interview, an infographic, or a podcast recording. By including a range of assignments, instructors can allow students to pursue projects in modes most comfortable for them and to develop a range of relevant technological skills.

- **Encouraging multiple means of participation.** UDL also recommends that we offer students multiple means of engagement in our courses as they may "lack the operational means to connect with their instructors and classmates" (Oswal & Melonçon, 2014, p. 289). Across my interviews, tools like discussion boards and specific LMS applications were discussed as cultivating a range of multimodal experiences. I encourage instructors to explore the applications available on their university's LMS and to contact their center for teaching and learning to learn more about such applications.

- **Complementing multimodality with accessible design.** Although multimodality allows students to engage in diverse ways, each digital tool may bring with it accessibility challenges (Walters, 2010). It is therefore important that instructors and/ or course designers attend to access. For example, auditory materials should be "accompanied by closed captioning, clearly thorough summaries, or . . . another text-based form" (Nielsen, 2016, p. 95), images should include "descriptive <alt> tags," PDFS should be "in readable (non-image) format," and audio files should complement "text-based lectures" (pp. 101–102). Such tactics ensure access and allow students to navigate the course in ways most comfortable for them.

These efforts can be used by instructors or course designers with attention to labor constraints. Specifically, they might incorporate projects that allow students to choose from a range of multimodal options rather than create multiple assignments that ask students to use specific media. For instance, students might respond to an assignment prompt by composing a traditional essay, a comic, a video, a podcast, a class syllabus, or an infographic. In this way, students can choose modes most accessible to them and build projects relevant to their professional goals.

    Exchanging capitalist productivity in digital courses for an anticapitalist politic. Based on the overarching influence of ableist productivity on students' agency in digital courses, I also recommend that instructors and/or course designers integrate disability justice by prioritizing class values beyond productivity. "Disability justice rejects impulsory able-bodiedness . . . and mandates for . . . productivity" (Hamraie & Fritsch, 2019, p. 22) because it recognizes disability as resistant to capitalist demands for a " 'normative' level of production" that presumes able bodyminds (Berne et al., 2018, p. 227). As such, disability justice understands human value beyond the limits of one's productive capacities. I thus recommend that instructors and designers of digital classes think critically about how different activities may privilege able-bodied student users. Specifically, I recommend the following practices:

- **Disrupting normalizing processes by complicating student dialogues.** When engaging in digital courses, instructors might disrupt conversations that promote productive standardization through responses that complicate student thinking. For example, instructors might ask in response to student posts, "How do your own experiences and embodied knowledge inform this claim? How might a person's perspective on this change depending on their experiences?" Such questions encourage students to consider the situated nature of knowledge construction and to explore the ideological forces that influence their personal beliefs.

- **Coupling standardized assessment with critical awareness.** When using LMS tools that quantify and rank participation, instructors might address the limits of such tools with students. Such critical discussions might be included in the introduction to each application and through subsequent emails, announcements, or responses. In addition, instructors might replace assignments that ask for standardized responses with prompts that encourage students to draw from their perspectives, positionalities, and backgrounds. Further, instructors might design assessment methods, like rubrics, to evaluate different contributions as equitably valuable. In this way, instructors and students can appreciate diverse knowledge-making strategies. Through such tactics, instructors can engage UDL by valuing "multiple understandings of the 'right' way to see, hear, think, and know" (Wilson, 2018). Likewise, such tactics can help students think critically about productive knowledge construction and engagement.

- **Repurposing normalizing technologies through communicative channels.** Since some LMS technologies and applications quantify student engagement, instructors can use such technologies to identify students, like Harry, who may be experiencing isolation or difficulty in digital courses. I recommend that instructors work with instructional designers to locate tools that may help to identify students with lower levels of participation so that they may support their access needs and/or connect them with on-campus resources like advising that can further assist them throughout the semester.

Collectively, many of these considerations can be made in a course's initial design. However, when engaging these strategies during the semester, instructors and/or digital class designers should do so in ways attuned to their unique labor constraints. For example, instructors might highlight a range of student responses to demonstrate the value of diverse contributions. Likewise, instructors might choose methods for engaging student feedback on technologies that complement their course loads and caps; for example, general surveys may be more manageable than individualized emails. Further, instructors might work with their teaching and learning centers to identify other ways to support student access throughout the semester. Through such methods, instructors and designers of digital class interfaces can communicate to students a value for learning beyond capitalist demands for able productivity.

Positioning students as leaders in course design. As my findings indicated, student participants grappled with agency in their online courses, particularly due to concerns with meeting instructor expectations. In addition, my findings indicated that although online learning may be accessible in many ways, unanticipated student access needs may arise throughout the semester. Because all individuals navigate "online technologies and pedagogies from an entirely different vantage point shaped by their social, physical, and educational experiences" (Oswal & Melonçon, 2017, p. 70), all student experiences with digital learning are unique to their dynamic intersectional positionalities. It is thus important that students are positioned as co-designers of digital learning spaces to share their insights from navigating digital courses. I recommend the following tactics to facilitate such leadership:

- **Offering students opportunities for providing ongoing feedback.** I recommend that instructors facilitate ongoing opportunities for students to provide feedback on their course experiences. UDL urges instructors to systematize feedback "at diverse times, and through diverse channels" (Price, 2011, p. 130). As previous TPC scholars have noted (Nielsen, 2018; Oswal & Melonçon, 2017), UDL pedagogies respond to students' embodied experiences. To foster such considerations, I recommend that instructors incorporate feedback methods throughout the semester, such as personalized email inquiries, anonymous class polls, informal office hours conversations, and/or surveys of varying lengths. Instructors should choose those methods that work best for their unique labor constraints and class needs. For example, questions could ask students to offer insight into the usability of the class design, their embodied experiences while navigating the class, their experiences engaging with different assignment types, and any life circumstances or access needs that might impact their class engagement. Instructors can then adjust their courses based on this feedback to ensure that their courses are accessible for a range of students.

- **Incorporating focus groups or semi-structured interviews.** Finally, instructors or course designers might incorporate focus groups and interviews after a course has ended to work with students to transform future courses (Bennett, 2022; Oswal & Melonçon, 2017). Should instructors wish to avoid potential bias in student responses, they might work with their teaching and learning center or with a colleague in conducting interviews or focus groups. I recognize that many of the insights provided to me by student interviews may not have been shared if I had been their instructor. Through such methods, instructors can validate the agency of all students in their classes, particularly their disabled students.

I encourage all stakeholders involved in designing online degree-granting programs to engage with some combination of these methods for engaging student feedback to ensure the ongoing accessibility of their digital courses.

Conclusion

The combination of a student-experience mindset with UDL and disability justice provides a valuable method through which instructors may reject ableist influences across digital courses and foster more equitable understandings of access that empower a range of student identities. Because access determines one's capacity to engage equitably as a citizen in public space, it is vital that we interrogate how class design may marginalize certain bodyminds. This chapter's case study was limited in that it revealed only three students' experiences with digital learning; future research might explore a broader range of student experiences to draw more extensive conclusions regarding digital course accessibility. As this chapter focused on disabled students' experiences, future research might examine the implications of digital interfaces for disabled instructors. Likewise, as this study revealed a potential connection between access limitations and high course loads or caps, future research might examine the impacts of labor constraints on accessible teaching. In addition, as this study critiqued the accessibility of standardized courses, future research might study student experiences across multiple sections of the same course to better understand the impacts of standardized digital instruction.

This chapter offers a starting point for integrating UX, UD, and disability justice in addressing localized concerns regarding digital class design. I encourage instructors and digital class designers to expand on offered insights in their own courses. As Dolmage (2017) has relayed, for access to be transformative, such "work must be change-enhancing, interactive, contextualized, [and] social" (p. 132). Through a combination of disability justice principles and methods grounded in UX and UD, instructors and designers of digital classes might position embodied differences like disability as resources for transformative change and work collaboratively with students to (re)design digital spaces that are more attuned to the needs of disabled bodyminds.

Appendix 3.A

Interview Protocol 1 Questions-History

1. Please tell me about your previous learning experiences (in college or prior to college). What experiences (if any) stand out to you? Please explain.

2. For students identifying as disabled only: On the previous survey, you identified as having a disability. Is there a metaphor you would use to describe your experience with your disability? Could you explain it to me? (For example: Writing an essay feels like running a marathon).

3. Have you registered for accommodations with the disability resource center? Could you please elaborate on why/why not?

    a. Possible follow up: Could you tell me about your previous experiences with accommodations in online or on-ground learning environments?

    b. Possible follow up: Could you tell me about a time where an accommodation significantly improved your learning experience?

    c. Possible follow up: Could you tell me about a time where an accommodation negatively impacted your learning experience?

4. Is there a metaphor you would use to describe your experience with online course interfaces? Can you explain it to me?

5. Do you receive any type of accommodations in this program? Did you receive any accommodations in high school?

    a. Possible follow up: Could you tell me about your previous experiences with accommodations in online or on-ground learning environments?

    b. Possible follow up: Could you tell me about a time where an accommodation significantly improved your learning experience?

    c. Possible follow up: Could you tell me about a time where an accommodation negatively impacted your learning experience?

6. Can you discuss your relationships with others at the university (faculty, peers, etc.)? Are there any external factors that impact your relationships with these individuals?

    a. Possible follow up: How do these factors impact your relationships?

    b. Possible follow up for students who identified as disabled: Does your disability impact your relationships with others (faculty, peers, etc.)? If yes, could you please describe how?

7. Could you tell me about your previous experiences with online education (prior to this class)?

8. Could you tell me about a time where you experienced difficulty in an academic or informal learning environment (online or on-ground)? What contributed to this difficulty?

9. Could you tell me about a time where you excelled in an academic or informal learning environment (online or on-ground)? What positively contributed to this experience?

10. Is there anything else that you want to share with me regarding your online or on-ground formal or informal educational experiences prior to this class?

Interview Protocol 2 Questions-Experience

1. Please tell me about what motivated you to pursue an online MA education.

2. Did you enroll in any other online classes this semester? Which ones/why?

3. Did you have accommodations during this course? Could you tell me about your experience with or without accommodations in this class?

   a. Possible follow up: What additional accommodations would have been useful for you in navigating this course? How might they have assisted you?

4. Could you tell me about something you discovered about yourself as a learner through your experience in this class?

   a. Could you elaborate upon any specific moment(s) that contributed to this learning?

   b. Possible follow up: Did the design/interface of the course contribute to, or interfere with, your self-growth as a learner? How?

5. Could you tell me about your experience engaging in this course? Do any experiences stand out to you?

    a. Possible follow up: Did any aspects of the Blackboard interface contribute to these experiences? If so, which ones?

    b. Possible follow up: Could you tell me about a specific experience in which you had some difficulty with some aspect of this online course? What contributed to this difficulty?

    c. Possible follow up: Could you tell me about a positive experience you had in this class that may not have been available to you in an on-ground classroom environment? What contributed to this positive experience?

6. Please describe a time when you experienced discomfort or uncertainty in relation to new learning in this class?

    a. Possible follow up: Please elaborate upon what, if anything, contributed to this discomfort.

    b. Please elaborate upon what, if anything, eased this discomfort.

7. Could you tell me about a particular experience (or set of experiences) that stand(s) out to you in relation to discussion in this online course?

    a. Possible follow up: Please elaborate upon what contributed to your desire/lack of desire to engage in these discussions.

    b. Possible follow up: Please explain what might have further supported your engagement in these discussions.

8. Is there anything else that you want to share with me about your experience in this class?

Notes

1. Disability studies understands disability as "a political and cultural identity" rather than a purely medical condition. Aiming to promote social transformation, the field prioritizes the experiences and perspectives of disabled individuals (Dolmage, 2017, p. 5).

2. This term challenges the separation of the body and mind by indicating their inherent connection (Price, 2014).

References

Bennett, K. C. (2022). Prioritizing access as a social justice concern: Advocating for ableism studies and disability justice in technical and professional communication. *IEEE Transactions on Professional Communication, 65(1)*, 226–240. https://doi.org/10.1109/TPC.2022.3140570

Bennett, K. C., & Hannah, M. A. (2022). Transforming the rights-based encounter: Disability rights, disability justice, and the ethics of access. *Journal of Business and Technical Communication, 36(3)*, 326–354.

Berne, P., Levins Morales, A., Langstaff, D., & Invalid, S. (2018). Ten principles of disability justice. *Women's Studies Quarterly, 46(1/2)*, 227–230. https://www.jstor.org/stable/10.2307/26421174

Cherney, J. L. (2019). *Ableist rhetoric: How we know, value, and see disability.* University of Pennsylvania Press.

Clinkenbeard, M. J. (2020). A posthuman approach to agency, disability, and technology in social interactions. *Technical Communication Quarterly, 29(2)*, 115–135. https://doi.org/10.1080/10572252.2019.1646319

Dolmage, J. T. (2017). *Academic ableism: Disability and higher education.* University of Michigan Press.

Greer, M., & Harris, H. S. (2018). User-centered design as a foundation for effective online writing instruction. *Computers and Composition 49*, 14–24 https://doi.org/10.1016/j.compcom.2018.05.006

Hamraie, A., & Fritsch, K. (2019). Crip technoscience manifesto. *Catalyst: Feminism, Theory, Technoscience, 5(1)*, 1–33. https://doi.org/10.28968/cftt.v5i1.29607

Hancock, D. R., & Algozzine, B. (2017). *Doing case study research: A practical guide for beginning researchers* (3rd ed.). Teachers College Press.

Hitt, A. (2018). Foregrounding accessibility through (inclusive) universal design in professional communication curricula. *Business and Professional Communication Quarterly, 8(1)*, 52–65. https://doi.org/10.1177/23299490617739884

ISO (International Organization for Standardization). (1998). *ISO 9241-11:1998 Ergonomic requirements for office work with visual display terminals (VDTs)— Part 11:* https://www.iso.org/standard/16883.html

ISO (International Organization for Standardization). (2014). ISO/IEC Guide 71:2014 Guide for addressing accessibility in standards. https://www.iso.org/standard/57385.html

Johnson, R. R. (1998). *User-centered technology: A rhetorical theory for computers and other mundane artifacts.* State University of New York Press.

Jung, K. E. (2003). Chronic illness and academic accommodation: Meeting disabled students' "unique needs" and preserving the institutional order of the university. *The Journal of Sociology and Social Welfare, 30*(1), 91–112. https://scholarworks.wmich.edu/jssw/vol30/iss1/6

Kent, M. (2015). Disability and elearning: Opportunities and barriers. *Disability Studies Quarterly, 35*(1). https://doi.org/10.18061/dsq.v35i1.3815

Lakshmi Piepzna-Samarasinha, L. (2018). *Care work: Dreaming disability justice.* Arsenal Pulp Press.

Melonçon, L. (2018). Orienting access in our business and professional communication classrooms. *Business and Professional Communication Quarterly, 81*(1), 34–51. https://doi.org/10.1177/2329490617739885

Nielsen, D. (2016). Can everybody read what's posted? Accessibility in the online classroom. In D. Ruefman & A. G. Scheg (Eds.), *Applied pedagogies: Strategies for online writing instruction* (pp. 90–105). Utah State University Press.

Oswal, S. K. (2019). Breaking the exclusionary boundary between user experience and access: Steps toward making UX inclusive of users with disabilities. In *SIGDOC '19: Proceedings of the 37th ACM International Conference on the Design of Communication,* 1–8.

Oswal, S. K., & Melonçon, L. (2014). Paying attention to accessibility when designing online courses in technical and professional communication. *Journal of Business and Technical Communication, 28*(3), 271–300. https://doi.org/10.1177/1050651914524780

Oswal, S. K., & Melonçon, L. (2017). Saying no to the checklist: Shifting from an ideology of normalcy to an ideology of inclusion in online writing instruction. *WPA: Writing Program Administration, 40*(3), 61–77.

Palmer, Z. B., Oswal, S. K., & Huntsman, S. (2019). Breaking the exclusionary boundary between user experience and access. In *Proceedings of the 37th ACM International Conference on the Design of Communication, 37,* 1–2. https://doi.org/10.1145/3328020.3353920

Petrie, H., & Bevan, N. (2009). The evaluation of accessibility, usability and user experience. In C. Stepanidis (Ed.), *The universal access handbook* (pp. 2–30). CRC Press. www.crcpress.com/product/isbn/9780805862805

Price, M. (2011). *Mad at school: Rhetorics of mental disability and academic life.* University of Michigan Press.

Price, M. (2014). The bodymind problem and the possibilities of pain. *Hypatia, 30*(1), 268–284.

Saldaña, J. (2016). *The coding manual for qualitative researchers.* Sage.

Seidman, I. (2013). *Interviewing as qualitative research: A guide for researchers in education & the social sciences.* Teachers College Press.

Spinuzzi, C. (2007). Accessibility scans and institutional activity: An activity theory analysis. *College English, 70*(2), 189–201.

Walters, S. (2010). Toward an accessible pedagogy: Dis/ability, multimodality, and universal design in the technical communication classroom. *Technical Communication Quarterly, 19*(4), 427–454. https://doi.org/10.10.1080/1057 22522010502090

Wilson, J. D. (2017). Reimagining disability and inclusive education through universal design for learning. *Disability Studies Quarterly, 37*(2). https://doi.org/10.18061/dsq.v37i2.5417

Wheeler, S. (2018). Harry Potter and the first order of business: Using simulation to teach social justice and disability ethics in business communication. *Business and Professional Communication Quarterly, 81(1),* 85–99. https://doi.org/10.1177/2329490617748691

Chapter 4

Preparing Professionals to Make Equitable Experiences of Visual Information for Users with Low Vision and Blindness

Philip B. Gallagher and Marci J. Gallagher

Since the early days of faculty teaching "graphical presentation" at Iowa State University, visual communication has become a part of technical communication (TC) (Connors, 2004, p. 14). In its infancy, visual pedagogy consisted mainly of emulating the forms of diagrams and data displays from science, engineering, and industry (Gallagher, 2019). Gradually, audience diversity for visual communication was recognized and nonexpert user needs were taken up out of necessity. The exigence for this shift in attention was the proliferation of technology throughout the world after World War II. The explosion of consumer technology meant that every new "machine needed a manual written for it" (Connors, 2004, p. 12), and those manuals had to be used by anyone who used the technology.

From 1950 through 1970, the accessibility needs of individuals with low vision, blindness, hearing loss, deafness, and limited mobility and cognition were invariably missing from the curriculum of TC. Technical artifacts were created with an ableist mindset regarding users. For technical communicators of the time, the absence of users with low vision and blindness, for example, meant that the usability of technical visual

105

information was limited to only sighted users who could interact with it. The poor attention to voices of low-vision and blind audience members by the field was a major barrier to creating equitable visual information access via empathy that can aid in document design that meets everyone's needs. That is, only by listening to users with low vision and blindness can technical communicators respond through empathy to make truly accessible visual designs.

Only after creation of the U.S. Access Board and ratification of the Rehabilitation Act of 1973—which contains Section 504 that protects individuals with disabilities from discrimination in education—and the passing of the Telecommunications Act of 1996, did visual communication instruction begin to take the accessibility needs of users seriously. In the mid-2000s, the U.S. Access Board extended consumer communications accessibility duties to include "telecommunications equipment, and . . . electronic and [digital] information technology" (United States Access Board, 2006). At this point, online information, technology, and communication industries began to seek professionals who had the skills to accommodate all user audiences, but filling positions has been slow.

Today, despite several decades of government accessibility initiatives, barriers to visual information for users with low vision or blindness remain. These barriers are especially prevalent in online documentation environments. Current TC pedagogy provides little guidance on how to address the needs of low-vision and blind users for multimodal communication. In fact, most textbooks, if they cover accessibility, provide only a topical look at addressing these users—usually talking about things like alt text and captions without much instruction on how to assess users' accessibility needs or expectations. Thereby, the pedagogy of these materials often fails to prepare novice professionals to support these users adequately.

The lack of an accessibility education that works is a problem now more than ever in TC. Most industries looking to hire communicators that are accessibility specialists cannot find qualified individuals (Alcántara, 2021). Many of these companies are creating technical documents for mass-market consumption, and the materials are often highly visual, especially when they are online (Cawley, 2017). Without adequately trained communicators who understand accessible design theories and practices, users with low or no vision are forced to confront documentation with formatting, symbols, pictures, tables, diagrams, maps, and other visual content that does not accommodate them. This means that a potential global audience of "2.2 billion people [who] have a vision impairment" may be trying to use information that is inaccessible (World Health Organization, 2019, p. 23). As a user group, then, this segment of the audience is expected to interact with online technical media, ranging from product summaries, instructions, descriptions, and recalls to proposals, reports, and white papers that often contain primary visual information. Primary visual information is graphic messaging that must be understood by the user to make the information actionable. To address this problem, we need to educate young technical professionals to design visual information using accessible forms based on information received from the users who need it.

Another industry pressure exacerbating the need for qualified communicators is the desire for contemporary technical artifacts to achieve cross-cultural communication through visuals (Wang, 2000). In this context, the primary visual information is understood as a "translation" of information across linguistic barriers, but not across "audiovisual" sensory barriers, wherein lies the problem (Universitat Autònoma de Barcelona, n.d.). The visuals aim to reduce the need for numerous iterations and to establish local communication (Lotito et al., 2013). Although this is beneficial to a company's bottom line, it may negatively impact users of their products and services who are low vision or blind. These users are disadvantaged by predominantly visual communications because, if there is no text associated with the information for access by screen readers or description services, these users cannot get the information, localized or otherwise. Thus, much of today's localized visual communications, according to analyst Andrew Martins (2019), are largely "inaccessible online" by this population. In response to these problems and pressures, our chapter offers a pedagogical approach to train technical communicators to create equitable user experiences for users with low or no vision. Our approach provides access-driven communication theories and practices and teaches procedures and tools to help communicators craft simple design elements like tags, alt text, and visual content descriptions meaningfully to boost the accessibility of visual information in digital documentation.

To define and support our TC pedagogy, our chapter opens with existing approaches to visual communication accessibility. Then, we provide a rationale for our approach based on studies of users with low vision and blindness (e.g., Power et al., 2012; Theofanos & Redish, 2003, 2005; Xie et al., 2015; Xie et al., 2020; Youngblood et al., 2018). From those studies, we identify these users' needs and provide a response leveraging common accessibility design elements—tags, alternative texts, and visual content descriptions. Afterward, weaving theory and practice, we share what topics are taught during our visual accessibility training. We discuss our access-first design approach and how it supports focus on accessibility needs and expectations; we cover how visual rhetorical theory supports interpretive analysis using systematized visual language and the identification of design choices using Kostelnick's (1989) matrix; we offer how design-thinking theory provides practice using empathy and human-centered, problem-solving procedures for accessible design work. Afterward, we explain Kostelnick's (1989) matrix as a tool to train visual interpretation and design and demonstrate the whole pedagogical approach deployed in a visual communication class. From that deployment, we share the pedagogical results and the most salient takeaways readers can use in their own classrooms. In conclusion, we provide readers with the benefits and limitations of our approach and call for continued research to meet the access needs of this important user group.

Literature Review

To start our research, we defined the pedagogical interests and objectives for our approach:

- use to create accessible visual information in digital documentation;
- address challenges, needs, and expectations of users with low and no vision; and
- focus on teaching communicators at the intersection of design, theory, and tools.

In response, we identified three distinct—but not mutually exclusive—pedagogical approaches based on technology tools and best practices (e.g., Roberts, 2006; Walters, 2010; Youngblood, 2012); ethics and user experience (UX) (e.g., Bennett & Hannah, 2022; Conway et al., 2020; Lancaster, 2018); and social justice (e.g., Bivens et al., 2020; Sonka et al., 2021).

Beginning with technology tools and best practices pedagogy for accessibility, it involves using conventional social and technological accoutrement for accessible design. According to Roberts (2006), teaching technical communicators how to help others overcome visual information barriers via technology is a core part of creating accessible digital documents—and rightfully so. However, pedagogues who take technology too far may lack knowledge of why an element is chosen or interpreted a certain way *in situ*. Additionally, educators using this pedagogy may emphasize the "generalizability" of accessible designs by teaching "universal design" (Walters, 2010, p. 432), "best coding practices" (Youngblood, 2012, p. 209), or "emulating" UX concepts (Youngblood, 2012, p. 222). Although these may have a place in the discussion of designing for access, overemphasis can reduce the effectiveness of visual communications made for unique situations, users, and needs.

Ethics and UX-based pedagogy for accessibility involves careful attention to user rights and satisfaction. According to Lancaster (2018), an ethic of care is necessary as part of the relationship between an information designer and users. This philosophy helps to improve communications to prevent negative consequences (p. 253). This pedagogical approach inspires action to design visual information for accessibility and its emphasis on satisfaction motivates our approach to address the challenges of users with low vision and blindness with satisfaction as a goal. We aim to make visual information interaction useful—a goal we share with Conway et al. (2020)—and a positive experience. However, this pedagogy often does not provide a well-defined procedure; therefore, we add design-thinking processes to scaffold accessible design work.

Last, pedagogy based on social justice for accessibility involves centering the right to equity and inclusion of users who are marginalized by ableist practices in TC education. This pedagogy helps marginalized users by maintaining the commitment of stakeholders "to creating user experiences centered on shared values of equity, sustainability, and usability" (Sonka et al., 2021, p. 264). Further, scholars push the pedagogical reach of a social justice initiative beyond the classroom to include accessibility and usability "knowledge and knowledge-making" by "crowdsourcing" the development of curriculum (Bivens et al., 2020, p. 70). In social justice pedagogy, marginalized users are the focus for transforming TC artifacts to make them fair, accessible, and representative forms of documentation. Our pedagogy, too, seeks to put marginalized low-vision and blind users at the center of attention, grounded in more practical, simple, actional theories and methods-based tools.

VISUAL COMMUNICATION PEDAGOGY: A USER NEEDS RATIONALE

According to users with low vision and blindness, much visual communication leaves a lot to be desired (Power et al., 2012; Theofanos & Redish, 2003, 2005). Despite initiatives by the U.S. government to make electronic media accessible, since 2015, "there have been over 240 online accessibility lawsuits" in telecommunications alone (Youngblood et al., 2018, p. 334). As a field that deals in the transaction of complex information, TC needs "to prepare students to understand what accessibility [and usability are] and how to make electronic media accessible [and usable]" (p. 334). To begin our journey to prepare students, we will provide data from five studies about low-vision and blind users of visual information and address how these studies establish rationale.

In the first study, Theofanos and Redish (2003) observed 16 blind users using the screen-reading software with which they were most familiar while completing tasks and providing feedback. First, users who rely on screen readers will "scan" information with their ears, like sighted users who scan text, to make decisions about information usefulness (Theofanos & Redish, 2003, pp. 3–4). Thus, information should be brief while offering necessary details. Further, users benefit when designers front-load vital information in a description because the user may listen to only the first few words (pp. 2–3). Second, users who rely on screen readers must divide their attention between the information that they are trying to access and the software that is helping them gain access: "always being in a 'help' system—having to split your cognitive energy between the task you are doing and how to use the system that is helping you" (p. 5). The authors have suggested that programmers should make software functions "mnemonic and intuitive" (p. 5). If we apply this to documentation, we can highlight the importance of creating memorable and direct descriptions.

In the second study, Theofanos and Redish (2005) also "observed low vision users as they worked with Web sites and assistive devices they typically use" (p. 9), including screen magnification and reading tools like JAWS, Window-Eyes, and ZoomText. The 10 users in the study were low-vision users, or individuals with vision not corrected by glasses or typical accessibility tools (p. 9). Their study participants were found to "customize many aspects of their screen, using the operating system, the browser, and screen-magnifying software . . . based on individual vision abilities and weaknesses" (p. 10). When observing participants' interactions with U.S. government sites, Theofanos and Redish (2005) found numerous challenges to low-vision users' access to government sites that included visuals. Participants had trouble "seeing" everything in documents, that items to the right-of-center or on the right side of the page were often missed, and that they needed to be able to modify colors to recognize and understand visual elements (pp. 13–14). From these observations, the authors realized that "the needs of low-vision users are too diverse for simple solutions to Web accessibility and usability" (p. 9), and they established six recommendations for document design: (1) do not rely on color to convey meaning or define sections, (2) outline tabs and sections with a black border so their purpose is clear without color, (3) do not use graphics as a replacement for text, (4) make text size responsive so that it can be adjusted, (5) ensure that the type of text allows it to be clearly enlarged, and (6) use sans serif fonts for low-vision users (pp. 14–17).

In the third study, 32 blind users were asked to complete a task-oriented procedure (Power et al., 2012), during which the researchers counted 1,383 user problems, most that involved visuals (p. 433). The procedures were tested on 16 websites that conformed to Web Content Accessibility Guidelines (WCAG), but the users often reported "no enhancements to multimedia content" on sites (p. 439). They also indicated that the meaning of content "was lost or modified by transformations" meant to improve usability and that tables and images lacked alternative text (p. 439).

In the fourth study, Youngblood et al. (2018) examined the legal and ethical exigency for designing media per accessibility needs and urged educators to teach their students to use accessibility media (p. 342) as well as highlighted web accessibility issues. They asked, "How can we make visual content accessible [and usable] to people who are blind and visually impaired?" (p. 338), noting that these users have significant visual description needs—needs that require what Newfield (2014) and Conway et al. (2020) called "transmodal translation" of the media. For example, users may require audio description (AD) of video content: that is, "narration added to the soundtrack to describe important visual details that cannot be understood from the main soundtrack alone . . . [that] provides information about actions, characters, scene changes, on-screen text, and other visual content" (American Council of the Blind, 2016). For technical communicators, much of today's communications involve video content for sharing information online; in these situations, preparing both written and audio descriptions of visual content provide access.

In the fifth study, users with low vision and blindness were found to experience additional problems with visual information (Xie et al., 2020) when navigating digital library spaces, which are virtual collections of online media organized to maximize access and usability for patrons. The researchers previously identified 17 physical and cognitive help-inducing challenges unique to this user group (Xie et al., 2015; Xie et al., 2018). In this follow-up study (Xie et al., 2020), 40 visually impaired users were observed in two scenarios: a 20-user control group using existing help features and a 20-user experimental group using new help features. Researchers examined accessibility criteria such as ease of use, usefulness, satisfaction, and usability metrics during help-seeking events of participants. They found several recurring access issues: locating items, making sense of visuals, understanding labels, understanding web functions/controls, not being able to access formats, and using search queries. Missing visual description and formatting information were significant barriers, and elements like tags, alternative texts, and visual descriptions were identified as necessary to reduce help-seeking events (Xie et al., 2020).

Observed collectively, the needs of users with low vision and blindness identified by these five studies (Power et al., 2012; Theofanos & Redish, 2003, 2005; Xie et al., 2020; Youngblood et al., 2018) to teach accessible visual communication and prepare students to produce tags, alternative texts, and visual/verbal content descriptions for technical visuals to meet the needs of these users. Therefore, our pedagogical approach needed to enable students to produce accessible visual media to

- help users interact with the whole document;

- make commonly positioned visuals obvious and useful;

- clarify purpose and function of visual information;

- support color modification and clearly align color descriptions with meaning;

- maintain information integrity if/when visually modified for access;

- include alternative forms of information for static and dynamic visual media; and

- use appropriate multimedia enhancements for visual information.

Hereafter, we describe our approach to teaching design elements, theories, and tools to TC students.

Our Pedagogical Approach

We spent one year piloting a pedagogical approach for teaching TC students how to create accessible visual document design. Our approach had three parts:

- essential access design elements;

- a set of actional theories; and

- Kostelnick's (1989) visual language matrix as a tool for working with visual language.

In this section, we explore each part of our pedagogy and intended outcomes of its use.

PART ONE: DESIGN ELEMENTS

Three essential elements of accessible digital visual design are tags, alternative texts, and visual content descriptions, and one or more of these elements are necessary for sharing visual information with low-vision or blind users (Power et al., 2012; Theofanos & Redish, 2003, 2005; Xie et al., 2020: Youngblood et al., 2018). Therefore, our pedagogical approach begins with these elements that improve visual accessibility. Each of these design elements has unique attributes that support visual information access in online technical artifacts and may be used to respond effectively to different needs of users with varied visual experiences.

Tags are digital labels that aid sorting and navigation of information within a document. Tags belong to two categories—keyword metatags and content tags. Keyword metatags are typically used within a digital document's code to guide search inquiries to relevant results. Technical professionals working with Search Engine Optimization carefully consider keyword metatags. Content tags are available to users to help them retrieve "other content within the system that contains the same tag" (Oregon State University: IT Department, n.d.). For example, content tags may appear in a blog to link readers to content with the same tag.

Both metatags and content tags are guided by a shared purpose: to help users find the content they seek. Per this purpose, our teaching interests lie with content tags and designer's ability to use them (and explain them) to increase the presence of graphic information, to group visuals together, and to link between written and visual content. The ability to design documents linking text and visual content is important to teach so designers can improve interactive accessible experiences of visual information.

The next essential access design element is alternative text (alt text). Alt text is a description of elements, including "images, image map hot-spots, audio, video, graphical buttons, applets, animations" (McEwan & Weerts, 2007, p. 2). The text interprets visual information into writing and thus is at once a literal alternative to the visual information, and, when done well, the necessary context for that information. Without alt text, visuals are inaccessible to users of screen readers. Teaching alt text requires instruction on what alt text is and how to write it with context, and the ethics and proper use of alt text is a design priority. This instruction helps technical professionals prepare to support users with low vision and blindness while they interact with an entire artifact, grasp the purpose and function of its visuals, and experience visual information with integrity.

The last essential design element is visual content descriptions, which create usable information from complicated visuals. Unlike alt texts, visual content descriptions are lengthy and used when images offer robust datasets or infosystems, such as to explain workflow diagrams, tables in spreadsheets, or maps. Well-written visual descriptions provide details about multifaceted primary visual information, sometimes from many perspectives that provide users with low vision and blindness access to primary visual data. Visual descriptions may come in different forms—written text or audio descriptions—depending on the medium of which they are a part. But, in any form, they act like "transmodal translations" (Conway, 2020, p. 69), taking information from static or dynamic visuals and reproducing it in another mode.

Further, visual descriptions for "complex images [that] typically need a long description for them to be fully understood" and must follow spe-cific practices (California State University: Northridge Universal Design Center, n.d.). These practices include using either a "Web Accessibility Initiative-Accessible Rich Internet Application (or WAI-ARIA) approach" or a similar page-linking method (California State University: Northridge Universal Design Center). The WAI-ARIA approach places long para-graph-form descriptions within a page, while the page-linking method connects visuals with an external page that contains the description. Whereas alt texts may be simple captions, visual content descriptions may require several paragraphs, pages, or minutes of audio presenting visual information, interpretation, and relationships. Our pedagogical approach provides students with experience using visual descriptions to address users' needs for functional complex information and for multimedia modification that supports equitable, satisfying information access.

PART TWO: RHETORIC AND DESIGN CONCEPTS

To create accessible visual design, we must teach technical professionals access-first-informed visual rhetoric and design. By doing so, we ensure that their work supports the equitable experience of low-vision and blind users of technical documents. We recognize that "technical communicators can remove [accessibility] barriers" to create a "more meaningful experience for people with disabilities" if they have the necessary education (Roberts, 2006, p. 14).

Access-First Design

Our concept of an access-first approach for technical communication is not dissimilar from Slatin's (2001) "access-first design." Slatin sought to create accessible experiences for his class websites using alt text for visual content (p. 76). Our pedagogy adds more design elements, rhetorical and design theories, and tools for improving "experience design" for visual documentation. We consider our endeavor like past Google CEO Eric Schmidt's (2010) call for "Mobile First" design. Schmidt's design idea centered on developing digital content for mobile users first, but our focus is on access first. This focus helps ensure that users have equitable experiences with documentation with any platform, device, or ability. Using an access-first design approach to inform instruction on visual rhetoric and analysis, design thinking, and the practice of using Kostelnick's (1989) visual language matrix gives our training an edge in TC situations.

Visual Rhetoric & Analysis

Teaching Kostelnick and Roberts' (2011) theory of visual rhetoric, we introduce learners to the concept of treating visual communication, like written and oral communication, as a rhetorical process and consider how they communicate with an audience for a purpose in a situation. Then, we discuss how interpreting a visual's message and evaluating its effectiveness relies on a process of gathering, questioning, recording, sorting, analyzing, and synthesizing information to understand and/or create visuals for those specific audiences, purposes, and situations. We stress that this process is used whether professionals are interpreting other's designs or creating their own. Learning and practicing the visual rhetorical process of analysis, we focus on how a visual responds to user expectations, achieves its goals, and works in finite locations to define a document's technical exigency as well as the role of the visual in relation to that motivation.

A complete rhetorical analysis of a visual communication by a designer creates a deep understanding of the visual's message in situ. With this knowledge, the designer can create visual access for users with low vision and blindness using tags, alt text, and visual content description elements. At this point, we discuss designing for user access first, and learners learn from users who have low or no vision to support access-first design.

Analysis pushes learners to think about their users' reasons for using a document and to think carefully about how physically, visually, aurally, and electronically those users encounter content. Learners consider and record what a sensory experience with the documentation is like for these users so they can discover the access-design elements that the documents should include. Thus, the access-first lens creates important learning opportunities for thinking about accessible design rhetorically.

Design Thinking

The practice of designing artifacts for specific users and situations is not new to TC. For more than 50 years, scholars have sought a process to help them use design to improve communication (e.g., Arnheim, 1969; Brumberger, 2004; Gallagher, 2020; Kress & van Leeuwen, 2006; McKim, 1980; Schriver, 1997; Williams, 2015). For our pedagogical approach, we start from a high-level view of a problem-solution situation to help learners think about designing visuals. They learn that they have primary visual information someone else needs—the problem—and that they must design documents with accessible visuals to effectively share that information—the solution. To support this design work, we teach design thinking theory and processes.

Design thinking was developed by IDEO, a multinational design firm known for innovative methods and client solutions (Brown, 2009). IDEO pioneered a humanistic design procedure to target users and professional problem-solution scenarios. We adopted IDEO's procedure because it is user-centered, and many existing UX and social justice pedagogies lack a clear, practicable process. For our pedagogical goals, we added the access-first design lens to focus on users with low vision and blindness. IDEO's design thinking is a framework for design that mirrors the writing process: understanding the problem and context, observing users, defining users' needs, ideating solutions in response to users, prototyping the best of those solutions, and testing the designs before implementation (Purdy, 2014, p. 627). Thus, design thinking's user devotion and empathy work with the access-first design lens.

As part of our training, we have our TC students use the first few steps of the design-thinking process. We simplified the four design-process steps into four questions that designers must answer prior to designing their visual information:

1. Understanding—"What do you understand about the challenges users with low vision and blindness may face when using your visual?" (List the challenges and strategize how to respond to each one.)

2. Empathizing—"What obstacles to accessing visual information would make me want to give up, cause me to get it wrong, or force me to ask for help?" (List obstacles and determine how to remove or surpass them.)

3. Defining—"Can you define a visual communication design that meets these users' needs and expectations that can inform your own design work?" (Record this definition and plan to make it actionable in design.)

4. Ideating—"What visual communication design elements and forms best respond to the definition I created based on these users' challenges, obstacles, needs, and expectations?" (Create several designs using these elements and forms that can be tested for accessibility prior to publishing directly with users or indirectly through grading criteria.)

Thus, our pedagogy works to combine access-first design with rhetorical analysis and design thinking theories and practices to form a solid foundation. We teach our students to not only analyze users with low vision and blindness to inform their design choices but to empathize with them and attend to their needs (Gallagher, 2020), thus creating balance (equilibriUX) between the designer and all users. In brief, learning these concepts and principles of rhetoric and design prepares our professionals for practicing accessible visual communication.

Part Three: The Visual Language Matrix

The practicable scaffolding for breaking down visual messages comes from training TC professionals to use visual language and a tool that supports systematic interpretation of visual communications: Kostelnick's (1989) visual language matrix. According to Kostelnick, teaching visual language and analysis prepares professionals who "need to know how to write and design documents consistent with the needs and expectations of their readers" (p. 77).

Training in Kostelnick and Roberts's (2011) visual language teaches designers how to communicate about visuals and how to use the design language of the "visual discourse communities" of professionals (p. 35). Visual language names the elements of information design—from the anatomy of fonts to the formatting of CSS—and supports analysis of those elements via the myriad conventions of professional documentation. When we teach this visual language, learners develop "familiarity with [how] to communicate [a visual's] intended meaning" (p. 35). Therefore, visual language enables learners to describe elements to audiences, to use them when designing, and to explain visual meaning from a granular to a macro view. For example, when we teach about visual "building blocks" (p. 82) of documents, they use visual elements from the very small (like fonts, kerning, and punctuation) to the very large (like featured images, navigation bars, and templates). They learn what roles these elements play in meaning making and how the textual, spatial, and graphic manipulation of these elements are interpreted situationally.

Learning the visual design elements and conventions in TC takes time, and when a designer breaks a particular convention, that choice may have different meanings in different situations. For this reason, visual language learning focuses on the attributes of technical information from an elemental view (i.e., a discussion of parts) and their interpretative value in specific situations. Targeted dissection helps the learner to study different documents by individual visual parts because "observing and analyzing documents can go a long way toward helping you acquire visual language skills" (Kostelnick & Roberts, 2011, p. 39) and toward sharing the meaning of visuals with those who need it.

As part of teaching visual language, we provide Kostelnick's (1989) visual language matrix as a tool to deconstruct visuals so learners practice identifying the parts and conventions of visual communication. They then better interpret the visuals' meaning in situ, piece by piece, for different users and needs. In this manner, Kostelnick's "model of visual communication" gives students a framework "to analyze and design technical documents as visual systems" of design elements (p. 77). The model consists of a matrix of 12 cells: four rows to present "levels of design" and three columns for "modes of coding" (pp. 81–83; see Table 4.1).

Table 4.1. Kostelnick's 12-Cell Visual Language Matrix

	Textual	Spatial	Graphic
Intra	1 variations in style, weight of letters, numbers; upper and lower case, italics	2 local spacing between textual units; size of letters, CPI, kerning; super- and subscript	3 punctuation; marks for emphasis; underscoring; iconic letter forms
Inter	4 levels of headings; numbers, letters, and symbols that signal textual structure; initial letters	5 variations in line endings, indentation; justified text; lists, matrices, tree configurations	6 bullets, icons; line work on tables; arrows, geometric forms on charts and diagrams
Extra	7 text that labels or describes pictures and data displays; legends, captions; numbers on graphs	8 plotting of data on X-Y axes; space between bars, lines; viewing angle, size of pictures; perspective	9 tone, texture on data displays (bars, pie charts); drawings, diagrams; detail in pictures, levels of realism
Supra	10 section titles and numbers; page headers, pagination; tabs that divide longer documents	11 arrangement of extra-textual designs in document layout; page breaks; size, shape, page thickness	12 marks, icons, color, line work, and logos that unify pages, screens, and sections of the document or website

Source: Created by the author.

The four levels of design—Intra, Inter, Extra, and Supra—are arranged from the smallest design elements to the largest, top to bottom, and the three types of visual coding—Textual, Spatial, and Graphic—are across the top from left to right. Each represents a category of visual encoding: Textual involves word forms, Spatial the different types of positive and negative space, and Graphic includes the many attributes and kinds of images. Each of the 12 cells provide descriptive lists of visuals that belong to the combination of level and code.

For example, at the *Extra* level of *Graphic* design, visual elements like use of shapes to show numbers in data displays, the amount of detail in a picture or a logo, the use of drawings or photographs, and other types of image effects are all identified as belonging to this combination of level and mode. These examples offer learners a guide to understand what elements to consider at different levels of a document's design and how these elements relate to each other across the codes. However, as Kostelnick and Roberts (2011) pointed out, the boundaries between adjacent cells are sometimes flexible. In other words, an element may seem to fit into more than one category. Kostelnick and Roberts have indicated that this placement issue is acceptable if the final position of the element aids the interpretation and creation of the visual information.

Learning the matrix results in designers who are suited to combining systematic visual analysis with access, rhetorical, and design concepts to generate rich visual translations and accessible materials. A designer must not describe every minute detail to a user, but they must be able to interpret the communicative value of the visual components holistically and formulate appropriate tags, texts, and descriptions that share all essential information. Therefore, this effort supports our equity-driven pedagogical approach by using visual language and the matrix to improve the development of effective accessible design elements and to share visual information accurately with users.

PEDAGOGY DEMONSTRATION

Using this pedagogical approach, the first author taught a visual communication course for undergraduate and graduate TC students. Students participated in design labs using Adobe software and in lectures, discussions, activities, and assignments driven by an access-first design initiative, visual rhetoric, and user-centered design. They learned about visual-design elements, accessibility, rhetoric and design-thinking theories, visual language, and use of Kostelnick's (1989) matrix. Additionally, they confronted ethical challenges involved in technical design work for clients. They learned the legal and moral imperatives behind meeting the needs of users with low vision and blindness in digital environments. Finally, they produced a visual rhetorical analysis report (VRAR) on a technical document with primary visual information and offered a redesign to meet users' needs and expectations. (An overview of the unit is provided in Table 4.2.)

Table 4.2. Visual Communication: Introduction to Accessible Design (General Overview)

Week	Topic(s)	Practice(s)
1	Visual Rhetoric & Analysis	Practicing visual rhetorical analysis
2	Users & Access First Design	Generating user personas
3	Design Thinking & Procedures	Exercising empathy and perception
4	Design Elements, Conventions, & Access	Evaluating designs for convention and access
5	Visual Language & Matrix Analysis	Applying Kostelnick's (1989) matrix to documents

Source: Created by the author.

Hereafter, we discuss the 5-week schedule, theories and practices, examples of coursework, and the major assignment for evaluation.

Week 1

During the first week of the unit, students were introduced to concepts of visual rhetoric and analysis. The class reviewed the classical and contemporary definitions of rhetoric, as well as the use of Aristotle's three rhetorical appeals. Then, these terms were recast to consider the persuasive nature of visuals, and why visual appeals are used to get the attention of users and to support logical, authoritative arguments in technical documents. Students then practiced visual rhetorical analysis in class by looking at how visual messages from sample professional and technical documents were used to communicate information effectively (or ineffectively) for specific users, purposes, and contexts.

Observing students' development during Week 1, the first author found that once students had a rhetorical background, they understood the need for a "communication's visual language [to] be tailored to its audience, [to] enable the communication to fulfill its purpose and be suitable for the context in which users actually interact with [it]" (Kostelnick & Roberts, 2011, p. 40). After students practiced visual rhetorical analysis, they were primed to look closer at users and the importance of access to visual messages.

Week 2

During the second week, students focused on users and access-first design. The class covered user-centered design considerations like how a user's background, education, knowledge, communities, and experiences shaped visual information needs and expectations. They studied how knowing user attributes can help to focus visual-design choices. Then, the core purpose of access-first design—to create visual communication focused on accessibility first to address potential user problems with visual communication—was discussed, including moral and ethical dimensions of accessibility for digital documentation. Students practiced creating user personas in groups to define the typical user for a popular citation website. Then, they considered how their user's experience of the site would be different if that user had low or no vision. From this new perspective, students discovered challenges that their users may face.

Week 3

During the third week, students' concept of design work was recast using designing thinking theory and design procedures. They considered the human-centered nature of design work and the need to adopt a problem-solution approach to addressing design needs, especially for visual communication. These concepts were necessary to move students to empathy and consider users' experiences of design, to encourage them to observe users, and to ingrain the importance of empathizing with users, which are essential to defining and responding to visual communication accessibility problems. The first author then introduced the steps of IDEO's design-thinking process by comparing it to the writing process they were already familiar with. To practice using the first few steps of the process students were asked to use empathy and record the challenges of users with low vision and blindness. They were given the actional questions from design thinking (earlier in this chapter) to answer while revisiting their work with the citation website from Week 2. Students then began to cover how human perception of visual information influences the interpretation of that information. They were given instruction on basic design principles (e.g., CRAP and Gestalt concepts) and were shown how these forms impact visual messages in context. The introduction to perception offered a segue to discussing relationships between the perception of design elements, conventions, and access to make visual information more accessible by being context aware.

Week 4

During the fourth week, students connected human perception and conventions to visual language and the use of access design elements—namely tags, alt text, and visual content descriptions to make visual information accessible for users. Additionally, students were asked to submit the technical document they wanted to analyze for their visual rhetorical analyses. Once approved, these documents were submitted to the class to practice evaluating samples with the instructor and to use Kostelnick's (1989) matrix to dissect the visual language. On their own, students practiced the first two stages of design thinking—observing users (directly or indirectly) and empathizing with their experiences. Students recorded the characteristics of all of the user experiences that they could gather and evaluated these experiences based on user access as a key factor. Using the third and fourth stages of design thinking—ideation and prototyping—students were asked to brainstorm and sketch accessible design elements for the primary visual information in their documents. At this point, they were ready to learn how to use the visual language matrix on their own for their analyses.

Week 5

Considering Kostelnick's (1989) matrix during Week 5, students used professional design language to structure and communicate their visual analyses. The matrix required students to dissect the visual language of a technical document systematically and to describe the visual design elements and conventions across the four levels and three codes. They evaluated how well the document's visual elements responded to the rhetorical situation and focused on how to make arguments about whether the visual language achieved its purpose by meeting the needs of its intended users.

As students prepared to start their analyses, they attended a workshop during which they used the matrix to break down their documents' visual design language, to work on intercepting and assessing the visual communication, and to ask any questions. The learners also finished the visual-language crash course for describing design elements and accessible choices for typography, textual fields, accessory visuals like charts, pictures, and icons, and whole document—"top-down design elements" (Kostelnick, 1989, p. 91), which allowed the professor to ensure that students had optimal scaffolding. At the end of Week 5, students had the "visual vocabulary" to express "design decisions [ranging from] using uppercase letters for a phrase to large-scale decisions about the size and shape of the communication" at their disposal for report writing based on their matrix analyses (p. 81).

Major Assignment

At the end of Week 5, students submitted the VRAR, which required (1) a rhetorical analysis of a document, (2) a study of the document's visual design using the matrix, (3) an interpretation of the visual messages and their effectiveness based on users, and (4) redesign recommendations that included access design elements. Students also were required to justify their access element inclusions using rationales based on low-vision and blind users' needs.

PEDAGOGY RESULTS

The first author evaluated student reports using a five-part rubric and the standard academic scale—grading for context, content, organization, style, and delivery. The author found that the majority (75% or more) of student reports

- articulated the results of their rhetorical analyses with competence;
- offered clear and complete descriptions of their document's visual language;
- assessed the effectiveness and accessibility of visual information in their documents;

- used content tags for their documents when they needed to group topics internally;

- contained alt text descriptions for each visual that they included in their own reports;

- crafted alt texts for visuals that contained visual language of the image, its message, and provided context based on the document;

- addressed and maintained the visual information integrity of documents; and

- redesigned complex visuals like data displays or tables following guidelines for visual content descriptions established in the "WAI-ARIA approach" (California State University, Northridge, n.d.).

Thus, student deliverables were favorably received, and these young professionals illustrated skills for using tags, alt text, and visual content descriptions to share complex technical visuals in-line with our pedagogical aspiration to help users with low vision and blindness.

Further, we compared overall performance data based on final grades for student deliverables from traditionally taught courses with those created from this pedagogical demonstration. On average, we found that performance evaluations done by the same faculty member increased 11% (a letter grade) between courses, which indicates that the new approach was at least better for preparing students for the VRAR, and that they performed well using the newly emphasized access design elements. Though this study presents preliminary findings, we are encouraged because the student work holds important information in visual form that must be transmediated for low-vision and blind users. Therefore, the improvement indicates that our approach may help students make a more equitable visual information experience for users, and our unique pedagogical combination of theories, elements, and tools shows promise to enhance information that is user centered over document centered.

Pedagogy Takeaways

Using this approach can help students learn to create visual information that supports a more equitable experience. By providing a complex way of thinking and a collection of processes and tools to make visuals accessible, the approach better prepares students for real world users of visual information and the challenges they may face. For example, as our students learned to assess visual messages rhetorically and to design empathically, they were given tools to put messages together with user needs at the center of interpreting and making visuals. Their practice making technical documentation accessible via simple access design elements integrated these elements into their repertoire and increased the usefulness of their documents for users with low and no vision. Overall, from what we observed, the new approach helped them learn that as technical writers they must maintain an object's visual messages through accessible design choices based on empathy with the audience and careful consideration of multisensorial needs.

In our observations and performance comparison, the young professionals showed that they were better able to create technical communications that preserve information integrity and make information easier to access for more users. For example, when students had to convert the information in tables using visual content descriptions, they practiced complex analytical skills to turn visual content into interpretations of trends and stories to give them meaning. This action not only made the information accessible but also more approachable for users. With our pedagogy, we aimed to help technical communicators make the best support possible for visual content. We now seek to pass along a few takeaways from our experience that may improve visual communication instruction in accessibility in any other technical communications course.

When using our pedagogical approach, consider these four takeaways from our experiences that may improve outcomes:

1. Teach tags, alt text, and visual content descriptions using demonstrations and redesign activities. (We recommend having students write a tag to connect common visuals in a textbook, having them write alt text for an image on a course website, and having them make a visual content description for an organizational chart.)

2. When teaching how to do visual rhetorical analysis and design thinking, provide step-by-step procedures students can follow. (We recommend numbering the parts of the rhetorical situation for analysis and having them compare/

contrast the writing process and design thinking process with guidance.)

3. If an instructor wants students to use Kostelnick's (1989) matrix, they must demonstrate it first. (We recommend using the matrix to dissect the visual levels and codes of a newsletter together to break the ice.)

4. The accessibility unit in our course demonstration can (and probably should) be expanded into a whole class. (We recommend making the VRAR into a capstone assignment and expanding each week into its own unit if you want to design a course around accessibility and visual communication.)

We hope these takeaways improve the experience of using this pedagogical approach in any classroom.

Conclusion

It is inspiring to consider the people with different levels of visual needs that this work may serve, including blind students we help through our course management systems, the low-vision patrons we guide through our libraries, and the users with visual signal-processing issues. As the world shrinks, our digital lives grow, and the visual information we use continually streams in endless iterations. Focusing attention on access, use, and recognition of all audiences and the value of including their voices is part of our calling for user-centered design advocacy.

This project has several limitations that should be addressed in future research. First, further comparison of designs made with and without our approach is necessary. This additional examination will help us look at how well the training works for the targeted user group, especially when other educators reproduce the instruction. Second, this current approach lacks user testing of artifacts produced in the classroom. A study that involves users with low vision and blindness should be conducted to determine if their experiences support the seeming usefulness of this approach. Third, although our teaching considers several professional fields and audiences, more field-specific data should be addressed. Future studies may help us find and fill these gaps and will allow us to refine the methodology to maximize the benefits of our pedagogy. Finally, we have not discussed all users; we must consider other users whose communication experience may be more equitable through continued research into improving information accessibility.

References

Alcántara, A. (2021, September 1). More companies are looking to hire accessibility specialists. *The Wall Street Journal*. https://www.wsj.com/articles/more-companies-are-looking-to-hire-accessibility-specialists-11630501200

American Council of the Blind. (2016). *The audio description project*. http://www.acb.org/adp/netflix.html

Arnheim, R. (1969). *Visual thinking*. University of California Press.

Bennett, K. C., & Hannah, M. A. (2022). Transforming the rights-based encounter: Disability rights, disability justice, and the ethics of access. *Journal of Business and Technical Communication, 36*(3), 326–354. https://doi.org/10.1177/10506519221087960

Bivens, K. M., Cole, K., & Heilig, L. (2020). The activist syllabus as technical communication and the technical communicator as curator of public intellectualism. *Technical Communication Quarterly, 29*(1), 70–89. https://doi.org/10.1080/10572252.2019.1635211

Brown, T. (2009). *Change by design*. HarperCollins.

Brumberger, E. (2004). The rhetoric of typography: Effects on reading time, reading comprehension, and perceptions of ethos. *Technical Communication, 51*(1), 13–24.

California State University, Northridge. (n.d.). *Best practices for accessible images*. Universal Design Center. https://www.csun.edu/universal-design-center/best-practices-accessible-images

Cawley, C. (2017, January 20). *The world is now more visual than ever before*. Tech.co. https://tech.co/news/world-now-visual-ever-2017-01

Connors, R. J. (2004). The rise of technical writing instruction in America. In J. Johnson-Eilola & S. A. Selber (Eds.), *Central works in technical communication* (pp. 3–19). Oxford University Press.

Conway, M., Oppegaard, B., & Hayes, T. (2020). Audio description: Making useful maps for blind and visually impaired people. *Technical Communication, 67*(2), 68–86.

Gallagher, P. B. (2019). Design: Changing zeitgeists, changing communication. In D. M. Baylen (Ed.), *Dreams and inspirations: The book of selected readings 2018* (pp. 194–205). International Visual Literacy Association Press.

Gallagher, P. B. (2020). *Technical communication as design: A design pedagogy study* (ID#8892). [Doctoral dissertation, Iowa State University]. ProQuest. https://doi.org/10.31274/etd-20200624-64

Kostelnick, C. (1989). Visual rhetoric: A reader-oriented approach to graphics and designs. *The Technical Writing Teacher, 16*(1), 77–89.

Kostelnick, C. & Roberts, D. D. (2011). *Designing visual language: Strategies for professional communicators* (2nd ed.). Longman.

Kress, G., & van Leeuwen, T. (2006). *Reading images: The grammar of visual design* (2nd ed.). Routledge.

Lancaster, A. (2018). Identifying risk communication deficiencies: Merging distributed usability, integrated scope, and ethics of care. *Technical Communication, 65*(3), 247–264. https://www.ingentaconnect.com/content/stc/tc/2018/00000065/00000003/art00003

Lotito, A., Spoto, G. L., Frisiello, A., Macchia, V., Bolognesi, T, & Ruà, F. (2013). Smart2poster. Bridging information and locality. In *Proceedings* of *the 15th International Conference on Human-Computer Interaction with Mobile Devices and Services* (MobileHCI '13). Association for Computing Machinery, New York, 582–587. https://doi.org/10.1145/2493190.2494440

Martins, A. (2019, August 16). Your business's website may be unusable to the blind. *Business News Daily.* https://www.businessnewsdaily.com/15264-is-your-website-inaccessible-to-the-visually-impaired.html

McEwan, T., & Weerts, B. (2007). Alt text and basic accessibility. In *Proceedings of the HCI 2007 The 21st British HCI Group Annual Conference.* University of Lancaster, UK (HCI), (1–4). https://www.scienceopen.com/hosted-document?doi=10.14236/ewic/HCI2007.64

McKim, R. H. (1980). *Experiences in visual thinking.* Brooks and Cole Publishers.

Newfield, D. (2014). Transformation, transduction and the transmodal moment. In C. Jewett (Ed.), *The Routledge handbook of multimodal analysis* (pp. 100–113). Routledge.

Oregon State University: IT Department (n.d.). *What are content tags used for?* Web Technology Training: OSU Drupal 7 Overview Technical Manual. https://webtech.training.oregonstate.edu/faq/what-are-content-tags-used

Power, C., Freire, A. P., Petrie, H., & Swallow, D. (2012). Guidelines are only half of the story: Accessibility problems encountered blind users on the web. In *CHI '12: Proceedings of the SIGCHI Conference on Human Factors in Computing Systems* (pp. 433–442). ACM Digital Library. https://dl.acm.org/doi/10.1145/2207676.2207736

Purdy, J. P. (2014). What can design thinking offer writing studies? *College Composition and Communication, 65*(1). 612–641.

Rehabilitation Act of 1973, 29 U.S.C. $35.104 et seq. (1973).

Roberts, L. E. (2006). Using an access-centered design to improve accessibility: A primer for technical communicators. *Technical Communication, 53*(1), 14–22.

Schmidt, E. [Google]. (2010, Feb. 18). Eric Schmidt at Mobile World Congress [Video]. *YouTube.* https://www.youtube.com/watch?v=YuqiE2lukDM

Schriver, K. A. (1997). *Dynamics in document design: Creating texts for readers.* John Wiley & Sons.

Slatin, J. M. (2001). The art of ALT: Toward a more accessible web. *Computers and Composition, 18*(1), 73–81.

Sonka, K., McArdle, C., & Potts, L. (2021). Finding a teaching Ally: Designing an accessibility-centered pedagogy. *IEEE Transactions on Professional Communication, 64*(3), 264–274. https://doi.org/10.1109/TPC.2021.3091190

Telecommunications Act of 1996, Pub. LA. No. 104-104, 110 Stat. 56 (1996). https://transition.fcc.gov/Reports/tcom1996.pdf

Theofanos, M. F., & Redish, J. C. (2003). Guidelines for accessible and usable websites: Observing users who work with screen readers. *Interactions, 10*(6), 36–51.

Theofanos, M. F., & Redish, J. C. (2005). Helping low-vision and other users with websites that meet their needs: Is one site for all feasible? *Technical Communication, 52*(1), 9–20.

United States Access Board. (2006). *Telecommunications Act accessibility guidelines; Electronic and information technology accessibility standards* [Proposed Rule]. Architectural and Transportation Barriers Compliance Board. https://www.federalregister.gov/documents/2006/07/06/E6-10562/telecommunications-act-accessibility-guidelines-electronic-and-information-technology-accessibility

Universitat Autònoma de Barcelona. (n.d.) TransMedia Catalonia. *Department of Translation and Interpreting and East Asian Studies.* https://grupsderecerca.uab.cat/transmedia

Walters, S. (2010). Toward an accessible pedagogy: Dis/ability, multimodality, and universal design in the technical communication classroom. *Technical Communication Quarterly, 19*(4), 427–454. https://doi.org/10.10.1080/1057 22522010502090

Wang, Q. (2000). A cross-cultural comparison of the use of graphics in scientific and technical communication. *Technical Communication, 47*(4), 553–560.

Williams, R. (2015). *The non-designer's design book* (4th ed.). Peachpit Press.

World Health Organization. (2019). *World Report on Vision.* W.H.O. Press. https://www.who.int/docs/default-source/documents/publications/world-vision-report-accessible.pdf

Xie, I., Babu, R., Joo, S., & Fuller, P. (2015). Using digital libraries non-visually: Understanding the help-seeking situations of blind users. *Information Research, 20*(2). http://informationr.net/ir/20-2/paper673.html#.YydZZuzMJro

Xie, I., Babu, R., Castillo, M. D., Lee, T. H., & Youi, S. (2018). Developing digital Library Design Guidelines to support Blind Users. In *Proceedings of the 20th International ACM SIGACCESS Conference on Computers and Accessibility.* (401–403). https://doi.org/10.1145/3234695.3241024

Xie, I., Babu, R., Lee, T. H., Castillo, M. D., You, S., & Hanlon, A. M. (2020). Enhancing usability of digital libraries: Designing help features to support blind and visually impaired users. *Information Processing & Management, 57*(3). https://doi.org/10.1016/j.ipm.2019.102110

Youngblood, S. A. (2012). Communicating web accessibility to the novice developer: From user experience to application. *Journal of Business and Technical Communication, 27*(2), 209–232. https://doi.org/10.1177/1050651912458924

Youngblood, N. E., Tirumala, L. N., & Galvez, R. A. (2018). Accessible media: The need to prepare students for creating accessible content. *Journalism & Mass Communication Educator, 73*(3), 334–345. https://doi.org/10.1177/1077695817714379

Chapter 5

Getting to the Heart of
User Experience and Usability in
Technical Communication Programs

K. Alex Ilyasova and Jamie May

User experience (UX) is deeply dependent on understanding users' needs, wants, behaviors, and the context in which they will engage with a product, service, and/or company. UX is about what users both think *and* feel, and both play a significant role in how users experience a product, service, and/or company. Additionally, users' experiences with a product, service, and/or company may change over time. As Babich (2020), an UX architect and writer for Adobe, explains, "When people start using a new product, they may have mixed feelings about it. However, as they become more familiar with it, they might easily change their minds" (p. 3). As a result, the criteria to evaluate users' experiences over time—value, function, usability, and general impression—may also change. UX, then, is an ongoing process and evolves as UX designers receive new feedback and data from users.

Several UX methods (e.g., journey mapping, empathy mapping, user testing, and persona development) rely on researchers' ability to empathize with and understand their users when collecting accurate data. Empathy is a skillset that involves being aware, communicating precisely, and regulating our own emotions, and it is an invaluable tool in the ongoing

UX process. Reviewing the literature on how empathy is practiced in UX reveals a variety of methods that *include* the use of empathy but do not teach *how* to develop empathy. As one example, from the industry training side, the Nielsen Norman Group offers the following practices on how to include empathy in UX (direct quote):

1. Use Qualitative Research Methods: qualitative methods, such as user interviews, cognitive mapping, and diary studies, allow us to dig into user behaviors, motivations, and concerns [. . .]. Practice empathy as you conduct research.

2. Recruit Diverse Users: make accessibility a part of your research plan. This approach allows you to test your assumptions and explore potential opportunities for improvement with actual end users.

3. Have Your Team Watch Research Sessions and See Real Users: when conducting research, invite all team members and key stakeholders to observe the sessions. Doing this vastly increases the potential of empathy, and the corresponding acceptance of research findings.

4. Use Videos of Users Whenever Presenting Research Findings to Stakeholders: Supplement your findings and recommendations with video clips showing how users actually perform the task. Not only will your findings be more compelling, but you'll also be building empathy towards your audience.

5. Make an Empathy Map: empathy maps capture users' emotions, hopes, and fears and distill your knowledge of the users into one place. An empathy map can help you discover gaps in your current knowledge and identify the types of research needed to address it.

6. Invest in a Diverse Team: Recruit team members with a variety of backgrounds and demographics. That will not guarantee empathy for users, but [it] will at least be a first step in the right direction.

7. Build Empathy into Your Design Guidelines: Within this diverse team, create protocols that encourage empathy. For example, Question Protocol uses consistent intention and

prioritization behind every question, rather than asking all possible questions to the user (some of which may make the user feel inadequate or uncomfortable). (Gibbons, 2019, pp. 3–6)

All these practices are valuable to the UX process, and they all advocate for us to "practice," "encourage," "build," and "increase" empathy at various points. However, the *how* is missing. The methods imply that by *doing*, we will develop, learn, assess, and improve empathy. For example, mapping exercises help map what users say, think, feel, and do (emotion map) or the phases, actions, thoughts, and mindsets/emotions of users (journey map). However, when a user identifies the feelings of anxiety, stress, or fear, it is not clear how we learn to empathize with these emotions in a way that leads UX practitioners to ask questions and to recognize the differences between these emotions and to more accurate and insightful data as a result. Without having a more precise emotional vocabulary and an understanding of emotions (e.g., the differences between anxiety, stress, and fear), researchers may find it more challenging developing questions or getting answers that are insightful. Additionally, building a diverse team does not "guarantee" empathy, as Gibbons stated. At some point we need to ask what can be done to find out why not. Such inquiry may help us develop and improve practices and methods that do increase empathy when we already have a diverse team or when we have gathered a diverse set of users.

Scholarship on technical communication (TC) curricula emphasizes including empathy in courses (especially for teaching UX design, writing, and research) and as it relates to meeting industry and employer expectations. For example, as Getto and Beecher (2016) pointed out, two of the most widely cited books at UX conferences are Unger and Chandler's *A Project Guide to UX: For User Experience Designers in the Field or in the Making* and Gray and colleagues' *Gamestorming: A Playbook for Innovators, Rulebreakers, and Changemakers* (p. 156). Each source provides learners with the methods that are most common to UX practitioners and for gaming projects. Additionally, both mention the importance of including empathy and accounting for the role "fuzzy goals" (such as emotions) have in the success, direction, and motivation of a project (Gray et al., 2010, p. 7). And both explain and offer exercises in the use of emotion maps. However, neither explain empathy, how to develop empathy, what empathy looks like in practice, nor the insight(s) empathy offers. Other texts are similar in scope, providing approaches, mentioning the need to account for the emotional needs of users but stopping short of actually explaining and/or teaching empathy (e.g., Colborne, 2011; Garrett, 2011; Krug, 2014; Lidwell et al., 2010; Rubin & Chisnell, 2008). Such practices, which involve enacting the values of UX and teaching the practice-level work, often do not make it into the classroom. We teach and advocate for thoughtful design and meeting users' needs through a process that requires empathy and compassion, but rarely formalize it in terms of articulated practices—that is, what do we actually teach when we state that empathy is needed, and what do we give the developmental time to in terms of applying and practicing such skills during the process of creation, revision and adaption, and collaboration. Like in industry, often what happens when we teach UX methods and research practices is that we teach the more tangible things like conducting user analysis, developing tailored questions, developing artifacts like storyboards, maps, personas, surveys, usability tests, and wireframes, and we tend to assume that these more tangible practices inherently teach, develop, and incorporate empathy, compassion, and emotion regulation and awareness.

Chong (2016) noted in her analysis of usability pedagogy in general that "[a]s a discipline that focuses on the needs of users, the literature in our field still, to some extent, stops short at addressing the 'rhetorical and epistemological nature' of usability 'practices' in our classrooms" (p. 21). She went on to state that "terms such as 'usability' and 'user-centered design' might sound quite simple [. . .], but [. . .] 'the study of users and their use of products is not straightforward'" (p. 21). Nowhere does that echo more than in the (lack of) teaching of empathy in UX. Recently, scholars and practitioners in UX and game design have been speaking to the practice of empathy: from not mistaking empathy maps for empathy (Siegel & Dray, 2019) to the effect of performative empathic embodiment that is possible in game design (Shultz Colby, 2022). Additionally, recent texts like *User-Experience Research: Discover What Customers Really Want* (2022) offer resources for foregrounding empathy in the process of UX and design thinking. The text is solely focused on the first stage of the five-stage design thinking model: empathize. As Gage and Murrell (2022) explained, "It's important to get this step right, because it is the foundation for everything that follows" (p. 10).

Lastly, research of the skills that UX practitioners need continues to show that soft skills like communication, empathy, and teamwork continue to encompass comments by employers (Lauer & Brumberger, 2016; Rosala & Krause, 2019; Rose et al., 2020). Although limited literature exists on teaching students how to practice empathy, as Cash (2018) explained, "A variety of situations positively show the obligation for technical communicators to understand the emotions and mental processes behind a user's experience and interaction with information" (p. 14).

Additionally, and arguably to our and our users' detriment, we tend to teach from the belief that we are mostly logical beings who occasionally become emotional, when the reality is more that we are emotional beings who are occasionally rational. Nowhere is this truer than in what appeals to us in design: "Beyond the design of an object, there is a personal component, . . . one that no designer or manufacturer can provide" (Norman, 2004, p. 6). Emotions are what move us, and emotions are *also* what move users toward or away from the products, processes, and information. Then we must explicitly include the teaching of empathy and emotion in the UX curriculum and develop explicit teaching practices around emotional literacy in TC programs. These practices would be the how, when, and why emotions and empathy move us and how they help us better understand agency in design.

Though a great need exists for our field to research pedagogical practices for teaching empathy (especially in UX), we aim to establish a lean baseline from which future pedagogical research on empathy might develop. In what follows, we discuss some of the research (albeit from other fields) on what empathy and emotion literacy are and what the roles of emotion awareness and regulation are. From this lean baseline, we offer key aspects that can be adopted in TC academic programs and specifically to the UX curriculum.

The goal of this chapter is to offer practical ways to teach empathy and to demonstrate what the research already confirms—that empathy is a teachable set of skills. We recognize this chapter is limited by our own anecdotal evidence and teaching experiences; offering generalizable teaching best practices or assessment outcomes based on pedagogical research is beyond the scope of this chapter. However, this may also serve as a call to conduct formal UX pedagogical studies and as a focus for future TC pedagogy research. As teachers who advocate and prioritize users' needs, we can incorporate empathy in our UX and TC curriculum to balance the rational and emotional sides of this work and understand better the experiences of our diverse users.

Rethinking Three Myths about Empathy

We start by addressing three myths about empathy. Addressing these myths can shift us into a more productive framework where we can better understand empathy as TC scholars and teachers who call for designers to consider it in UX.

The first myth, often mentioned by students when asked what empathy looks like, is "walking in someone else's shoes." In other words, empathy is the ability to imagine what it is like to be someone else, to be "them." This is putting yourself in "their" situation and taking a walk around to see what that experience "feels" like. Rethinking this myth, Waytz has noted, "When trying to empathize, it's generally better to talk to people about their experiences than to imagine how they might be feeling" (as cited in Goleman, 2017, p. 113). As he goes on to explain, talking to people may seem simplistic, but it is more accurate, and

> A recent study bears this out. Participants were asked how capable they thought blind people were of working and living independently. But before answering the question, some were asked to complete difficult physical tasks while wearing a blindfold. Those who had done the blindness simulation judged blind people to be much less capable. That's because the exercise led them to ask, "What would it be like if *I* were blind [. . .]" rather than "What is it like for *a blind person* to be blind?" (p. 114)

On the surface, this seems like a reasonable approach to gaining a better understanding of another person's experience. However, the main problem with this is that we ultimately *cannot* put on someone else's shoes and experience *how they* have experienced something. What we end up doing is understanding someone else's experience through our own lens, through *our* understanding of *our* world. In other words, it is *our* take on *their* experience. Brown (2021) commented on this notion: "We need to dispel the myth that empathy is "walking in someone else's shoes." Rather than walking in your shoes, I need to learn how to listen to the story you tell about what it's like in your shoes *and* believe you even when it doesn't match my experiences" (p. 123). Learning to listen to and believe others' experiences is particularly useful when working with diverse users who may have different experiences with the same event, practice, or people.

As UX practitioners and teachers, we consistently remind our students that they are not the user. Breaking down and rethinking this myth with students can go a long way in teaching them the skills to learn *how* not to be the user and what it means to empathize and listen to someone else's experience in the process.

The second myth about empathy is the assumption that empathy means feeling the same feelings as the other person. This may seem redundant to the first myth, but there is a useful difference. Myth one is about sharing the experience, doing the same thing they did, and understanding what is involved in that experience. Myth two is about the ability to *feel* the same feelings about said experience. Although these two myths are related, it is useful to separate them to understand the nuances of empathy. The main problem with this second myth is the assumption that empathy can be reduced to a single attribute—the ability to feel the same feelings, and if we do not *feel* the same way, we are not really empathizing. Empathy, however, can be felt or expressed in three different ways:

1. Cognitive empathy: the ability to understand another person's perspective

2. Affective empathy: the ability to feel what someone else feels

3. Empathic concern: the ability to sense what another person needs from you (Goleman, 2017, p. 4).

Cognitive empathy is a skill that enables people to explain themselves in meaningful ways and requires them to *think* about feelings rather than to *feel* them directly (Goleman, 2017, p. 4). It is an outgrowth of self-awareness and curiosity, where we learn to monitor or get curious about our own thoughts and feelings and apply that same level of thought and curiosity to the thoughts and feelings of someone else. This skill is sometimes called *perspective taking*.

Affective empathy is what students tend to think of when we discuss empathy—the ability to feel what someone else is feeling. It is a skill that is useful in group dynamics, mentoring, and managing people. It is the ability to tune into the emotional states of others and make a connection. Goleman (2017) stated, "It requires the capacity to access two types of attention: attention on what we are feeling, and awareness and attention on the other person's face, voice, and other external signs of emotion" (p. 6). In other words, affective empathy requires understanding our own feelings while also relating to the feelings and external signs of emotion of others. It's an emotional attunement.

Lastly, *empathic concern*, the ability to sense what another person needs, is what we want from a project manager, group leader, our boss, our physician, and our spouse. It implies that we care and are concerned for that person and are willing to act. One neural theory, as Goleman (2017) discussed, explains that this response triggers both the amygdala's ability to sense danger and the prefrontal cortex's release of oxytocin, the chemical for caring. The implication, as Goleman explained, "is a double-edged feeling. We intuitively experience the distress of another as our own. But in deciding whether we will meet that person's needs, we deliberately weigh how much we value his or her well-being" (p. 8). This has the potential to overwhelm or numb us. Getting this mix right requires practice and awareness so that our feelings do not over-whelm us and cause suffering, as in the case of burnout or compassion fatigue, or feeling so anxious about people and their circumstances that it leads us to protect or to numb our feelings and lose touch with empathy.

Depending on the situation and the person, the relationship between the people involved, as well as our own general mood, energy levels at the time, and so on, we can empathize in one or more of the ways described. For students, challenging this second myth, that empathy is not singular and different ways to empathize exist, may be a relief. Rethinking empathy provides students the opportunity to recognize that they can empathize in one or more of these ways and to consider how to develop empathy skills more explicitly.

The last myth that comes up with students is about language: the difference between sympathy and empathy and whether empathy is a skill that can be used to hurt people. Sympathy and empathy are not synonyms. Sympathy is actually closer to pity in meaning than it is to empathy. It creates disconnection and a safe distance from which to convey one's sentiment. As Brown (2021) explained, sympathy is about "not me" while adding "but I do feel sorry for you" (p. 124). At best it can be an empathic miss "when we share our experiences with people whose iden-tities afford them more physical, social, and emotional safety" (p. 125). The following example helps clarify the difference further: "'I'm imagining not being called on in a meeting and I don't think it's so painful,' says the man to the woman who is often overlooked, or the white [*sic*] woman to the Black woman who is routinely ignored—versus hearing what that experience felt like for the person sharing the story and responding by saying, 'Feeling invisible is painful' " (p. 125). Sympathy is at a distance. We can feel for the person but do not relate to the situation or feelings. By contrast, empathy involves us; we relate to the other person's experience and feelings. Defining these terms and using them correctly matters if we want to teach empathy and want to include diverse users.

Regarding the second part of this—whether empathy can be used to hurt people—language matters as well. We get the question and have a discussion with students on the difference between empathy and manipulation. This is where we, as technical communicators, scholars, and teachers have another opportunity to illustrate how words and language matters. Empathy is about making a connection and trying to understand the experiences and feelings of another person. If someone is doing this to leverage the emotions of another person, it would not be considered empathy: it is manipulation or exploitation. As Brown (2021) stated, "Language matters. Don't use a word that has an almost universally positive connotation to describe a dangerous behavior that is hurtful. Use a word that issues a caution and demands accountability" (p. 124). The following example from the business management industry helps clarify this further: "Scott Sullivan, WorldCom's CFO, told his staff that recent and rapid decline in revenue warranted that all of them treat the situation like an aircraft carrier and they needed to perform account adjustments so that the 'planes' (the investors, employees, and pensioners who were counting on them) could safely 'land' with their financial security intact" (adapted from Baker, 2017, p. 576). The staff complied despite their misgivings and otherwise ethical records. The CFO committed fraud and manipulated his staff's emotions to do it—their feelings of loyalty to a highly regarded company (the only Fortune 500 company in the state); their respect for the opinions of their CFO; their fear of losing their jobs, reputations, and wealth; and their concern over the fate of their coworkers and community (Baker, 2017). Empathy, instead, would have built on the staff's loyalty, respect, and concern to help identify the root cause of the revenue declines and explore numerous options to address the source of the problem without violating ethical or legal codes. Manipulation uses others' emotions to avoid accountability and responsibility, whereas empathy builds on the feelings of others to create connection and understanding.

Empathy and Making Connections:
Practical Ways to Teach Empathy

In addressing these three myths about empathy, one aspect that is clearer now is that empathy is about connecting with another person by sharing or understanding the feeling and then reflecting that understanding back to the person. With that in mind, empathy is not specifically about the experience but rather about connecting with what someone is feeling about the experience. This requires paying attention to another person, and it also involves "connect[ing] to your own experience in a 'thinking' way that creates emotional resonance: *Oh, yeah. I know that feeling*" (Brown, 2021, p. 128).

In teaching empathy, four useful attributes can be used: perspective taking, staying out of judgment, recognizing emotions, and communicating our understanding about the emotion (Wiseman, 1996, pp. 1162–1167). These attributes are also applicable to UX. In this section, we discuss the skillsets that we can teach to develop the last two attributes—recognizing emotions and communicating our understanding about the emotion.

Research shows that empathy can be taught (Brackett, 2019; Brown, 2021; David, 2016b; Goleman, 2017). As Fuller et al. (2021) summarized, "Didactical approaches such as role-playing could be particularly useful, allowing students to practice improvisation, manage their own emotions, and respond verbally and nonverbally in different contexts" (p. 5). Other teaching strategies from business and medical schools include emotional vocabulary development, service learning, journaling, and case analysis.

Several interconnected methods and tools can help students develop empathy and specifically the three skills discussed earlier (emotion awareness, labeling, and regulation). In the next section, we discuss three approaches to help TC students develop empathy: (1) emotional vocabulary lists and emotion mapping, (2) mindfulness, and (3) reflective writing and storytelling. For ease, the first one is geared toward raising awareness and developing emotional literacy, and the last two are geared toward developing emotion regulation. However, all are interconnected and help with all three aspects of empathy.

Vocabulary Lists and Mapping

Vocabulary lists and mapping can help with learning the language of emotion and with paying attention to emotion. There are a variety of lists and maps available online that can be used to help students develop their awareness and emotional vocabulary, such as the examples below (Tables 5.1 and 5.2 and Figure 5.1). Using these lists and maps can be done in a few different ways. The first involves a three-step process that includes

1. identifying the accurate emotion being experienced;

2. identifying how intense the feeling is; and

3. describing and communicating the feeling as exactly as possible.

One place to start with students is at the beginning of class: to give them an opportunity every class meeting for an emotion check in or "meter reading." This can be done individually, in pairs, or in small groups depending on the size of the class. For example, at the start of a class, give the students 5–10 minutes to check in with themselves and then share that information with the class, with the other student, or within the small group. The goals are to get students to define what they are feeling at the start of class by writing things down. They identify (1) high or low energy, (2) the category or group of emotions that tend to apply to that energy level, (3) two to three emotions from that category or group, and (4) the reasons, triggers, and/or events that have contributed to those feelings. After the fifth week of class, we also asked students to identify how they want to feel (if applicable) and what they can do to move in that direction.

Table 5.1. A List of Emotions by Categories

Angry	Sad	Stressed	Hurt	Happy
Grumpy	Disappointed	Afraid	Jealous	Thankful
Frustrated	Mournful	Anxious	Betrayed	Trusting
Annoyed	Regretful	Vulnerable	Isolated	Comfortable
Defensive	Depressed	Confused	Shocked	Content
Spiteful	Paralyzed	Bewildered	Deprived	Excited
Offended	Dismayed	Cautious	Lonely	Relieved
Irritated	Disillusioned	Worried	Guilty	Relaxed

Source: Adapted from David (2016a).

Table 5.2. Vocabulary of Emotions/Feelings

	Happiness	Caring	Depression	Inadequate	Fear	Confusion	Hurt	Anger	Loneliness	Remorse
Strong	Delighted	Adoring	Alienated	Blemished	Appalled	Baffled	Abused	Affronted	Abandoned	Abashed
	Ebullient	Ardent	Barren	Broken	Desperate	Befuddled	Aching	Belligerent	Black	Debased
	Ecstatic	Cherishing	Beaten	Crippled	Distressed	Chaotic	Anguished	Bitter	Cut off	Delinquent
	Elated	Compassionate	Bleak	Damaged	Frightened	Confounded	Crushed	Burned up	Deserted	Depraved
	Energetic	Crazy about	Dejected	Feeble	Horrified	Confuse	Degraded	Enraged	Destroyed	Disgraced
	Enthusiastic	Devoted	Depressed	Finished	Intimidated	Flustered	Destroyed	Fuming	Empty	Evil
	Excited	Doting	Desolate	Flawed	Panicky	Rattled	Devastated	Furious	Forsaken	Exposed
	Exhilarated	Fervent	Despondent	Helpless	Paralyzed	Reeling	Discarded	Heated	Isolate	Humiliated
	Overjoyed	Idolizing	Dismal	Impotent	Petrified	Shocked	Disgraced	Incensing	Marooned	Judged
	Thrilled	Infatuated	Empty	Inferior	Shocked	Shook up	Forsaken	Infuriated	Neglected	Mortified
	Tickled pink	Passionate	Gloomy	Invalid	Terrified	Speechless	Humiliated	Outraged	Ostracized	Shamed
	Turned on	Wild about	Grieved	Powerless	Terror-stricken	Startled	Mocked	Provoked	Outcast	Sinful
	Vibrant	Worshipful	Grim	Useless	Wrecked	Stumped	Punished	Seething	Rejected	Wicked
	Zippy	Zealous	Hopeless	Washed up		Taken-aback	Rejected	Stormiest	Shunned	Wrong
			In despair	Whipped		Thrown	Ridiculed	Truculent		
			Woeful	Worthless		Trapped	Ruined	Vengeful		
			Worried	Zero			Scorned	Vindictive		

	Happiness	Caring	Depression	Inadequate	Fear	Confusion	Hurt	Anger	Loneliness	Remorse
Medium	Aglow	Admiring	Awful	Ailing	Afraid	Adrift	Belittled	Aggravated	Alienated	Apologetic
	Buoyant	Affectionate	Blue	Defeated	Alarmed	Ambivalent	Cheapened	Annoyed	Alone	Ashamed
	Cheerful	Attached	Crestfallen	Deficient	Apprehensive	Bewildered	Criticized	Antagonistic	Apart	Contrite
	Elevate	Fond	Demoralized	Dopey	Awkward	Puzzled	Damaged	Crabby	Cheerless	Crestfallen
	Gleeful	Fond of	Devalued	Feeble	Defensive	Blurred	Depreciate	Cranky	Companion-less	Culpable
	Happy	Huggy	Discouraged	Helpless	Fearful	Disconnected	Devalued	Exasperated	Dejected	Demeaned
	In high-spirits	Kind	Dispirited	Impaired	Fidgety	Disordered	Discredited	Fuming	Despondent	Down-hearted
	Jovial	Kind-hearted	Distressed	Imperfect	Fretful	Disorganized	Distressed	Grouchy	Estranged	Fluster
	Light-hearted	Loving	Downcast	Incapable	Jumpy	Disquieted	Impaired	Hostile	Excluded	Guilty
	Lively	Partial	Down-hearted	Incompetent	Nervous	Disturbed	Injured	Ill-tempered	Left out	Penitent
	Merry	Soft on	Fed up	Incomplete	Scared	Dizzy	Maligned	Indignant	Lonely	Regretful
	Riding high	Sympathetic	Lost	Ineffective	Shaky	Foggy	Marred	Irate	Oppressed	Remorseful
	Sparkling	Tender	Melancholy	Inept	Skittish	Frozen	Miffed	Irritated	Uncherished	Repentant
	Up	Trusting	Miserable	Insignificant	Spineless	Frustrated	Mistreated	Offended		Shame faced
		Warm-hearted	Regretful	Lacking	Taut	Misled	Resentful	Ratty		Sorrowful
			Rotten	Lame	Threatened	Mistaken	Tortured	Resentful		Sorry
			Sorrowful	Overwhelmed	Troubled	Misunderstood	Troubled	Sore		
			Tearful	Small	Wired	Mixed up	Wounded	Spiteful		
			Upset	Substandard		Perplexed		Testy		
			Weepy	Unimportant		Troubled		Ticked off		

continued on next page

Table 5.2. Continued.

	Happiness	Caring	Depression	Inadequate	Fear	Confusion	Hurt	Anger	Loneliness	Remorse
Light	Contented	Appreciative	Blah	Dry	Anxious	Distracted	Annoyed	Bugged	Blue	Bashful
	Cool	Attentive	Disappointed	Incomplete	Careful	Uncertain	Let down	Chagrined	Detached	Blushing
	Fine	Considerate	Down	Meager	Cautious	Uncomfortable	Minimized	Dismayed	Discouraged	Chagrined
	Finial	Friendly	Funk	Puny	Disquieted	Undecided	Neglected	Galled	Distant	Chastened
	Gratified	Interested in	Glum	Tenuous	Goosebumpy	Unsettled	Put away	Grim	Insulated	Embarrassed
	Keen	Kind	Low	Tiny	Shy	Unsure	Put down	Impatient	Melancholy	Hesitant
	Pleasant	Like	Moody	Uncertain	Tense		Rueful	Irked	Remote	Humble
	Pleased	Respectful	Morose	Unconvincing	Timid		Tender	Petulant	Separate	Meek
	Satisfied	Thoughtful	Somber	Unsure	Uneasy		Touched	Resentful	Withdrawn	Sheepish
	Serene	Tolerant	Subdued	Weak	Unsure		Unhappy	Sullen		
	Sunny	Warm forward	Uncomfortable	Wishful	Watchful		Used	Uptight		
		Yielding	Unhappy		Worried					

Source: Adapted from Drummond (2022). Permission to use under Creative Commons.

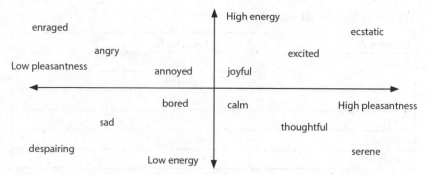

Figure 5.1. Emotion Meter Map. *Source*: Adapted from Yale Center for Emotional Intelligence (Brackett, 2020).

To help students identify emotion(s), it may help to start with the physical sensations and work their way to a specific emotion. For example, noticing their breathing, heart rate, and even how fast or slow they are blinking can help identify whether they are excited or anxious, content, or bored. Another way to start is with an emotion meter map, as shown in Figure 5.1. In the emotion meter map, students first identify feelings of pleasantness (x-axis) and energy level (y-axis), and then use whatever quadrant they land in to choose the emotion(s) they may be experiencing (Brackett, 2020). Next, intensity can be labeled as least intense, middle-level intense, and high-level intense. Defining these as a class exercise might be useful as students learn to pay attention and become more aware of emotion(s). Lastly, the physical sensations and level of intensity can be used with the third step—describing and communicating the feeling with as much detail as possible. Using lists like those in Tables 5.1 and 5.2 is useful in providing options (see Table 5.1) and raising awareness about intensity (see Table 5.2).

Another way to use the lists and maps involves identifying the patterns and progressions of emotion. As Caruso and Salovey (2004) explained, emotions do have rules and certain patterns and progressions (p. 124). Practicing identifying the patterns and progressions can help with communicating emotions and anticipating and understanding what might be felt next. For example (using Table 5.3),

1. Take some of the emotion words under "Anger" from one of the lists, e.g., annoyed, furious, enraged, irritated, frustrated, upset.

Table 5.3. Emotional Management

Column A	Column B
Irritated	Irritated
Furious	Annoyed
Annoyed	Frustrated
Upset	Upset
Enraged	Furious
Frustrated	Enraged

Source: Adapted from Caruso and Salovey (2004).

2. Create two lists of these words: one that is jumbled (Column A) and one that is in sensible pattern (Column B).

3. Have students identify which one is in sensible order and why.

Emotion experts would likely say Column B (Table 5.3) is a better illustration of how emotions progress. This has to do with identifying the increasing levels of intensity and, if the emotion or situation continues, the escalating progression of feelings building. As Caruso and Salovey (2004) stated, "If you know how emotions work and how they don't work, you can learn how to predict the future—at least [understand] how someone might feel if a certain event occurred, how a client will react to your proposal, or how you will feel if you take a new role" (p. 130).

One application of this work could be in the creation of a personal journey map. Since students will hopefully be more in tune with their own emotions after the check-in at the beginning of class, we can then ask them to interact with an interface to recognize their own emotions when completing a difficult task. A journey map (Howard, 2014) is a document that describes and illustrates the process a user goes through to complete another process, and we often focus journey maps on users or composites of users. In this case, we would ask students to complete a journey map about themselves.

One task that has worked in getting students to empathize with users who are completing a difficult task is having students work through the instructions on learning how to knit using only video instructions available on YouTube. This is typically a skill that is relatively new to most students and proves complicated and arduous if you aren't familiar with the process. The exercise would aim to document the emotional journey of learning this task, including the "pain points" or emotional meltdowns. By having students examine their own experiences first, rather than those of others, they are given opportunities to recognize how emotions are part of the process, understand how their reactions are caused by those emotions, and eventually, how they might better understand how our users feel in completing a similar unfamiliar or difficult task.

Using the tables provided can help students to better describe the emotions they encounter when working with a document or interface. Having students create tables, like Table 5.3, can also help them see patterns and progressions to emotions in particular to the process of learning to knit, and ideally, help them identify how to anticipate what might be felt next by someone in their position and if that is something they want to avoid or encourage. Lastly, such an exercise, and specifically this task of learning to knit, tends to avoid the "putting yourself in their shoes" myth by helping students experience what it is like to complete a difficult or unfamiliar task rather than simply relying on hearing from experienced knitters' histories or what someone else might explain to them about their experiences. Creating a journey map, explicitly focusing on emotions and using emotional mapping and vocabulary, will provide students with an opportunity to examine and explain their own emotions so that they can be better aware of others' emotions.

MINDFULNESS

Demystifying this practice and situating it in research may seem to be effective first steps for compelling teachers and students to incorporate it. As David (2016b) plainly explained, mindfulness is "a technique for paying attention, on purpose and without judgment" (p. 97). Arguably this is exactly what we want to do in TC and UX research, with diverse users, and in diverse teams. Recent research on mindfulness shows that practicing mindfulness keeps us from being distracted, helps us focus, and also increases competence. Additionally, a regular mindfulness practice reduces stress and improves memory, creativity, and mood. Regarding emotions, mindfulness allows us to focus on our emotions and pay attention to what is going on inside and around us, noting on how we feel rather than ignoring our feelings, getting entangled in them, or just going along with the program and numbing out. For example, "When you're mindful of your anger, you can observe it with greater sensitivity, focus, and emotional clarity, perhaps discovering that your 'anger' is really sadness or fear" (p. 97).

However, as David (2016b) explained, being mindful and developing a calm awareness of *just being* does not come easily to everyone. In a world full of distraction and competition for our attention, simply being quiet with our thoughts can be miserable for some of us: "A series of studies at Harvard and the University of Virginia put this idea to the test. Psychologist Timothy Wilson and colleagues asked participants to sit alone with their thoughts for a period of ten minutes. Most of the subjects were miserable. Some went so far as to choose the option of giving themselves a mild electric shock rather than simply sitting there and being present" (p. 99). Mindfulness can get us more comfortable with this inner world and create the space necessary to be more reflective on our thoughts, actions, and, importantly, our emotions. In other words, mindfulness can help with emotion regulation—the ability to have any number of troubling thoughts and emotions *and* still manage to act in a way that serves how we most want to behave.

Lastly, developing a practice of calm when sitting with emotions—staying open and nonjudgmental—leads to a natural partnership with curiosity and to creative solutions, another skill we value in UX (David, 2016b). Here we have adapted a few mindfulness exercises from David (2016b); each one has an example assignment for either TC and/or UX application that can develop mindfulness and be incorporated into a particular unit. For each there is a practice and a reflective part to the assignments:

1. **Breath work:** Start with a full minute (add more if you can) and do nothing but focus on your breathing—breathing in slowly and breathing out slowly. It might be helpful to count the in-breath and out-breath (say, for a count of four each to start). The goal is not to empty the mind of thoughts but rather to monitor the thoughts and emotions and just let them be. Each time a thought pops up, notice it, and focus back on your breathing without judging or berating yourself.

   a. **TC unit on workplace documents:** Have students meditate and focus on their breathing when writing a resume and cover letter. After each section in the two documents, for example after the summary of skills section in the resume or the introductory paragraph in the cover letter, have students meditate and try to focus on their breathing for one minute. After each minute of meditating, have them note the thoughts and feelings that came up using the comments feature in Word. Students might be dealing with emotions like fear or rejection, self-doubt, or anxiety. An additional aspect is to have them note the physical sensations like their rate of breathing, tightness in their chest, sweating that may also be happening, which can help them see the connections between specific emotions and physical sensations.

b. **UX unit on journey maps:** Have students pause, meditate for one minute, when they hit a "pain point" or feel an emotional meltdown coming. Having them focus on their breathing, for example, can help them regulate their growing frustration. Having them repeat this multiple times during the journey map process can also help develop their awareness that emotions are fleeting if we give them the space to dissipate.

2. **Mindful observation:** Pick an object in your immediate environment to focus on for a full minute: for example, a pen or pencil, your computer, your shoe. Really look at it and try to isolate and identify its various aspects, features, and dimensions—color, texture, size, movement, and so on.

a. **UX unit on think-aloud protocol:** This could be a two-part exercise for students to empathize with testers during this type of user testing. This can be done in pairs or in small groups of three to four depending on time. The first part would have the students record themselves mindfully observing an object of their choosing, for example, those listed above and thinking out loud about their observations. The second part would have students mindfully observing an object of their choosing again, performing the think-aloud protocol, and having other students observe. After they have completed both tasks, give students three to five minutes to write and reflect on how they felt performing the tasks, and what, if any difference, it made to have someone watching and listening. They can listen to the recording as well to connect with how it feels to hear themselves speak and be recorded. Afterward, in pairs or as a group, they can discuss those reflections and how that might shape how they engage with testers during think-aloud protocol testing.

3. **Rework a routine:** Pick something that you do every day and take for granted—making coffee or brushing your teeth. Next time you do it, focus on each step and action, each element of sight, sound, texture, smell, etc. Try to be fully aware.

   a. **TC unit on instructions:** This can be used to ready students to be fully aware, before starting on creating instruction sets and/or performing the activities that support it. This could take the form of a small reflective write-up of the exercise or a short (2- to 3-minute) recording that has students reflect on how it felt and how well they maintained focus on each step and action, and element of a task.

   b. **UX unit on empathy maps:** Have students fill out the emotion map template for the routine task, documenting the various elements—sees, thinks, says, does, feels, hears. Like with the instructions, have them provide a short reflection on how it felt and how well they maintained focus on each step and action, and element of a task.

4. **Listening:** Pick a piece of music and really tune in—using headphones would be ideal. Try to identify different aspects of rhythm, melody, structure, voices, instruments, etc. (David, 2016b, pp. 101–102).

   a. **UX unit on think-aloud protocol and TC units on instructions:** If you had students record their observations in the think-aloud protocols and/or for the instruction unit to ready them for work, you can use these to have students practice this listening exercise. Like with the other exercises, the reflective part of this assignment would be to discuss how it felt and how well they were able to listen to the recording.

The goal with mindfulness exercises is not to get "good" at it. All of these exercises at some point may trigger resistance in students' minds, distract their thoughts, or cause the urge to do something more "useful." The goal, rather, is to develop the ability to pay attention to those distracting thoughts and urges *and* return to the practice (i.e., counting our breath, observing the object, paying attention to the routine, listening). This goal is worth explaining to and reminding students of throughout the various units, as well as prior to having them complete these specific TC and UX exercises. That is how we develop mindfulness—focus, calm, and nonjudgment, not in the absence of distraction but in the midst of troubling and distracting thoughts and emotions.

REFLECTIVE WRITING AND STORYTELLING

Writing or journaling about emotions is not new. We know that journaling about the events, people, and feelings we encounter can have a positive effect on our mental and emotional well-being. As David (2016b) stated, "We can say with confidence that showing up and applying words to emotion is a tremendously helpful way to deal with stress, anxiety and loss" (p. 92). Using such writing exercises can help develop empathy as well; this is why we include reflective assignments to mindfulness exercises, for example. As students get more comfortable with reflective writing, adding the following specific guidelines to those writing assignments can assist them in developing empathy and better regulation. Have students include

- Pleasant[1] emotion words frequently, because we tend to pay less attention to these;
- Unpleasant emotion words moderately, to help avoid feeding a particularly intense emotion;
- Casual words and phrases like "caused me to" and "led me," to help identify stimulus; and
- Insight words and phrases like "understand" and "realize" to help understand the emotion, stimulus, and response (p. 136).

The following are guidelines for emotions-based journal writing that can lead students to develop empathy and emotion regulation.

1. Write about an event that explores emotions *and* thoughts.

2. Write for 20 minutes per day for three days.

3. Write without stopping.

4. Write without thinking about what to say or how to say it.

5. Include positive and negative emotion words, and casual and insightful phrases.

Students may be asked to write for three days while they try to learn a new task: for example, building on the knitting exercise from the journey maps and having students learn knitting over the course of three days or having them learn a software feature, like pivot tables in Excel over that period. Such journal writing can be written to a specific person (e.g., another user or colleague) or to themselves to help examine and explain how the student felt and what they experienced. Such writing activities also provide opportunities in class to build emotional literacy by getting students to discuss and define pleasant and unpleasant emotion words and additional casual and insight words and phrases they can and/or are using in their journals.

An additional exercise that can be used is storytelling, which is a "a powerful tool in [UX to] help you understand users and their experiences better" (Quesenbery & Brooks, 2010). Unlike traditional storytelling involving exposition, rising action, climax, falling action, conclusion, storytelling in TC/UX classrooms (and later workplaces) could involve creating scenarios of emotion progression. This exercise would be useful to follow the emotion progression exercise discussed above. With emotion storytelling, we start with how someone is feeling as a baseline, and then predict possible reactions to various aspects from an event. Caruso and Salovey (2004) offered the following three steps. First, start by listing three emotions in progression (almost any emotional pattern can be created):

1. You are wondering (baseline).

2. You are anxious.

3. You are surprised.

4. You are feeling shocked.

5. You are feeling devastated.

Using these three steps, students write a story that follows the progression of emotion. In other words, they tell a story that, as reasonably as possible, illustrates how someone can feel each of the emotions. Here is one story based on list of emotions above: "I was sitting at my desk, *wondering* how the job I applied for would change my life. I started feeling *anxious* about when I would hear from the company. I was *surprised* when I heard the email ding and there was an email from them. But when I read that I didn't get the job, I was *shocked*. I was *devastated* to realize that I would not be able to change things right now" (adapted from Caruso, 2004, p. 128). Once the stories are written, students might discuss whether each story was easy to create and/or if the series of events and emotions make sense and why. TC instructors can add more emotions to the list and ask how much harder or easier it would be to tell a story that made sense. "This is not a creative writing exercise; you have to tell a story that would make sense to you and to most other people. Sure, anything is possible, but the less an emotional progression follows the emotional rules, the less sense our story makes" (Caruso & Salovey, 2004, p. 129). The goal with this kind of storytelling exercise is to understand how emotions change and transition, to identify the underlying causes of emotions, and to gain insights or data from these predictions on how to then manage/regulate emotions and respond to a specific situation or event.

A clear connection to UX research practices and storytelling is the development of storyboards. Storyboards are similar to journey maps in that they indicate the steps in a process. However, whereas a journey map might present the "big picture" of a user's experience, a storyboard breaks down steps in a process into more detail, providing specific description of a user's experience, including emotional reactions (Shi et al., 2020). By incorporating storyboarding as an activity, the teacher can ask students to create a storyboard focusing on the nuanced emotions of a user during observation, potentially using think-aloud protocol, as described in the previous section regarding mindfulness.

As mentioned earlier, emotions contain valuable information or data about people, situations, and events. Such data when understood can lead to empathy and help us understand and connect with another person over a shared feeling or understanding of the feeling. The vocabulary lists and mapping, mindfulness, and reflective writing and storytelling are a few ways to develop the skills inherent in empathy. Arguably these exercises and techniques ask us as technical communication scholars and teachers who consistently call for being empathic in UX design to take a few steps back and show students *how* to be empathic. Each of these exercises or techniques can be used in our TC and/or UX curriculum, asking students to make use of vocabulary lists to mapping exercises, storytelling to personas, and all three to how they listen and collaborate when in groups and/or with users that evoke emotion. Additionally, making room for reflection in our classrooms is another aspect that can be built into our current practices.

Empathy and Emotion Literacy: Understanding Different Cultures, Different Users

We would be remiss if we did not address one additional aspect that empathy helps develop: understanding differences in cultures and in users. In the 1970s, Ekman (2003) made the case that all human faces express six "basic emotions" in much the same way across life experiences and cultures:

- Happiness
- Sadness
- Anger
- Fear
- Surprise
- Disgust

However, although we may all facially express these basic emotions in more or less the same way, it is a stretch to conclude that we all experience these emotions the same way. For example, "When schoolchildren in the United States are asked to draw a happy face, it sports a huge smile. When Asian children are given the same task, the smiles are smaller. This doesn't mean they are less happy than their American counterparts, only that perhaps they experience and express their happiness differently" (Brackett, 2019, p. 59). Recent research "suggests that as many as twenty-two emotions are recognized in the face at the above-chance levels (that is, above 50/50) across different cultures" (p. 81).

In addition to understanding facial expressions, accurately reading emotion involves interpreting vocal tones, body language, cultural influences, differences in personality, relationship quality, and even accounting for the opinions of others. To make this even more complex, other prejudices exist like gender stereotypes and implicit racial biases that influence how we read emotions. In other words, neat categories of emotion do not exist. No wonder we are all prone to being misinterpreted and to misunderstanding the emotional states of others. Lastly, evidence suggests that we are actually getting worse in reading emotions. As we spend more and more time communicating through electronic screens and less time face-to-face (or even voice-to-ear), we get less and less practice reading nonverbal cues (Brackett, 2019). Maybe all of this is why we tend to assume this skillset is beyond our teaching abilities, or that we are either innately born to be good at it, or we are not. We argue that all of these factors are why it *is* important to actually teach empathy and emotion literacy because it is not innate and we need to develop it like other skills that we *do* teach.

Empathy and emotion literacy are interconnected and need to develop together. Having a more precise emotional vocabulary (emotion literacy) allows us to process and understand emotions and experiences better (empathy). As Brackett (2019) explained, "Preschoolers have one word for angry: mad. Older children in schools where we work learn to make fine distinctions, using concepts such as annoyed, aggravated, irritated, livid, and enraged" (p. 59). If all we have is the word *mad* or *angry*, it suggests that not only can we not discern the differences between those other words—*annoyed, aggravated, irritated, livid,* and *enraged,* we also can't understand them either. Brown (2021) explained that "over the course of five years of collecting and surveying more than seven thousand people on 'all the emotions that they could recognize and name as they were experiencing them [. . .]' the average number of emotions named across the surveys was three. The emotions were happy, sad, and angry" (p. xxi). Both Brackett and Brown agree to the effect of this limited vocabulary—a limited experience of our world and the world of others. Language is what helps us make meaning: to learn and to gain more self-awareness. Having access to the right words can lead to options for understanding ourselves and others and, if needed, finding solutions to our and others' experiences. In UX, when we do not have the language to talk about what our users are experiencing, our ability to make sense of what is happening to them is limited. This ability to recognize and label emotions is often referred to as *emotional granularity*.

Consequently, developing empathy and emotion literacy can also help us better understand other cultures. There are many languages that have words for emotions and experiences that do not easily translate into English. For example, as Evans (2001) wrote in *Emotion: The Science of Sentiment*, after a friend expressed how glad he was that Evans joined their band: "I can still remember vividly the intense reaction that comment produced in me. A warm wave spread outwards and upwards from my stomach, rapidly enveloping the whole of my upper chest. It was kind of joy, but unlike any moment of joy I had felt before. It was a feeling of acceptance, of belonging, of being valued by a group or people whom I was proud to call my friends" (p. 1). We can probably relate to this feeling, if not that particular moment, because it seems to be what "millions of football fans and religious worshippers seem to feel something similar every weekend" (p. 2). However, we don't have a specific word for it in English. In Japan, it seems the word for this feeling is *amae*. It means "comfort in another person's complete acceptance" (p. 2). So why is there no word for this in English? As Evans explained, "The different ways in which various languages carve up the world reflect different cultural needs. Perhaps the Japanese need a word for *amae* because the emotion it designates accords with the fundamental values of Japanese culture. Unlike the situation in the English-speaking world, which prizes independence, self-assertion, and autonomy, in Japan it is often more important to fit in with others and live in harmonious groups. *Amae* is an emotion that helps people to comply with these values" (p. 3). The fact that in English we do not have a specific word for this feeling does not point out some fundamental difference between people of the US and Japan. The lack of a word does not prevent us from *feeling* that emotion. Rather, learning and understanding what words do or do not exist in other languages for emotions, feelings, and experiences can help us understand what diverse users might value and what their needs are, thus giving us opportunities to connect and understand each other better. Developing competency here means expanding our vocabulary and learning to get more and more granular about our basic emotions. For example, if we start with anger, getting granular can look like the process in Figure 5.2.

We can and often do feel more than one emotion. Learning to get granular about this leads to a better understanding of the data that "angry" is trying to tell us. For example, determining if we are angry and hurt, angry and threatened, or angry and frustrated helps us "be angry with the

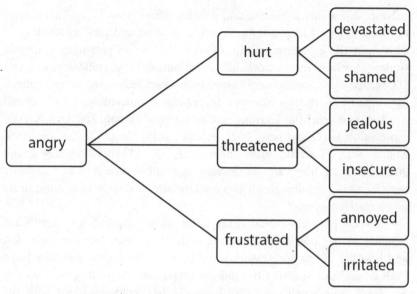

Figure 5.2. Example of Emotional Granularity. *Source*: Created by the author.

right person, to the right degree, at the right time, for the right purpose, and in the right way" that simply being "angry" does not (Aristotle, n.d.).

Connecting all this back to empathy, it is easier to see the components that make empathy possible and teachable. In order to connect with another person over a shared feeling or understanding of the feeling, and then be able to reflect that understanding back to them, we need (1) self-awareness or the skill to notice when emotions are happening; (2) language, to name what we are experiencing or seeing; and (3) the ability to regulate our feelings, so as to not either become overwhelmed or numb to the point where we cannot communicate or reflect our understanding to the other person. The exercises described here can help promote those requirements.

Conclusion

Being educated about empathy leads to a powerful skill set. As Battarbee et al. (2015) explained, "research shows that when we are empathetic, we enhance our ability to receive and process information [. . .]. This type of thinking helps us to put information in context and pick up contextual cues from the environment" (p. 61). In addition to preparing empathic students for their future workplace situations, writing, collaborations, and UX-related work, we also might emphasize newer technological applications for empathy work. For example, increasing advancements in emotional AI, VR, and machine learning makes developing empathy in UX even more valuable (Grandinetti, 2021; Hocutt, 2018; Hocutt & Ranade, 2019; Miller, 2007; Pavliscak, 2019; Tham et al., 2022; Verhulsdonck & Tham, 2022). After all, how can you develop algorithms, data sets, and usability tests for those technologies if you are unaware or unable to communicate or regulate emotions?

Developing emotional skills inherent in empathy has significant benefits. In our post-COVID world, with what seems like more division and hostility in our social worlds and politics, developing empathy helps both us and our students beyond the classroom. According to researchers Zurek and Scheithauer (2017, pp. 57–68), empathy helps with the following aspects:

- Interpersonal decision-making;
- Ethical decision making and moral judgments;
- Short-term subjective well-being;
- Relational bonds;
- Allowing people to better understand how others see them; and
- Prosocial and altruistic behavior.

These additional upsides prove useful if we consider usability as well because they can help with the ongoing part of the development of a design, users, and the rights and interests of users as an ethical responsibility of the designer.

And as we and our users spend more of our emotional life online, empathy practices can help us as developers, practitioners, and humans better balance and understand those demands and needs, especially the unreasonable demands on our time and emotional well-being. Lastly, our students will face new workplace challenges where empathy applies; "Consumers' relationships with—and experiences of—companies are changing, as well. Businesses worldwide are being held increasingly accountable for the long-term social and environmental impacts. This is driving many to adopt new policies and practices around energy conservation, sourcing, production, and sustainability" (Battarbee et al., 2015, p. 62). Companies are being held accountable for the work environment they create—both in terms of the physical spaces of work and the humane aspects of culture, inclusion, and diversity. Empathy can be a powerful force for all of that, and TC and UX practitioners can choose to gain and develop empathy in our practices as a small way to contribute and prioritize those developments.

Note

1. The terms have changed from "positive" and "negative," used by Caruso and Salovey (2004), to "pleasant" and "unpleasant" to avoid judging emotions in the process of writing the emotions journal.

References

Aristotle. (n.d.). *Nicomachean ethics.* http://classics.mit.edu/Aristotle/nicomachaen.html

Babich, N. (2020, November 24). *What you should know about user experience design.* IX Ideas Adobe. https://xd.adobe.com/ideas/career-tips/what-is-ux-design

Baker, D. F. (2017). Teaching empathy and ethical decision making in business schools. *Journal of Management Education, 41*(4), 575–598.

Battarbee, K. J., Fulton Suri, J., & Gibbs Howard, S. (2015). Empathy on the edge: Scaling and sustaining a human-centered approach to innovation. *Rotman Management Magazine, University of Toronto, 2015*(3), 61–65.

Brackett, M. (2019). *Permission to feel.* Cedadon Books.

Brackett, M. (2020, January 19). *The color of our emotions.* https://www.marc-brackett.com/the-colors-of-our-emotions

Brown, B. (2021). *Atlas of the heart: Mapping meaningful connection and the language of human experience.* Random House.

Caruso, D. R. & Salovey, P. (2004). *The emotionally intelligent manager: How to develop and use the four key emotional skills of leadership.* Jossey-Bass.

Cash, B. L. (2018). Beyond audience analysis: Conceptualizing empathy for technical communication. *Cornerstone: A Collection of Scholarly and Creative Works for Minnesota State University,* 2–22.

Chong, F. (2016). Pedagogy of usability: An analysis of technical communication textbooks, anthologies, and course syllabi and descriptions. *Technical Communication Quarterly, 25*(1), 12–28.

Colborne, G. (2011). *Simple and usable web, mobile, and interaction design.* New Riders.

David, S. (2016a, November 10). 3 ways to better understand your emotions. *Harvard Business.* https://hbr.org/2016/11/3-ways-to-better-understand-your-emotions

David, S. (2016b). *Emotional agility: Get unstuck, embrace change, and thrive in work and life.* Avery.

Drummond, T. (2022, June 11). *Emotion/feeling vocabulary.* https://tomdrummond.com/leading-and-caring-for-children/emotion-vocabulary

Ekman, P. (2003). *Emotions revealed: Recognizing faces and feelings to improve communication and emotional life.* Times Books.

Evans, D. (2001). *Emotion: The science of sentiment.* Oxford University Press.

Fuller, M., Kamans, E., van Vuuren, M., Wolfensberger, M., & de Jong, M. D. T. (2021). Conceptualizing empathy competence: A professional communication perspective. *Journal of Business and Technical Communication, 35*(3), 1–36.

Gage, M., & Murrell, S. (2022). *User experience research: Discover what customers really want.* Wiley.

Garrett, J. J. (2011). *The elements of user experience* (2nd ed.). New Riders.

Getto, G., & Beecher, F. (2016). Toward a model of UX education: Training UX designers within the academy. *IEEE Transactions on Professional Communication, 59*(2), 153–164.

Gibbons, S. (2019, April 21). *Sympathy vs. empathy in UX.* NN/g Nielsen Norman Group. https://www.nngroup.com/articles/sympathy-vs-empathy-ux

Goleman, D. (2017). *Empathy: HBR emotional intelligence series.* Harvard Business Review Press.

Grandinetti, J. (2021). Examining embedded apparatuses of AI in Facebook and TikTok. *AI & Society.* https://doi.org/10.1007/s00146-021-01270-5

Gray, D., Brown, S., & Macanufo, J. (2010). *Gamestorming: A playbook for innovators, rulebreakers, and changemakers.* O'Reilly.

Hocutt, D. L. (2018). Algorithms as information brokers: Visualizing rhetorical agency in platform activities. *Present Tense, 6*(3). http://www.presenttense-journal.org/?s=Algorithms+as+information+brokers

Hocutt, D. L., & Ranade, N. (2019). Google Analytics and its exclusions. *Digital Rhetoric Collaborative Blog Carnival 16.* https://www.digitalrhetoriccollab-orative.org/2019/12/19/google-analytics-and-its-exclusions

Howard, T. (2014). Journey mapping: A brief overview. *Communication Design Quarterly, 2*(3), 10–13.

Krug, S. (2014). *Don't make me think, revisited: A common sense approach to web and mobile usability.* New Riders.

Lauer, C., & Brumberger, E. (2016). Technical communication as user experience in a broadening industry landscape. *Technical Communication Quarterly, 63*(3), 248–264.

Lidwell, W., Holden, K., & Butler, J. (2010). *Universal principles of design, revised and updated: 125 ways to enhance usability, influence perception, increase appeal, make better design decisions, and teach through design.* Rockport Publishers.

Miller, C. R. (2007). What can automation tell us about agency? *Rhetoric Society Quarterly, 37*(2), 137–157. http://doi.org/10.1080/02773940601021197

Norman, D. A. (2004). *Emotional design: Why we love (or hate) everyday things.* Basic Books.

Pavliscak, P. (2019). *Emotionally intelligent design: Rethinking how we create products.* O'Reilly.

Quesenbery, W., & Brooks, K. (2010). *Storytelling for user experience: Crafting stories for better design.* Rosenfeld Media.

Rosala, M., & Krause, R. (2019). User experience careers: What a career in UX looks like today. https://www.nngroup.com/reports/user-experience-careers

Rose, E. J., Putnam, C., & MacDonald, C. M. (2020). Preparing future UX professionals: Human skills, technical skills, and dispositions. *SIGDOC '20,* October 3–4, 2020, Denton, TX, USA.

Rubin, J., & Chisnell, D. (2008). *Handbook of usability testing: How to plan, design, and conduct effective tests* (2nd ed.). Wiley.

Shi, Y., Cao, N., Ma, X., Chen, S., & Liu, P. (2020, April). EmoG: Supporting the sketching of emotional expressions for storyboarding. In *Proceedings of the 2020 CHI Conference on Human Factors in Computing Systems* (pp. 1–12). ACM. http://doi.org/10.1145/3313831.3376520

Shultz Colby, R. (2022). Embodying empathy: Using game design as a maker pedagogy to teach design thinking. *Technical Communication Quarterly.* Advanced online publication. https://doi.org/10.1080/10572252.2022.2077453

Siegel, D., & Dray. S. (2019). The map is not the territory: Empathy in design. *Interactions,* March–April, 82–85.

Tham, J., Howard, T., & Verhulsdonck, G. (2022). Extending design thinking, content strategy, and artificial intelligence into technical communication and user experience design programs: Further pedagogical implications. *Journal of Technical Writing and Communication.* Advanced online publication. https://doi.org/10.1177%2F00472816211072533

Verhulsdonck, G., & Tham, J. (2022). Tactical (dis)connection in smart cities: Postconnectivist technical communication for a datafied world. *Technical Communication Quarterly.* Advance online publication. https://doi.org/10.1080/10572252.2021.2024606

Waytz, A. (2017). The limits of empathy: It's exhausting. In D. Goleman (Ed.) *Empathy: HBR emotional intelligence series* (pp. 97–116). Harvard Business Review Press.

Wiseman, T. (1996). A concept analysis of empathy. *Journal of Advanced Nursing, 23*(6), 1162–1167.

Zurek, P. P., & Scheithauer, H. (2017). Towards a more precise conceptualization of empathy: An integrative review of literature on definitions, associated functions, and developmental trajectories. *International Journal of Developmental Science, 11,* 57–68.

Part Two

Rhetoric of Health and Medicine Topics

Several of the authors in this collection are researching how to create balance in the UX methods and processes that they use in creating effective medical and health communication. In the last few years, we have seen innovation in the ways that designers, researchers, and scholars consider mental health intervention, community support systems, patient-centered discourse, and user-centered tools. Even more exciting is that the context goes beyond health and medical communication to consider intercultural communication and design—beyond the United States to Nepal and China. Additionally, their research expands to specialized environments that address multiple purposes and yet maintain balance (that *equilibriUX*)—specifically in end-of-life documentation across the United States. Another study focuses on Audio Descriptions created in collaboration between blind and Deafblind or low-vision users and those who are able to view media to enhance social inclusion and mental health. And the final chapter analyzes materials for college athletes and their coaches who together are addressing eating disorders. The diversity of topics, users, and collaborations in these five chapters is exciting.

Rhetoric of Health and Medicine Topics

Chapter 6

User Localization and mHealth App Design in the Global South Context

KESHAB RAJ ACHARYA

The growing use of mobile health (mHealth) applications or apps to access information and stay current with advancements in the healthcare industry is vital. The U.S. Food and Drug Administration (FDA, 2019) and the Pew Research Center's Internet and American Life Project (2012) recognized the rapid expansion and broad applicability of mobile apps for health information and management. To ensure high-quality mHealth interventions, applications, and overall access for multicultural users, including healthcare practitioners, it is crucial to promote, implement, and assess health-related technologies like mHealth apps in today's globalized society. Research has also shown that mHealth apps can assist users not only in managing chronic diseases and fitness and wellness (Bervell & Al-Samarraie, 2019; Henriquez-Camacho et al., 2014) but also in offering quality medical treatment and improving service productivity (Helmi et al., 2019).

Scholars have investigated several areas of mHealth use in local and global contexts that we can examine broadly for effectiveness (in the sense of a user's ability to complete a task in a given context) and efficiency (in the sense of a user's ability to complete a task with speed and accuracy): clinical decision-making (Adepoju et al., 2017); chronic-disease management (Henriquez-Camacho et al., 2014; Welhausen & Bivens, 2021); medical education and training (Sezgin et al., 2017); and fitness and wellness (Liew et al., 2019). As mobile devices become more convenient with app software capabilities, mHealth apps become a powerful tool for transforming global healthcare delivery. As usability is a prerequisite for the success of mHealth interventions, the design of mHealth apps must take into account the value of a culturally localized user experience (UX) that prioritizes the "use practices of individual local users and values their efforts of user localization" (Sun, 2020, p. 7).

Because users' willingness to try mHealth apps has increased globally (Vaghefi & Tulu, 2019), it is important to address users' expectations and reduce localized usability barriers to promote empowerment and equity, especially in the low- and middle-income Global South (GS) context where health and medical resources are limited. As such, ensuring that these apps are designed in ways that support GS users' knowledge of use, their local conditions and practices, as well as the nature of work they are engaged in, remains critical. Recognizing the importance of and need for bringing uniformity to care-related activities and practices, medical and health researchers have assessed the effectiveness and efficiency of mHealth apps in healthcare delivery in the GS (e.g., Bhatta et al., 2015; Osei & Mashamba-Thompson, 2021; Pokhrel et al., 2021). Technical communication (TC) scholars have also observed how users interact with such apps, attempting to identify users' functional and productive usability concerns (Welhausen & Bivens, 2021), implying that such tools should be designed to meet the UX needs of multiple stakeholders in a complex healthcare system (Kirkscey, 2020). UX research is thus a critical component for the success of mHealth technology in both national and international healthcare environments.

Recently, mHealth apps designed in the high-income industrialized Global North (GN), such as Medscape, Epocrates, PEPID, and UpToDate, are becoming increasingly popular among GS users. These apps are used by healthcare practitioners around the world for different purposes, including making health-related decisions, accessing pharmacological and disease pathology information, studying the etiology of common presentations, and much more (e.g., Acharya, 2022b; Roth, 2014). However, the ways GS healthcare practitioners perceive and experience these emergent digital spaces created in the GN cultural context have not been well explored in the TC field. More specifically, research on the extent to which GN mHealth apps are designed from the perspective of localized usability in the GS context is scarce in the field. The idea of localized usability is to maximize the effectiveness and efficiency of technical products or tools (such as mHealth apps, medical devices, and print or online care-related documents) by considering users' local practices, cultural values, behaviors, and meanings in the context of use.

This chapter reports a case study that investigates Nepalese health-care practitioners' perceptions of and experiences with the Medscape app, which was designed in the GN cultural context to provide medical and healthcare information as well as continuing education for physicians and health professionals worldwide. Semi-structured interview questions were used to collect data in the form of written responses that I call "user narratives." One of the findings suggests that northern designers of mHealth apps should consider developing native apps, which directly interact with the mobile device hardware, for southern users, as well as organizing or managing content in ways that are easy to use and access in a critical care situation. Findings also suggest that while developing mHealth apps for global use, GN designers need to pay heightened attention to GS users' local knowledge making practices, cultural values, behaviors, and meanings as legitimate grounding for enhancing user agency through localized UX design and practices.

Overview of the Medscape App

The Medscape app—a free point-of-care resource—was created in the United States by WebMD Corporation for both iOS and Android device users. It provides a wide range of medical and health information, such as details on medications, medical conditions, treatments, and the most recent clinical guidelines. In addition to featuring the drug database, blogs, and a directory of pharmacies and physicians (Lazakidou & Iliopoulou, 2012), the app also includes tools, such as drug interaction checker, pill identifier, and calculators. More so, the app allows users, especially healthcare practitioners, to follow the customized activity list, stay current in practice, and earn continuing medical education credit (WebMD LLC, 2022). Users can connect with a network of over 250K healthcare practitioners worldwide via Medscape Consult—an online community—to exchange cases, ask questions, and benefit from their knowledge and experiences (Helmi et al., 2019).

The app's drug information includes over 8,500 monographs on prescription, over-the-counter, and herbal/supplemental medications, as well as information from the First DataBank drug library (WebMD LLC,

2022). In addition to the drug class and name, the app also displays pregnancy safety information (Apidi et al., 2017). Its drug reference tool, which includes current prescription and safety information, allows users to learn about recent international medical and health issues. After completing a brief registration process, users can access the entire network of websites and services through the app. As of this writing, the app has been downloaded more than five million times in the Google Play Store alone (Google Play, 2022). Its version (10.2 at the time of writing and 10.1.1 in April 2022; Apple Store Preview, 2022) and database are also updated on a regular basis (Ming et al., 2016).

This app was chosen for this case study because it (1) was developed in the GN, (2) is widely used by Nepalese healthcare professionals (Chaudhary et al., 2019; Pokhrel, 2021), and (3) is free for both iOS and Android device users. As demonstrated by the author's recent survey results, the app is also popular among Nepalese medical students seeking evidence-based information for their medical education and training (Acharya, 2022b). Also, healthcare practitioners in countries like Nepal can use the app to stay updated with recent medical research articles that are combined into a single, easy-to-use platform. The drug reference and interaction checker alone make it an invaluable source for daily clinical use in resource-constrained healthcare settings.

Literature Review

As the TC field increasingly focuses on global audiences, a number of scholars have advocated for localized usability, emphasizing that a product designed for one culture must be revised or customized to fit the user of another (e.g., Acharya, 2019; Dorpenyo, 2020; Gonzales & Zantjer, 2015; Gu & Yu, 2016). Localization, as defined by Hoft (1995), is "the process of creating or adapting an information product for use in a specific target country or specific target market" (p. 11). In user localization, a technical product or tool (such as an mHealth app, a medical device, or documentation) is adapted to specific cultural and technical environments (Suchman, 2002; Sun, 2012; Yunker, 2003). Localizing technical tools by merely adding, altering, or modifying some aspects to meet the target culture in which these tools are utilized without incorporating local practices, behaviors, and values into the design is insufficient for user localization. Sun has addressed this concern by pointing out the gap caused by the existence of two levels of localization: user localization that occurs at the user's site and developer localization that takes place at the developer's site. This distinction clarifies how developer localization falls short of meeting the social affordances of technology in specific context of use. By social affordances, Sun means "the affordances on the activity level that emerge from use interactions in the sociocultural and historical context" (p. 79). Neglecting social affordance in design impacts UX of the product and can limit its adoption in the target culture.

Broadly speaking, user localization should result in an understanding of users' local practices and activities in the target context. However, localization practices are negatively impacted by the lack of a broader understanding of culture, which is the most important factor to consider during the design process (Sun, 2012). Scholars have developed contemporary models for understanding the importance and values of culturally localizing products (Batova & Clark, 2015; Gonzales & Zantjer, 2015; Sun, 2020). For instance, while studying the biometric verification system's usability in Ghana's 2012 general elections, Dorpenyo (2020) found how users "reconfigure the original intent of a technology" (p. 146) and developed the idea of "subversive localization," where users are determined to adapt technologies to fit local exigencies (p. 146). Likewise, through several case studies involving social media technology design in the global context, Sun (2020) illustrated the importance of local users' adaptations of social media tools in improving their localized usability for empowerment and engagement. In another study conducted to examine the functions of mobile text messaging in China, Sun (2012) demonstrated how local users' adaptations of a technology (i.e., text-messaging) can be useful in improving product usability for specific target users in the specific target context.

When designing health-related tools, including mHealth apps, for use in contexts other than the contexts of production, designers should consider how such tools support users to perform desired tasks to accomplish desired goals in their particular local context of use. But, in many cases, performance gains are frequently obstructed by users' unwillingness to accept and use the tools because they fail to meet users' expectations and needs (Vaghefi & Tulu, 2019). These concerns can be addressed by considering high-quality measures as key determinants of user acceptance and adoption at a global level. Such measures include, but are not limited to,

- Perceived ease of use: The degree to which an individual believes that using a specific tool will be easy (Davis, 1989);

- Perceived usefulness: The degree to which the user believes that using the tool will enhance their performance on the job (Davis, 1989);

- Perceived credibility: The degree to which the user of the interface believes the information provided to be trustworthy and persuasive (Fogg, 2003); and

- User motivation: The innate desire to perform an action because it is rewarding (Fogg, 2003).

Additionally, designers should consider incorporating a wider variety of contextually situated sociotechnical components, such as users' local practices, cultural values, behaviors, and meanings in the development and UX evaluation of health-related tools for culturally diverse health-care practitioners in the increasingly globalized world (Acharya, 2019). The failure to consider sociocultural factors in the design process might result in a disconnect between developer localization and user localization.

As our field "recognizes a change in *how* and *why* and for *whom* we localize" (Lancaster & King, 2022, p. 2, original emphasis), the gap between product designers and users can be bridged by understanding the user's local practices and activities in the target culture, easing tensions between developer localization and user localization. Considering these perspectives, the sustained interest in user-centered design has shifted the focus from usability that is associated with "do-goals" (accomplishments with a product) to UX that prioritizes "be-goals" (feelings about a product; Hassenzahl, 2008) to culturally localized UX that emphasizes "sociocultural-goals" (meanings of a product in relation to improving users' lives; Acharya, 2022a). In response to this shift in focus, system designers need to orient their attention toward culturally localized/context-specific UX design for promoting user agency and achieving balance between user localization and UX, or what Lancaster and King (2024) call equi-libriUX, in the design of user interfaces. Given this shift of sociocultural and context-specific turns of usability, the localized usability of health technology can be best determined in consultation with potential users as co-designers.

Co-designing with target users, who can offer valuable user inputs in negotiating design requirements, is one way to meet user needs and expectations in the locally situated cultural context (Dorpenyo, 2020; Rose, 2016). Collaboration is, of course, not new in TC scholarship or practice. Rose's (2016) notion of "design as advocacy," Agboka's (2013) idea of "participatory localization," and Spinuzzi et al.'s (2019) concept of "coworking" communicate the need to collaborate with users or nonexperts. As global markets for healthcare technology continue to grow, mHealth tools need to be developed in collaboration with target users to cater to culturally localized UX practices, behaviors, and meanings. TC scholars have made significant contributions to this sector by employing UX-driven design approaches for the study and development of effective and efficient care-related tools in local and international contexts (Arduser, 2018; Breuch et al., 2016; St.Amant, 2021). For instance, St.Amant (2017a) proposed a script-prototype approach for understanding and addressing the audience's communication expectations in different medical and healthcare settings. Gonzales et al. (2022) advocated for community-based UX collaboration, particularly with Indigenous communities, who actually possess the cultural and linguistic expertise required to create localized health materials that can better serve and represent the world's increasing diversity of people, languages, and cultures. Similarly, Sun (2020) expanded on her *culturally localized user experience* (CLUE)" approach as "*culturally localized user engagement and empowerment* (CLUEE)—the CLUE^2 (CLUE-squared) approach" for designing empowering technology with empowering use practices (p. 6, emphasis in original). Thus, culturally sustaining localized UX approaches are critical in the design of mHealth technology to ensure improved healthcare quality and safety in the GS context.

Method

The primary aim of this study was to investigate the perceptions and experiences of Nepalese healthcare practitioners regarding the use of mHealth technology developed in the GN cultural context for global use. More specifically, the goal was to generate ideas that could inform the design of northern mHealth apps, such as the Medscape app, from a localized usability perspective for multicultural, multilingual healthcare practitioners in the GS. Accordingly, I conducted a case study of the Medscape app for a number of reasons. For instance, as case studies are bound by time and activity (Stake, 1995), they allow researchers to study one or more units of study in depth (Creswell, 2009) and researchers can gather data using methods like interviews and surveys. Similarly, case studies are useful and valuable in developing a thorough understanding of specific people, problems, or situations (Patton, 2014). Furthermore, case studies have demonstrated value and effectiveness for understanding complex systems and embodying UX during the development and testing of healthcare tools (Arduser, 2018; Kennedy, 2018; Kirkscey, 2020).

For the purposes of this IRB approved study, an interview-based case study was conducted to elicit responses from a specific pool of Medscape app users, especially healthcare practitioners, in Nepal. In TC research and scholarship, interviewing key informants has also been proven an effective data collection method (Melonçon & St.Amant, 2018). However, when conducting this type of research in international contexts, researchers must be cautious of cultural differences that can influence the sampling procedure. Many methods of data collection that were developed in the West may not be effective in non-Western contexts. In some Asian cultures, for instance, asking someone face to face about the design of a tool or product and saying something negative about it is considered impolite (Smith et al., 2007). Though previous research has suggested that the mode of interviewing has no effect on participant responses (Sturges & Hanrahan, 2004), careful consideration should be given to which interview method (face-to-face, online, or written) is more appropriate and will facilitate similar quality of data collected via in-person or face-to-face online semi-structured interviews. To engage potential participants in a way that "provided them with a sense of agency and influence over the research process" (Rose, 2016, p. 434), participants were asked about their preferred way of participation: traditional face-to-face, online, or written interviews. All favored the writing option.

Participants who met the following criteria were included in this study:

- user of the Medscape app for at least a month;

- healthcare practitioner in Nepal; and

- proficient in reading, writing, and speaking English.

I utilized asynchronous email interviews in which participants were asked to respond to the same open-ended interview questions as in other types of interviews to generate in-depth and rich qualitative data (Dahlin, 2021). Unlike synchronous email interviews in which both interviewer and interviewee are online simultaneously and questions are asked sequentially in real-time, asynchronous interviews offer participants greater flexibility in answering questions (Amri et al., 2021). The timeline also allows the participant to reflect on the interview questions as well as have a better chance of owning their narratives (Dahlin, 2021). Additionally, the scheduling advantage of asynchronous email interviewing increases participation and access among potential respondents (Fritz & Vandermause, 2018; Hawkins, 2018).

Responses as data were collected via email for this case study. A total of 12 participants (n=12), ranging in age from 21 to 39 years, took part in this study. Among them were physicians, residents, and interns from different hospital departments, including emergency, gastroenterology, pediatrics, anesthesiology, and obstetrics. Individual interviews took place over a 4- to 8-week period of iterative email exchanges. Based on the initial user narratives, probing interview questions included more in-depth inquiries to elicit further details and reflect on emerging themes. Interview questions included details about Nepalese healthcare practitioners' experiences with the case app (i.e., the Medscape app), its usefulness for clinical examinations and medical education as well as training, their overall perception of the app in the GN healthcare context, and design suggestions for effectiveness and efficiency. In addition, the interviews focused on usability-related issues that participants did or did not encounter when using the case app in their locally situated cultural context.

Informed by user localization (Sun, 2020) and grounded theory (Corbin & Strauss, 2015) and following standard qualitative research coding techniques (Coffey & Atkinson, 1996), interview data were coded to identify thematic patterns in the data, patterns common across participants' experiences with the case app. To derive themes, three stages of data coding were performed: open coding, axial coding, and selective coding. While the open coding process allowed me to break down, examine, compare, conceptualize, and categorize user narratives as data, the axial coding process was employed to select core categories, validate the relationships between them, and fill in those that needed further refinement and development. Axial coding was followed by selective coding, the third stage of analysis, when I began to establish a connection between the categories derived from selective coding. At this stage, I was able to select and integrate thematic categories into cohesive and meaningful expressions. Following the formation of thematic categories, each user narrative was reviewed and analyzed in light of its broader implications. The next section provides a subset of findings concerning the effectiveness and efficiency of the case app in the GS healthcare context.

Results and Discussion

Qualitative analysis of user narratives yielded several results. Primarily, narratives revealed four major factors that influenced users' perceptions of and experiences with the case app in the GS context. These factors, as discussed below in relation to localized usability and UX improvement for effectiveness and efficiency, are categorized into four themes: perceived ease of use, perceived usefulness, perceived credibility, and user motivation.

PERCEIVED EASE OF USE

Perceived ease of use, defined as "the degree to which a person believes that using a particular system would be free of effort" (Davis, 1989, p. 320), is a key factor in users' acceptance of information systems. Individuals will be more willing to learn about a system if it is relatively easy to use in their locally situated cultural environments (Acharya, 2019). In the context of this study, perceived ease of use refers to the extent to which users believe that their continued use of the case app is easy. In the user narratives, many comments invoked the issue related to perceived ease of use in relation to increasing user agency in the GS healthcare context, explaining how using the case app is sometimes a challenge to obtain desired information quickly and easily: "I remember being posted in the emergency room and a patient with brucellosis arrived. At the time, someone asked me to check the number of weeks doxycycline should be given, but I could not search for it because the internet wasn't working. There are times when I really think the app with offline content would help." Providing information with graphical and diagrammatic representations also helps users to access information quickly and easily. As indicated by one user narrative respondent, "italicizing, underlining, or bolding key terms or concepts" would reduce the cognitive load on users' tasks.

To improve the case app's perceived ease of use in the GS healthcare context, one user narrative also mentioned the need for incorporating highly descriptive images or pictures into the app along with step-by-step instructions to navigate needed sources quickly and easily: "To make the [app] more usable, I believe the management or clinical sections should be more graphical and diagrammatical in nature; charts and graphs could be more helpful and take less time to read. A step-by-step approach would simplify our task and reduce confusion during treatment procedure, though this may not be possible in all cases." User interaction with mHealth apps can be improved with a clean and simple interface (Welhausen & Bivens,

2021). Excerpts from the user narratives demonstrate how irrelevant content and the frequent need for login information interfere with users' perceptions of the tool's ease of use in other cultures: "The hamburger menu is on the homepage and doesn't contain much useful information. It would help a lot in the drugs and condition section for easy access and save time in Wikimed. Removing news from the homepage will make the UI [user interface] clean." "One of the most difficult aspects of using this app is that I have to log in with my id/password every time I want to use it, which I frequently forget." As mHealth apps gain popularity among healthcare practitioners by providing them with a convenient and timely opportunity to obtain health knowledge and self-management (Wang & Qi, 2021), practitioners should be able to navigate information with ease of use.

Regardless of how important and valuable the mHealth app is, the determining factor of the app's localized usability is the perceived ease of use in the context where the app is adopted and utilized. For instance, GS healthcare practitioners may find northern mHealth apps difficult to use if they have no sidebars or limited subheadings in the top-down menu, as mentioned in the user narrative: "The navigation portion of this app is difficult with no sidebars and limited subheadings in the drop-down menu also force me to scroll through for a while, and I have to look intently to catch the content of my interest." Respondents also reported the lack of "topic-wise division" and "guidelines for different conditions." One user narrative emphasized the importance of "using the right color combination, rather than the stereotypical black and white colors, to make the app easier to use and more appealing," and suggested "adding features like night or dark mode to adjust the brightness according to the user's preference and ease."

Throughout these user narratives, respondents expressed the importance of and need for designing GN mHealth apps that are easy to use and convenient for GS users to manage health-related behaviors and activities. Furthermore, these narratives suggest that GN mHealth apps should allow GS users, especially GS healthcare practitioners, not only to seek answers to health-related issues but also make the process of navigating them as simple as possible.

PERCEIVED USEFULNESS

Perceived usefulness refers to "the degree to which a person believes that using a particular system would enhance his or her job performance"

(Davis 1989, p. 320). An application perceived to be more useful than another is more likely to be accepted by users (Schnall et al., 2015). The case app's perceived usefulness is primarily demonstrated by its ability to deliver pertinent and updated clinical information, as well as by allowing healthcare practitioners for professional networking and patient counseling. Some user narrative respondents also mentioned how they built a network by sharing information about the usefulness of the case app and recommending it to others as an essential source for healthcare practitioners.

> I have always recommended the Medscape app to all my friends and especially juniors because there was someone who rightly suggested that I use this app and this has been a great help and I would like to pass [this information] on to my juniors and friends. There are not many apps that actually provide medical-related stuff [for] free.
>
> In fact, I think almost everyone in our [institution] has this app on their cell phones. One of the reasons I recommend this app is because it simplifies pharmacology. In day-to-day practice, it is impossible to memorize all the dosages and side effects of the drugs. Also, patient counseling is very easy after getting to know about the disease and side effects of the [prescribed] drugs.

Respondents also noted the case app's perceived usefulness for learning more about what is currently going on in the medical industry and sharing information with others. For example, one user narrative respondent stated, "The medical field is a forever evolving field and depending just upon books would never suffice. I suggest this app as a guide to my colleagues and juniors." Another user narrative respondent mentioned that mHealth tools such as the case app should always provide "high-quality information" to enhance perceived usefulness and "important findings in images should always be labeled by providing a short explanation of each."

Highlighting the perceived usefulness of available content in the splash screen, one user narrative respondent also stated that the startup screen "should be clean and irrelevant content such as adverts should be avoided from the screen." Another one suggested adding features "to notify users regarding the current updates and recent advances in the medical field." Incorporating more design features that provide additional access to revisiting chosen topics might boost perceived usefulness, as mentioned in the user narratives: "There should be a feature to make notes so that the article and note appear simultaneously while revisiting any topic." Adding features that address specific user concerns, especially those of GS healthcare practitioners, can increase the perceived usefulness of such apps in the GS context. In other words, the presence or absence of a single design feature can have a significant impact on the perceived usefulness of a mHealth app in a different cultural context.

Because GS countries like Nepal lack widespread high-bandwidth internet infrastructure, many app users experience connectivity issues. So, if users can access downloaded content while offline, they are more likely to use the app frequently, as reflected in a user narrative: "Offline access is a must especially in Southeast Asian nations like Nepal where data costs are extremely high and [users] do not have access to Wi-Fi everywhere." To address such concerns, GN designers should consider developing native mHealth apps that can be installed directly on the device, run entirely on a single processing unit, and store data on the device (Budiu, 2016). By using such apps, users can obtain health-related information on demand without time or location restrictions, especially in nations with expensive data costs and weak internet infrastructure. One user narrative respondent mentioned that stand-alone mHealth apps are also "useful for patient counseling after learning about the disease and side effects of the drugs."

Perceived usefulness of the case app included greater self-awareness of one's knowledge gap, ability to calculate dosages based on weight, ability to explore pathophysiology and clinical features, current patient treatment procedures, and desire to customize app features to suit individual needs in the context of use. User narrative respondents also expressed greater confidence in their ability to provide high-quality healthcare services in the workplace after utilizing the app: "I use the app specially to look up the details of treatment and management of the cases that I am assigned to in my clinical postings. Besides this, I use the app if I have to revise recently learned topics during my break. The apps allow me to have a sneak peek into the recent advances in our field. This also bolsters my confidence in what I have studied and learned." These user narratives provide insight into building mHealth apps by recognizing what constitutes perceived usefulness across GS cultures. In addition to focusing on effectiveness and efficiency for promoting beneficial behaviors via mHealth apps, GN designers should take into account the perceived usefulness of such apps to enhance localized usability. As Mirel (2004) has argued, "Separating usefulness from ease of use and focusing primarily on the latter produces a product of incomplete usability" (p. 32).

PERCEIVED CREDIBILITY

Perceived credibility refers to the extent to which the user considers information sources in human-computer interfaces to be believable and trustworthy (Fogg, 2003). Users' perceptions of credibility affect their intention to alter their attitudes based on information presented in the context of cultural differences and diversity (St.Amant, 2017b). Because mHealth apps offer a wealth of health-related information, users may be exposed to ambiguous or controversial messages (Wang & Qi, 2021). Consequently, perceived credibility is a crucial concept with regard to considering the reliability of information presented in user interfaces. As such, the present study focused on exploring which source credibility was meaningful in the context of the case app adoption by GS healthcare practitioners and how their perceptions of perceived credibility influenced user localization.

Despite some respondents' concerns about information reliability, the majority of them expressed trust in the app because of the content being referenced: "Initially, because I lacked clinical exposure, I was unsure whether I could trust the information provided by the app. As the day passed along with my clinical exposures in the rounds and having read through clinical books, pdfs, and journal articles, I did not find contrasting differences in the information provided in the app. Also, the contents of the app are adequately referenced." Users' perceived credibility can be increased by providing up-to-date information, and sources such as articles and references should be carefully proofread and edited to ensure that information in mHealth tools like the case app and trade literature (i.e., books, book chapters) is consistent. Excerpts from user narratives echo similar sentiments:

> Mostly, I trust the information provided by the Medscape app. But sometimes, due to the mismatch of information in the app and textbooks, I get confused. Also, I have noticed some mistakes when reading the articles included in the app.

> I do trust the app's contents. However, it is actually not possible all the time. There are times when the information provided in sources such as medical books differ from the information

provided in the [case] app. At that time, I find it better to believe the way it is written in those sources.

These narratives support Peng et al.'s (2016) findings that some users will not solely rely on a single source to evaluate app credibility. By checking with other sources, such as trade literature, users will judge the source's reliability and trustworthiness.

Users normally perceive friends' recommendations for adopting care-related tools as highly trustworthy in the GS context (Peng et al., 2016). Additionally, users would recommend such tools as credible sources to their colleagues, as these excerpts from the user narratives explain:

> I would recommend this [case] app to all my friends and others. In fact, I have already done so for more than 10 juniors. Almost everyone in our [institution] uses this app and finds it very helpful and credible.

> Yes, I would definitely recommend this app to my friend. This app has helped me a lot to gain abundant knowledge related to disease conditions, drugs, etc. So, I also want my friends and other medical practitioners to take advantage of this app that contains reliable medical and clinical sources.

The system or tool's level of usage and the sources used to substantiate the information are other factors that impact perceived credibility, especially in the low- and middle-income GS context. This sentiment was also reflected in the user narratives: "The major reason for using the Medscape app is that it is such a large network and the app is used by millions of people all over the world. Credible references are also added to other stuff. This is what builds the trust in me toward this application." Thus, considering that health information has a great impact on health and wellness, especially in the GS healthcare environment, source credibility of mHealth apps is a concern for healthcare practitioners. Research-based references, recommendations, and usage level are factors that influence the perceived credibility of mHealth apps in the GS.

User Motivation

To be motivated means to be moved to action. Unmotivated individuals thus lack inspiration or impetus to act, whereas motivated individuals are energized or activated toward a goal. Motivators are extrinsic (external) and intrinsic (internal) factors that influence whether or not individuals, for instance, use mHealth apps (Peng et al., 2016). Whereas extrinsic motivators, in this study's context, are guided by social factors that influence the adoption of the case app in a social system, intrinsic motivators are internal desires or "energizing forces that arise directly from an activity or situation" such as certain features in the case app that prompt the user to engage with the tool (Fogg, 2003, p. 204). Interpersonal communication channels (face-to-face or word-of-mouth communication) are important external motivators for implementing health-related tools in GS cultural contexts. Examples from the user narratives show how the case app users were motivated to adopt the case app for educational and training purposes:

> I found out about the app from my friends as well as my seniors who used it and then suggested it to me as I made my way around med school, trying to simplify my learning methods and make them as effective as I could.

> I learned about this app from my friends when I was a first-year medical student. Since then, I have been using the app and it has been quite useful. My colleagues and I discuss a lot of topics together while using the information provided in the application.

The user's positionality in the network of advice-seeking and advice-giving relationships that binds a social system—an organization, community, or virtual network—can also predict users' motivation to use innovations or tools (Dearing & Cox, 2018). As inferred in the user narratives, when healthcare practitioners hear about the tool from others within their professional community or observe it being used by colleagues or seniors, they are motivated to use it for a variety of reasons:

> The first time I started using mobile apps was during my second year of medical school. The Medscape app was quite popular at that time in our college as it provides all the details of drugs, dosages, and complete information about diseases, starting from the etiology to the management. We used to have clinical postings of 3 hours every Wednesday, and one of our interns back then had suggested using the app to get

through important differentials quickly and learn the etiology of common presentations.

As these narratives indicate, one of the external motivators was seeing other people using the case app and recommending it to others for beneficial behaviors, which are, in this context, associated with clinical practices and improving and maintaining wellness by supporting users' personal needs, expectations, values, beliefs, and practices. Respondents also mentioned that using mHealth tools such as the case app is not just for diagnosing illnesses, updating best practices, and writing prescriptions; they are also "quite fun to use and play with."

Respondents discussed how internal motivators can impact case app adoption across cultures. Users are more likely to adopt health-related tools that are tailored to their preferences and provide personalized guidance (Wang & Qi, 2021). While one user narrative respondent recommended providing "hyperlinks for medical terminology and conditions," another suggested to include flowcharts "for disease management or patient approach." Two user narratives indicated the importance of customization features in motivating and engaging users in health-related activities:

> When I use a mobile application, I want it to be customizable to a large extent, which is not the case with the Medscape app. I think they should include a creative team and enhance the options for customization of their app for users like us. We should be able to change the layout of the app based on our personal preferences.

> The app should allow users for more customization so that it can be individualized by the users. I would like this option for theming and navigation.

Most user narratives cited medical education and training as factors influencing their use of the case app. One user narrative respondent reported adding other motivating features such as "a feature of posting one clinical question as an MCQ [multiple choice question] daily." Doing so will motivate healthcare practitioners to "check the app at least once a day." Another respondent wished to have a user comment section so that "app designers might get feedback from users." In the user narratives, respondents also noted features for inclusion that would allow them to "join online webinars and medical information sharing networks" locally and globally.

As these user narratives suggest, many case app users, especially GS healthcare practitioners, are motivated to engage in health-related activities if the apps are designed from users' perspectives. Needs and motivations differ among users according to their local logics, practices, behaviors, and meanings (Sun, 2020). Understanding user needs and motivations for technology adoption in diverse cultures might aid designers in developing mHealth tools with localized usability. The implication is that localization should emphasize user motivation and professional engagement, including social, cultural, and environmental factors. In short, the better northern app developers know *when*, *how*, *why*, and *where* GS healthcare practitioners use mHealth apps, the more effectively they can design the apps for user motivation and engagement.

Thus, in examining the Medscape app as a case from localized usability perspectives in this chapter, findings provide insight into building GN mHealth apps by recognizing what constitutes perceived ease of use, perceived usefulness, perceived credibility, and user motivation (see Table 6.1).

By considering the items in the description section as illustrated in Table 6.1, GN developers can address the concern related to how the technology design process can create a "disconnect between action and meaning" in the GS healthcare environment (Sun, 2012, p. 8). Essentially, in designing mHealth technology for worldwide use, northern designers should orient their attention toward extending localized UX practices and principles to increase user agency, support cultural differences and diversity, and expand high-quality health deliverables for GS healthcare practitioners.

Conclusion

Given the global interconnectedness and interdependence, northern designers of mHealth technology must consider the context of use of their products, such as mHealth apps, in GS healthcare settings. Such consideration can allow them to design, conceptualize, compromise, or localize the products for effectiveness and efficiency in GS users' local contexts. User localization work must rethink the cultural and context-specific usability that leads to the production, adaptation, and diffusion of mHealth technology across cultures. This rethinking must start with acknowledging local

Table 6.1. Influencing Factors with Descriptions Based on a Qualitative Analysis of the User Narratives on the Case App in the GS Context

Influencing Factors	Description
Perceived Ease of Use	The app is designed with less sophisticated functions The app's signup process is quick and easy Key terms or concepts are bolded, italicized, or underlined The downloading process is easy and efficient Information is simplified and well organized The user is not required to memorize information in order to perform the same task, such as login information Users can access information needed, including offline content, with speed and satisfaction
Perceived Usefulness	High-quality information and details of treatment procedures are provided Features to notify users about recent advances in the relevant field are available The displayed information is accompanied by explanations and assessments The app works standalone so that users can confirm the disease and side effects of the drug anytime, anywhere Drug brand names, as well as prices based on the countries where the app is used, are included The app is free of irrelevant information, such as advertisements
Perceived Credibility	Content is adequate, accurate, and reliable Information is updated regularly Users can double-check content to ensure its reliability and trustworthiness before downloading The topic of discussion includes accurate references and research-based, scholarly sources The app provides high-quality images and photos that are clearly labeled for better comprehension Information is uploaded by reliable healthcare agencies, institutions, or organizations
User Motivation	The app includes motivating features to encourage users to complete additional tasks, such as interactive questions and quizzes at the end of each topic Users feel a sense of adventure while navigating information The app contains sketch-like figures, animations, and/or flowcharts to approach patient condition Hyperlinks for medical terminology and condition are included Users can customize the layout of the app and create shortcuts for better performance User comment features are available The app allows users to join online webinars and medical information sharing networks both locally and globally

Source: Created by the author.

knowledge-making practices, cultural values, behaviors, and meanings as legitimate grounding for enhancing change through UX design and practices. The development of health-related tools such as mHealth apps with rich functionality for GS healthcare practitioners entails GN designers implementing sustainable UX-driven design approaches as well as more effective usability heuristics for extending localized usability to promote user agency and diversity. As for mHealth app design for multilingual, multicultural GS healthcare practitioners in my case study, the research findings are relatively rich, including perceived ease of use, perceived usefulness, perceived credibility, and user motivation. However, when applied beyond one GS country where the research was conducted, such a focus can make it difficult to judge the accuracy or generalizability of this study's findings.

Though email interviews can generate data that is both necessary and sufficient for some significant research tasks and should thus be included in the qualitative researcher's methodological toolbox (Dahlin, 2021), they do have some limitations. For instance, because email interviews do not always provide the same kind of information as face-to-face interviews, relying on written responses to understand users' behaviors, practices, and experiences may pose limitations. At the same time, those without reliable internet access and computer proficiency are likely to be excluded from participation because both are prerequisites for email interviews. Additionally, the user narratives in my dataset may not reflect how users, especially healthcare practitioners, in other low- and middle-income GS countries are experiencing the app. Thus, to design more effective and efficient care-related tools, further research on the perspectives and experiences of healthcare practitioners from other GS countries is needed. An extensive qualitative study of usability testing or user observations at sight would contribute to a more holistic view of the localized usability of mHealth technology design for GS healthcare practitioners.

References

Acharya, K. R. (2019). Usability for social justice: Exploring the implementation of localization usability in Global North technology in the context of a Global South's country. *Journal of Technical Writing and Communication, 49*(1), 6–32. https://doi.org/10.1177/0047281617735842

Acharya, K. R. (2022a). Promoting social justice through usability in technical communication: An integrative literature review. *Technical Communication*, 69(1), 6–26 https://doi.org/10.55177/tc584938

Acharya, K. R. (2022b). Exploring localized usability implementation in mHealth app design for healthcare practitioners in the Global South context: A case study. *Technical Communication*, 69(4), 44–63.

Adepoju, I.-O. O., Albersen, B. J. A., De Brouwere, V., van Roosmalen, J., & Zweekhorst, M. (2017). mHealth for clinical decision-making in sub-Saharan Africa: A scoping review. *JMIR mHealth and uHealth*, 5(3). https://doi.org/10.2196/mhealth.7185

Agboka, G. Y. (2013). Participatory localization: A social justice approach to navigating unenfranchised/disenfranchised cultural sites. *Technical Communication Quarterly*, 22(1), 28–49. https://doi.org/10.1080/10572252.2013.730966

Amri, M., Angelakis, C., & Logan, D. (2021). Utilizing asynchronous email interviews for health research: Overview of benefits and drawbacks. *BMC Research Notes*, 14(1), 1–5. https://doi.org/10.1186/s13104-021-05547-2

Apidi, N. A., Murugiah, M. K., Muthuveloo, R., Soh, Y. C., Caruso, V., Patel, R., & Ming, L. C. (2017). Mobile medical applications for dosage recommendation, drug adverse reaction, and drug interaction: Review and comparison. *Therapeutic Innovation & Regulatory Science*, 51(4), 480–485. https://doi.org/10.1177/2168479017696266

Apple Store Preview. (2022). Medscape. What's new. Version history. https://apps.apple.com/us/app/medscape/id321367289

Arduser, L. (2018). Impatient patients: A DIY usability approach in diabetes wearable technologies. *Communication Design Quarterly*, 5(4), 31–39. https://doi.org/10.1145/3188387.3188390

Batova, T., & Clark, D. (2015). The complexities of globalized content management. *Journal of Business and Technical Communication*, 29(2), 221–235. https://doi.org/10.1177/1050651914562472

Bervell, B., & Al-Samarraie, H. (2019). A comparative review of mobile health and electronic health utilization in sub-Saharan African countries. *Social Science & Medicine*, 232, 1–16. https://doi.org/10.1016/j.socscimed.2019.04.024

Bhatta, R., Aryal, K., & Ellingsen, G. (2015). Opportunities and challenges of a rural-telemedicine program in Nepal. *Journal of Nepal Health Research Council*, 13(30), 149–153.

Breuch, L., Bakke, A., Thomas-Pollei, K., Mackey, L., & Weinert, C. (2016). Toward audience involvement: Extending audiences of written physician notes in a hospital setting. *Written Communication*, 33(4), 418–451. https://doi.org/10.1177/0741088316668517

Budiu, R. (2016). Mobile: Native apps, web apps, and hybrid apps. http://www.nngroup.com/articles/mobile-native-apps

Chaudhary, R., Rabin, B., & Masum, P. (2019). Attitudes about the use of smart-phones in medical education and practice in emergency department of tertiary care hospital. *Journal of Health & Medical Informatics, 10*(5), 1–6.

Coffey, A., & Atkinson, P. (1996). *Making sense of qualitative data: Complementary research strategies.* Sage.

Corbin, J, & Strauss, A. (2015). *Basics of qualitative research: Techniques and procedures for developing grounded theory* (4th ed.). Sage.

Creswell, J. (2009). *Research design: Qualitative, quantitative and mixed methods approaches* (3rd ed.). Sage.

Dahlin, E. (2021). Email interviews: A guide to research design and implementation. *International Journal of Qualitative Methods, 20,* 1–10. https://doi.org/10.1177/16094069211025453

Davis, F. D. (1989). Perceived usefulness, perceived ease of use, and user acceptance of information technology. *MIS Quarterly, 13*(3), 319–340. https://doi.org/10.2307/249008

Dearing, J. W., & Cox, J. G. (2018). Diffusion of innovations theory, principles, and practice. *Health Affairs, 37*(2), 183–190. https://doi.org/10.1377/hlthaff.2017.1104

Dorpenyo, I. K. (2020). *User localization strategies in the face of technological breakdown: Biometric in Ghana's elections.* Palgrave Macmillan.

FDA. (2019). Policy for device software functions and mobile medical applications: Guidance for industry and FDA staff. https://www.fda.gov/regulatory-information/search-fda-guidance-documents/policy-device-software-functions-and-mobile-medical-applications

Fogg, B. J. (2003). *Persuasive technology: Using computers to change what we think and do.* Morgan Kaufmann.

Fritz, R. L., & Vandermause, R. (2018). Data collection via in-depth email interviewing: Lessons from the field. *Qualitative Health Research, 28*(10), 1640–1649. https://doi.org/10.1177/1049732316689067

Gonzales, L., & Zantjer, R. (2015). Translation as a user-localization practice. *Technical Communication, 62*(4), 271–284.

Gonzales, L., Lewy, R., Cuevas, E. H., & Ajiataz, V. L. G. (2022). (Re)designing technical documentation about COVID-19 with and for Indigenous communities in Gainesville, Florida, Oaxaca de Juárez, Mexico, and Quetzaltenango, Guatemala. *IEEE Transactions on Professional Communication, 65*(1), 34–49. https://doi.org/10.1109/TPC.2022.3140568

Google Play. (2022). Medscape. https://play.google.com/store/apps/details?id=com.medscape.android

Gu, B., & Yu, M. (2016). East meets west on flat design: Convergence and divergence in Chinese and American user interface design. *Technical Communication, 63*(3), 231–247.

Hassenzahl, M. (2008, September). User experience (UX) towards an experiential perspective on product quality. In *Proceedings of the 20th Conference on l'Interaction Homme-Machine*, 11–15. ACM. https://doi.org/10.1145/1512714.1512717

Hawkins, J. E. (2018). The practical utility and suitability of email interviews in qualitative research. *The Qualitative Report, 23*(2), 493–501.

Helmi, R. A. A., Thillaynadarajan, K., Jamal, A., & Fatima, M. A. (2019). Health adviser application for Android. *International Journal of Medical Toxicology & Legal Medicine, 22*(1&2), 125–129. https://doi.org/10.5958/0974-4614.2019.00028.7

Henriquez-Camacho, C., Losa, J., Miranda, J. J., & Cheyne, N. E. (2014). Addressing healthy aging populations in developing countries: Unlocking the opportunity of eHealth and mHealth. *Emerging Themes in Epidemiology, 11*(1). https://doi.org/10.1186/s12982-014-0021-4

Hoft, N. L. (1995). *International technical communication: How to export information about high technology.* John Wiley & Sons.

Internet and American Life Project. (2012). *Mobile Health 2012: Half of smartphone owners use their devices to get health information and one-fifth of smartphone owners have health apps.* Pew Research Center. https://www.pewresearch.org/internet/2012/11/08/main-findings-6

Kennedy, K. (2018). Designing for human-machine collaboration: Smart hearing aids as wearable technologies. *Communication Design Quarterly, 5*(4), 40–51. https://doi.org/10.1145/3188387.3188391

Kirkscey, R. (2020). mHealth apps for older adults: A method for development and user experience design evaluation. *Journal of Technical Writing and Communication, 51*(2), 199–217. https://doi.org/10.1177/0047281620907939

Lancaster, A., & King, C. S. T. (2022). Editorial: Localized usability and agency in design: Whose voice are we advocating? [Special issue]. *Technical Communication, 69*(4), 1–6.

Lancaster, A., & King, C. S. T. (2024). *EquilibriUX*: Designing for balance and user experience. In A. Lancaster & C. S. T. King (Eds.), *Amplifying voices in UX: Balancing design and user needs in technical communication.* State University of New York Press.

Lazakidou, A., & Iliopoulou, D. (2012). Useful applications of computers and smart mobile technologies in the health sector. *Journal of Applied Medical Sciences, 1*(1), 27–60.

Liew, M. S., Zhang, J., See, J., & Ong, Y. L. (2019). Usability challenges for health and wellness mobile apps: Mixed-methods study among mHealth experts and consumers. *JMIR mHealth and uHealth, 7*(1). https://doi.org/10.2196/12160

Melonçon, L., & St.Amant, K. (2018). Empirical research in technical and professional communication: A 5-year examination of research methods and a call for research sustainability. *Journal of Technical Writing and Communication, 49*(2), 128–155. https://doi.org/10.1177/0047281618764611

Ming, L. C., Hameed, M. A., Lee, D. D., Apidi, N. A., Lai, P. S. M., Hadi, M. A., Al-Worafi, Y. M. A., & Khan, T. M. (2016). Use of medical mobile applications among hospital pharmacists in Malaysia. *Therapeutic Innovation & Regulatory Science, 50*(4), 419–426. https://doi.org/10.1177/2168479015624732

Mirel, B. (2004). *Interaction design for complex problem solving: Developing useful and usable software.* Elsevier.

Osei, E., & Mashamba-Thompson, T. P. (2021). Mobile health applications for disease screening and treatment support in low-and middle-income countries: A narrative review. *Heliyon, 7*(3). https://doi.org/10.1016/j.heliyon.2021.e06639

Patton, M. Q. (2014). *Qualitative research & evaluation methods: Integrating theory and practice* (4th ed.). Sage.

Peng, W., Kanthawala, S., Yuan, S., & Hussain, S. A. (2016). A qualitative study of user perceptions of mobile health apps. *BMC Public Health, 16*(1), 1–11. https://doi.org/10.1186/s12889-016-3808-0

Pokhrel, P., Karmacharya, R., Taylor Salisbury, T., Carswell, K., Kohrt, B. A., Jordans, M. J. D., Lempp, H., Thornicroft, G., & Luitel, N. P. (2021). Perception of healthcare workers on mobile app-based clinical guideline for the detection and treatment of mental health problems in primary care: A qualitative study in Nepal. *BMC Medical Informatics and Decision Making, 21*(1), 1–12. https://doi.org/10.1186/s12911-021-01386-0

Rose, E. J. (2016). Design as advocacy: Using a human-centered approach to investigate the needs of vulnerable populations. *Journal of Technical Writing and Communication, 46*(4), 427–445. https://doi.org/10.1177/0047281616653494

Roth, V. J. (2014). The mHealth conundrum: Smartphones & mobile apps—how much FDA medical device regulation is required? *North Carolina Journal of Law & Technology, 15*(3), 359–429.

Schnall, R., Higgins, T., Brown, W., Carballo-Dieguez, A., & Bakken, S. (2015). Trust, perceived risk, perceived ease of use and perceived usefulness as factors related to mHealth technology use. *Studies in Health Technology and Informatics, 216*, 467–471.

Sezgin, E., Özkan-Yildirim, S., & Yildirim, S. (2017). Investigation of physicians' awareness and use of mHealth apps: A mixed method study. *Health Policy and Technology, 6*(3), 251–267. https://doi.org/10.1016/j.hlpt.2017.07.007

Smith, A., Joshi, A., Liu, Z., Bannon, L., Gulliksen, J., & Li, C. (2007). Institutionalizing HCI in Asia. In *Proceedings of the 11th IFIP TC 13 International Conference on Human-computer Interaction*, 85–99. Rio de Janeiro: Springer. https://doi.org/10.1007/978-3-540-74800-7

Spinuzzi, C., Bodrožić, Z., Scaratti, G., & Ivaldi, S. (2019). "Coworking is about community": but what is "community" in coworking? *Journal of Business and Technical Communication, 33*(2), 112–140. https://doi.org/10.1177/10506 51918816357

St.Amant, K. (2017a). Of scripts and prototypes: A two-part approach to user experience design for international contexts. *Technical Communication, 64*(2), 113–125.

St.Amant, K. (2017b). Mapping the context of care: An approach to patient-centered design in international contexts. *Connexions: International Professional Communication Journal, 5*(1), 109–124.

St.Amant, K. (2021). Creating scripts for crisis communication: COVID-19 and beyond. *Journal of Business and Technical Communication, 35*(1), 126–133. https://doi.org/10.1177/1050651920959191

Stake, R. E. (1995). *The art of case study research.* Sage.

Sturges, J. E. & Hanrahan, K. J. (2004). Comparing telephone and face-to-face qualitative interviewing: A research note. *Qualitative Research, 4*(1), 107–118. https://doi.org/10.1177/1468794104041110

Suchman, L. (2002). Located accountabilities in technology production. *Scandinavian Journal of Information Systems, 14*(2), 91–105.

Sun, H. (2012). *Cross-cultural technology design: Creating culture-sensitive technology for local users.* Oxford University Press.

Sun, H. (2020). *Global social media design: Bridging differences across cultures.* Oxford University Press.

Vaghefi, I., & Tulu, B. (2019). The continued use of mobile health apps: Insights from a longitudinal study. *JMIR mHealth and uHealth, 7*(8). https://doi.org/10.2196/12983

Wang, C., & Qi, H. (2021). Influencing factors of acceptance and use behavior of mobile health application users: Systematic review. *Healthcare, 9*(3). https://doi.org/10.3390/healthcare9030357

WebMD LLC. (2022). *Medscape* (10.2) [Mobile app]. App Store. https://www.medscape.com/public/medscapeapp

Welhausen, C. A., & Bivens, K. M. (2021). mHealth apps and usability: Using user-generated content to explore users' experiences with a civilian first responder app. *Technical Communication, 68*(3), 97–112.

Yunker, J. (2003). *Beyond borders: Web globalization strategies.* New Riders.

Chapter 7

How Do You Want to Live . . . or Die?

A Case Study Examining Advance Directive Forms and User Advocacy

Felicia Chong and Tammy Rice-Bailey

As the topic of rhetoric in health and medicine continues to expand in the field of Technical and Professional Communication (TPC), many scholars (e.g., Melançon, 2017; St.Amant, 2017, 2021) have advocated for applying usability approaches to healthcare contexts. Usability approaches have been applied to some medical consent forms. For example, informed consent forms for weight-loss surgery were identified as "technical rhetorics" intended to persuade patients to undergo the surgery, rather than innocuous questionnaires intended for information gathering (Frost & Eble, 2015). However, few discussions in the field of rhetoric and technical communication (TC) focus on the specific topic of death and dying. These include Kopelson's (2019) study on medical doctors' epideictic rhetoric of how to die, along with Keränen's (2007) analysis of the patient preferences worksheet within the context of code status rhetoric at a hospital.

As a field, we are beginning to understand the rhetorical nature of some medical forms, but we know little about advance directives (ADs) and user experiences with them. ADs are "legal documents that allow patients to put their healthcare wishes in writing, or to appoint someone they trust to make decisions for them, if they become incapacitated" (Miller, 2017, para. 5). As Dresser and Astrow (1998) argued, these forms are generic and sometimes unhelpful because the patient's treatment is still largely dependent on the critical care staff's interpretations of the forms and the patient's desire to live. Yet, in 2022, ADs are still being distributed widely without much redesign. TC scholars are well positioned to research issues of user advocacy and experience in the design of these forms.

Although ADs are meant to allow patients to express their wishes, the high education level to read them, inconsistency of the form, and the patient's potential unfamiliarity with the terms and processes likely negatively impact the patient's agency and their ability to advocate for themselves. This chapter examined readability and design issues in ADs. We present a case study of ADs from 34 states in the United States to establish advocacy for those who are facing potential end-of-life choices and care. Our overarching research question is: How are ADs meeting criteria for comprehension and how might this complicate end-of-life decisions made by terminally ill patients?

Literature Review

Although previous research on ADs is limited in the field of TPC, medical researchers, including physicians and nurses, have been studying the effects and usage of state ADs. A content analysis of 50 state ADs (Gunter-Hunt et al., 2002) found a large variability among the forms and advocated for "national dialogue to standardize some provisions of AD documents" (p. 51). A historical overview of ADs (Miller, 2017) argued that nurses need to better understand the AD laws in their states because of how ADs vary per state. Understanding the limitations of ADs, a computer-based decision aid has been created to help individuals with the process of advance care planning (Levi & Green, 2010).

FEDERAL REQUIREMENTS

Under the current federal law, an AD is defined as "a written instruction, such as a living will or durable power of attorney for healthcare, recognized under state law (whether statutory or as recognized by the courts of the state), relating to the provision of healthcare when the individual is incapacitated" (Public Health, Subpart I Advanced Directives, 1992). Although "advance directive" and "living will" are sometimes used interchangeably,

ADs are more comprehensive and may include forms such as living will and durable power of attorney for health, whereas a living will usually is limited to life-sustaining medical treatments that the patient would or would not prefer (Gunter-Hunt at al., 2002).

Initially, ADs were a response to people's fears that they would be kept alive by costly artificial means without regard to the quality of their life. Newman Giger et al. (2006) explained that some feared that life-extending technology might potentially increase their suffering without actually prolonging their life. Callahan (1988) additionally noted that many of these individuals wanted the ability to be self-determining. In response to these concerns, in 1990, the U.S. legislature implemented the Patient Self Determination Act (PSDA).

The goal of the PSDA is to put patients in control of medical treatment decisions, and it outlines requirements for healthcare agencies to follow for ADs. Under this law, healthcare facilities and agencies (e.g., hospitals and nursing homes) that receive Medicare and Medicaid are required to do the following (direct quote):

1. inform patients of their rights under State law to make decisions concerning their medical care;

2. periodically inquire as to whether a patient executed an advanced directive and document the patient's wishes regarding their medical care;

3. not discriminate against persons who have executed an advance directive;

4. ensure that legally valid advance directives and documented medical care wishes are implemented to the extent permitted by State law; and

5. provide educational programs for staff, patients, and the community on ethical issues concerning patient self-determination and advance directives. (Patient Self Determination Act, 1990)

STATE LAW REQUIREMENTS

In addition to the federal laws governing AD forms, state laws also exist, as the bill allows individual states to decide this right for their residents.

Therefore, each state has its own policies and processes. For example, Idaho gives the option to register on its state registry website (Idaho Department of Health and Welfare, n.d.), and three states—Massachusetts ("Massachusetts law," n.d.); Michigan ("Power of attorney," n.d.); and New York (New York State Department of Health, n.d.)—do not have living will statutes. Similarly, each state varies in terms of its process for executing the AD: Many states require two witnesses, some require two witnesses and a notary, and few do not require any witnesses.

Each state also varies in terms of the content of the ADs. For example, some states include both living will and durable power of attorney for healthcare in their AD forms, whereas others include only the living will. Some states allow patients to be specific in their end-of-life wishes such as disposal of body and anatomical gifts, while others may exclude these options.

COMMON RECOGNIZED PROBLEMS WITH/CRITICISMS OF ADVANCE DIRECTIVES

Previous studies of ADs identified problems with the forms and levied various criticisms of them. These problems (addressed below) include the following:

- **Low completion rates.** Although federal law mandates that healthcare providers and institutions provide patients with ADs, most patients do not complete them (Miller, 2017). Yadav et al. (2017) found that approximately one in three U.S. adults completes any type of AD for end-of-life care; they speculated that healthcare providers are not promoting the use of ADs due to a lack of trust in their effectiveness. Furthermore, legal processes, such as requiring two witnesses and a notary, could create barriers for executing the ADs.

- **Stress of the situation.** When patients are in the middle of a healthcare crisis, they are not in "an ideal state of mind to make competent decisions" and in these circumstances, "it is no longer an advance directive, but a crisis directive" (Miller, 2017, para. 2). In such cases, it may be difficult for patients to understand and complete the ADs, even with the assistance of nurses.

- **Lack of knowledge or ability to predict.** Patients may find it challenging to accurately predict the future medical decisions they would make, lack sufficient knowledge and experience to know what they are making decisions about, and overestimate the impact that a particular disability would have on their life (Levi & Green, 2010). Yadav et al. (2017) found that it is especially difficult to get healthy people to make these future predictions.

- **Limitations of scope and specificity.** Many ADs are limited, both in the scope of their questions and the space in which they provide the patient to respond. These limitations complicate the task of translating ADs into appropriate end-of-life medical decisions (Levi & Green, 2010).

- **Healthcare agent's/proxy's lack of understanding.** Although many ADs allow the patient to designate or name a healthcare agent or proxy, Shalowitz et al.'s (2006) meta-analysis shows that these agents or surrogates predicted patients' treatment preferences with only 68% accuracy and that "neither patient designation of surrogates nor prior discussion of patients' treatment preferences improved surrogates' predictive accuracy" (p. 493). Spalding's (2021) recent literature review found that "surrogates vary in the extent to which they accurately predict patients' preferences for medical treatment" (p. 20). Hence, unsurprisingly, Gunter-Hunt et al. (2002) emphasized the importance for the patient to ensure that their proxy or agent understands their preferences for care.

- **Inconsistent use of terms.** Although government agencies, such as the National Institute of Aging (NIA), have attempted to include information on ADs, the guidelines that they provide are still limited. For example, the NIA provides only two to three paragraphs explaining CPR, ventilator use, artificial nutrition, and comfort care (National Institute on Aging, n.d.). Miller (2017) also pointed out the varying definitions of incapacitation, which can make it challenging for patients to plan accordingly.

ADs are used in many states, but the construct of the forms, the communication around the patient wishes, and the explanations of the ADs are scarcely explored topics in the field of professional and technical communication. Our research aims to address this gap in our field's literature.

Methods

Our case study employed an exploratory, mixed methods approach. We collected 34 state ADs and conducted both quantitative and qualitative content analysis to determine the readability scores for these forms, the types of content that exist in the forms, and the design of the content both in terms of language and formatting.

DATA COLLECTION

As stated earlier, individual states decide the requirements for ADs. Some states have official or "standard" forms that their residents must use, whereas others allow hospital systems or healthcare facilities to create their own versions. We systematically searched for ADs of each of the 50 states and saved the PDF version of the ADs that appeared on their state government website. We located the "standard" AD or living will forms for 34 states by searching on Google in April 2022, using keywords that included the name of the state, along with "official," "state" and "advance directives," and "living will."

The remaining 16 states, which we excluded from this case study, either did not have the form listed on the state government website or did not require an official AD form or only had a power of attorney or healthcare proxy form. These states are Colorado, Illinois, Indiana, Iowa, Massachusetts, Mississippi, Nebraska, Nevada, New Hampshire, New York, Rhode Island, South Dakota, Vermont, Washington, West Virginia, and Wyoming. For some of these states, we found forms that look similar to "standard" AD or living will forms, but because the forms were not listed or located on the state's official website or were listed on a nonprofit organization's website (with a .org extension), we excluded them from this case study.

Although some patients may be provided with ADs and assistance in completing them in a hospital or healthcare facility, others may seek these forms on their own to prepare for future medical decisions. Given that the impetus for the case study is to determine how the document's content and formatting would impact user or patient comprehension and readability, we determined that the typical patient who is looking to complete these forms most likely would consider their state's website as a first resource due to its ethos and relevancy. Furthermore, there are nonprofit websites (e.g., caringinfo.org) that claim to list each state's living wills or AD forms, but they often use a generic template for each state, with the name of the state as the only difference. In other words, what they are showing is not what we would consider the "standard" form found on the official state government or legislature website.

Also, some states may publish a "standard" AD form on their state government websites, but they still allow individual healthcare facilities to use their own forms. For example, Michigan does not have a state-mandated form, so each hospital system (e.g., Ascension, Beaumont, and Henry Ford) developed and distributed its own. However, on the Michigan state government website, there is a version that Michigan residents can use ("Power of attorney," n.d.), and that is the version that we chose for analysis instead of the ones from each hospital system.

Furthermore, some of the states that we excluded from this analysis may have a state-approved form, but that form does not appear on its state government website. For example, Colorado residents can complete the "Declaration as to Medical or Surgical Treatment" form, which is explained in the state's house bill (C.R.S. 15-18-104), but we were unable to retrieve the fillable form on the Colorado state government website through our Google search. Therefore, Colorado's form was not collected or analyzed. Another reason for exclusion was if a form was only a living will or power of attorney form. For instance, Illinois, which has a one-page living will form that gives the patient only the option to indicate on its state government website that they want to die naturally, and Nebraska, which has a two-page power of attorney form that the patient can only indicate what power they want to give to their attorney, were excluded from our analysis.

DATA ANALYSIS

This section presents a quantitative and qualitative content analysis of state ADs. The quantitative content analysis is a readability analysis. The qualitative analysis considers the content, language or constructions, and formatting of the forms.

Quantitative Content Analysis: Readability Analysis

Readability scores are used widely to measure the readability of patient education materials or written health information (Bange et al., 2019; Hadden et al., 2017; Jindal & MacDermid, 2017; O' Sullivan et al., 2020; Roberts et al.,

2016). As rhetorically minded technical communicators, we understand the limitations of these quantitative readability formulas, as cautioned by Jindal and MacDermid (2017); for this reason, the readability analysis is only one of the methods we used to assess the ADs that we collected.

To check the readability scores of the forms that we collected, we converted the PDF version of the ADs into Microsoft Word files. Then, we ran the readability test in Microsoft Word to determine the scores for the Flesch Reading Ease (FRE), the Flesch-Kincaid Grade Level (FKGL), and the percentage of passive sentences. Microsoft's website provides an explanation of how the scores or numbers are formulated or determined (Microsoft, n.d.). Essentially, the higher the FRE number is, the easier it is to understand the document. For FKGL, the higher the number is, the higher the grade level is. The score for the passive sentences is calculated based on the percentage of passive sentences, so the higher the number is, the more passive sentences the document contains.

Food and Drug Administration guidelines for the Institutional Review Board state that consent forms should be written at or below the eighth-grade level (U.S. Food and Drug Administration, 2014). Previous research (e.g., Bothun et al., 2021; Cherla et al., 2012; O'Sullivan et al., 2020) on readability of consent forms and patient education materials went further by arguing that the forms' reading level should not be higher than a sixth- or seventh-grade level to ensure that they are written in language understandable to the human subject. Although ADs are not regulated by the institutional review board (as consent forms are), both ADs and medical consent forms are intended for the general public with varying reading or literacy skills; therefore, we argue that ADs should also be written at or below the eighth-grade level to accommodate the low adult literacy rates in the United States (U.S. Department of Education, 2019).

Because some of the ADs have explanatory or ancillary materials that typically appear at the beginning of the document, we separated the scores based on the type of the content that we found. For example, the Alaska document contains both the form (for patients to complete) and the explanatory materials at the beginning of the document. These materials might include content such as state statutes or law, explanations of each part of the form, instructions for the patients (e.g., where to file the form), or a glossary.

We ran the readability tests twice for these types of documents—once with the explanatory materials included and the second time with the form itself (without the explanatory materials)—to see how the scores would differ.

Qualitative Inductive Content Analysis

We examined the content in three separate qualitative analyses with an inductive approach. For the first analysis, we examined the form's content. Both of us acted as the coders for this case study. We began the open-coding process by individually reviewing the forms to develop a list of themes or codes in a Word document. These themes include proxy, agent, conservator, or attorney of power, mental health treatments, pregnancy, donation/anatomical gifts, disposal of body, and life-sustaining treatments. Then, we reviewed our codes to ensure that we have an exhaustive list of themes and their definitions. Using the "master" codebook, we reviewed each form and coded all the forms on a spreadsheet together to reach agreement on our coding. This iterative process helps to ensure the quality of our analysis. Once the coding was complete, we tallied the number of occurrences for each code.

For the second analysis, we reviewed each form for language or constructions that are known to impact reading comprehension based on the 2011 version of the Federal Plain Language Guidelines (FPLG), which were developed to help improve communication from the federal government to the public. For example, active voice is preferred over passive voice (p. 20), "I" pronoun is preferred over "you" when referring to an action that the user is taking (p. 31), and double negatives should be avoided (p. 54).

For the third analysis, we looked at the formatting of the forms. More specifically, we looked at headings; alignment (justified, left-aligned, or centered); and instruction types ("choose one" and/or "check all that apply," open/fillable textbox and cross or strike out, fill in the blank). We also looked at how the text is formatted (e.g., in red or all capital letters). These elements were also chosen for analysis based on the FPLG.

Findings

In this section, we explain our findings from both the quantitative and qualitative content analysis. We also provide examples when applicable.

Overall Readability

Virginia has the lowest FRE score of 31.3 (making it most difficult to read), whereas Alabama has the highest FRE score of 70.3 (easiest to read; see Table 7.1). The average FRE score for 34 ADs is 47.27, which

Table 7.1. Readability Test Scores of Advance Directive Forms

	Readability (forms only)			Readability (entire document, with explanatory/ancillary materials)		
	Flesch Reading Ease	Flesch-Kincaid Grade Level	Passive Sentences	Flesch Reading Ease	Flesch-Kincaid Grade Level	Passive Sentences
Alabama	70.3	7.9	17.8			
Alaska	54.7	9.7	13.7	54.2	10.2	13.6
Arizona	56.9	9.4	13.4	51	10.6	14.7
Arkansas	54.1	9.7	21.6			
California	38.7	14	20.5	46.5	12	15.2
Connecticut	41.9	13.5	27.2			
Delaware	33.5	16.1	25.3	38.3	14.8	22.2
Florida	36.3	12.9	24.1	45.9	11.7	20.7
Georgia	51.1	11.6	18.7	40.6	13.7	19.1
Hawaii	51.4	9.8	7.1	59.8	8.4	6.5
Idaho	41.7	11.6	25.6			
Kansas	34.2	13.5	31.8			
Kentucky	36.3	12.8	36	52.9	10.3	25.7
Louisiana	48	11.2	19.3	48.1	10.9	20.1
Maine	49.1	11.7	14.8			
Maryland	55.6	9.6	7.2	56.8	9.3	12.4
Michigan	44.8	11.7	18.1	55	9.3	10.2

	Readability (forms only)			Readability (entire document, with explanatory/ancillary materials)		
Minnesota	67.9	8.2	26.3			
Missouri	41.5	13.1	20.6			
Montana	53.7	9.9	21			
New Jersey	37.4	13.9	37.5			
New Mexico	49.4	11.6	16.6	56.7	10.1	15
North Carolina	51.8	11.3	15.9			
North Dakota	60.7	9.3	14.4			
Ohio	45.1	11.5	16.5	44.8	11.5	20
Oklahoma	32.8	14.9	48.2			
Oregon	59	9.5	10			
Pennsylvania	45.1	13	23	51.9	10.9	21.8
South Carolina	38	13.7	25.9			
Tennessee	54.1	10.3	18.6			
Texas	38.7	13	32.6			
Utah	57.8	9.5	5.8	63.9	8.6	11.8
Virginia	31.3	14.4	14.8			
Wisconsin	44.2	12.4	38.4	33.3	14.9	41.4

Source: Created by the author.

is considered low or difficult to read. According to Kher et al.'s (2017) readability assessment of online patient education material, FRE scores of 60–69 are equivalent to U.S. education level of eighth grade and U.S. Department of Health and Human Services readability rating of "average." Following this guideline, we found that only three of 34 states (9%) have scores at 60 or higher.

Virginia has the highest FKGL score of 14.4 (making it the most difficult to read), whereas Alabama has the lowest FKGL score of 7.9 (easiest to read). The average FKGL score is 11.65, or eleventh-grade reading level, which is too high, considering that medical documents should be written at or below eighth-grade reading level. Using eighth grade as the baseline, we found that almost all, or 32 out of 34 states (94%) exceeded that reading level.

Oklahoma has the highest percentage of passive sentences at 48.20%, whereas Hawaii has the lowest percentage at 7.10%. The average percentage of passive sentences is 21%. Active voice is considered easier to read than passive voice, so this average percentage is still too high for eighth-grade reading level.

Sixteen states include explanatory/ancillary information on their ADs. When comparing the average scores with this information, the average FRE score is up slightly at 49.98. The average FKGL score is similar at 11.08, whereas the percentage of passive sentences goes slightly down to 18.15%.

FORMATTING

Some ADs are basic two-page formats (e.g., Arkansas and Tennessee), and others are text-heavy with explanatory/ancillary information, spanning more than 20 pages (e.g., Arizona, Georgia, and Louisiana). Some contain helpful headings and subheadings, and others are more free-form.

The ADs are formatted in numerous ways, with varying attention paid to headings, text justification, and sustained use of all capital letters. Thirty-one of 34 states (91.18%) contain visible headings that separate each section. Eleven of the 34 states (32.35%) use justified alignment to varying extent. For example, Arizona uses justified alignment throughout the entire 29-page document (Arizona Attorney General, n.d.), whereas Georgia uses it intermittently throughout the explanatory information part of the form (Georgia Department of Health and Human Services, 2010). Louisiana includes the following language in red and all capital letters throughout the ancillary information: "IF YOU DO NOT UNDERSTAND

OR HAVE QUESTIONS ABOUT THE USE OF THIS FORM, CONTACT AN ATTORNEY, HEALTH-CARE PROVIDER, OR SOCIAL WORKER" (Louisiana Elder Law Task Force, 2018).

COMPREHENSION

This section explains elements that affect (either positively or negatively) a patient's comprehension, including inclusion of a glossary or definition of terms and use of double negative language.

Glossary or Definition of Terms

Seventeen ADs included explanatory/ancillary information, ranging from one to 15 pages. Six of these states also include a glossary or list of legal or medical terms. More specifically, three (Arizona, Georgia, and Wisconsin) of the six states incorporate the definitions into the front matter (with explanatory/ancillary information). Two states embed the definitions into the form itself. On the Delaware form, patients read about the differences between "permanently unconscious," "terminal condition," and "serious illness or frailty" right before they decide which option they choose if they experience the condition (Committee on Law and the Elderly of the Delaware Bar Association, n.d.). Similarly, Ohio provides the definition of terms in both sections of its AD form: The healthcare power of attorney form and the living will declaration form (Department of Rehabilitation and Correction, 2021). Louisiana is the last and only state that includes the glossary in the back matter (at the end of the document; Louisiana Elder Law Task Force, 2018).

Double Negative Language

Because ADs are meant to present the patient's dying wishes, they frequently contain language related to wanting or not wanting various interventions. Thus, negative language such as "do not" and "except" commonly appear in the forms. For example,

- California's form says, "I do not want my life to be prolonged if (1) I have an incurable and irreversible condition that will result in my death within a relatively short time" (Office of the Attorney General, n.d.).

- Louisiana's form says, ". . . that all life-sustaining procedures, except nutrition and hydration, be withheld or withdrawn so that food and water can be administered invasively" (Louisiana Elder Law Task Force, 2018).

- Minnesota's form says, "If I DO NOT want my healthcare agent to have a power listed above in (A) through (D) OR if I want to LIMIT any power in (A) through (D), I MUST say that here" (Office of Minnesota Attorney General, n.d.).

Negative language, and even more so double negative language, can be confusing. For example, page 7 of the Oregon form contains an example of negative language paired with a positive action. The form states, "I do not want life-sustaining procedures if I cannot be supported and be able to engage in the following ways" (Office of the State Public Health Director, n.d.). It then lists various options, including "express my needs," "be free from long-term severe pain and suffering," and "know who I am and who I am with." This stem phrase is confusing because it combines a negative ("I do not . . .") with both another negative ("if I cannot . . .") and a positive ("and be able to . . .") (Office of the State Public Health Director, n.d.).

Similarly, Idaho's form includes the following statement: "If my death is imminent, I do not want any artificial life-sustaining medical treatment, care or procedures except for artificial nutrition and hydration as follows" (Idaho Healthcare Directive Registry, n.d.). It then lists various options, including,

- only artificial hydration;

- only artificial nutrition; and

- both artificial hydration and nutrition.

If patients do not read this question carefully, they may inadvertently select an option they do *not* want. The problem, again, is that the first part of the statement is a negative (*I do not want*), but the second part ("*except for*") negates the negative (creating a positive).

INSTRUCTION TYPES

This section explains the different types of instructions that are found on ADs. They may contain one or more of the following instructions: Choose one, check all that apply, cross out what does not apply, open ended, and fill-in-the-blank.

Choose One, Check All That Apply, Cross Outs, and Open Text Boxes

The ADs vary on what they instruct the patient to do. Almost all the ADs (33 of 34, or 97.06%) have checkboxes that ask the patient to pick one option (first type of instruction). Most forms (30 of 34, or 88.24%) also include open text boxes (second type of instruction), which allow the patients to fill in an area of the form with their own words. Many forms (23 of 34, or 67.65%) have checkboxes and ask the patients to check "all that apply" (third type of instruction). A fourth instruction, found on only three forms (8.82%), directs patients to cross out directives that they do not want. The Virginia form uses all four of these instruction types.

Open-ended and Fill-in-the-blank Questions

Five ADs (Hawaii, Minnesota, Montana, North Dakota, and Oregon) include targeted, open-ended, or fill-in-the-blank questions about ethical, spiritual, or religious values. For example, Oregon asks, "Do you have spiritual or religious beliefs you want your healthcare representative and those taking care of you to know?" (Office of the State Public Health Director, n.d.).

Four ADs (Maryland, Minnesota, North Dakota, and Oregon) include questions that prompt the patients to provide their goals, fears, or beliefs. For example, Maryland provides an open textbox that corresponds with this statement: "I want to say something about my goals and values, and especially what's most important to me during the last part of my life" (State of Maryland Office of the Attorney General, 2022).

KEY ISSUES

This section explains the major issues that appear on ADs, including questions about nutrition, hydration, alternative treatments, serious illness, terminal illness and frailty, mental health and pregnancy, anatomical gifts, and disposal of body and autopsy.

NUTRITION, HYDRATION, AND ALTERNATIVE TREATMENTS

ADs vary on the way they ask patients to indicate their wishes regarding life-sustaining nutrition, hydration, and alternative treatments. Twenty-eight of the 34 states (82.35%) include options for artificial nutrition, hydration, or tube feeding on their ADs. For example, Idaho asks patients to pick if they want artificial nutrition, hydration, or both. Kansas is either all in or all out: They ask patients to sign a living will declaration that states, "If at any time I should have an incurable injury, disease, or illness certified to be a terminal condition by two physicians who have personally examined me . . . that I be permitted to die naturally with only the administration of medication or the performance of any medical procedure deemed necessary to provide me with comfort care" (Adult Development and Aging, n.d.).

Alternative treatments also vary widely among the 34 states, and they include CPR/respiratory/breathing machines (13, 38.23%); pain-relief medications (8, 23.53%); antibiotics (7, 20.59%); surgery (5, 14.71%); dialysis (4, 11.76%); blood transfusions (3, 8.82%); and chemotherapy (3, 8.82%).

Serious Illness, Terminal Illness, and Frailty

Several ADs differentiate the treatment options that patients may select by how their medical condition is categorized. Some ADs define the different terms; others do not. For instance, Delaware differentiates "terminal illness" from "permanent unconsciousness" and "serious illness and frailty" and provides separate definitions of each (Delaware Health and Social Services, n.d.). Similarly, Maryland separates "terminal illness" from "persistent vegetative state" and "end-stage conditions" but only defines the latter two categories and does not define "terminal illness" (State of Maryland Office of the Attorney General, 2022). Tennessee differentiates between "completely dependent" and "permanently unconscious" and provides definitions of each (Tennessee State Government, n.d.). New Jersey includes a definition of "brain-dead" and considers it to be the legal standard for the declaration of death (New Jersey Commission, n.d.).

Mental Health and Pregnancy

Few ADs ask patients to indicate their wishes regarding mental health treatment(s) and in the event that they are pregnant. Only three states (Alaska, Arizona, and Michigan) allow patients to indicate their desires related to mental health treatment (such as electroshock treatment). However, none of the three forms that contain information about mental health treatments are the same: Alaska calls out specific mental health treatments and asks patients to indicate which one they consent to; Michigan allows the patient to authorize their patient advocate to make mental health decisions; and Arizona includes an open text box with the following explanation: "Durable Mental health treatments that I expressly DO NOT AUTHORIZE if I am unable to make decisions for myself: (Explain or write in 'None')" (Arizona Attorney General, n.d.).

Only seven ADs have places to indicate the patients' wishes if they are pregnant, and three of the seven states that address the pregnancy issue do not allow the patients additional options. For example,

- Alabama's form states, "If I am pregnant, or if I become pregnant, the choices I have made on this form will not be followed until after the birth of the baby" (Alabama Public Health, n.d.).

- Arizona's form states, "Regardless of any other directions I have given in this Living Will, if I am known to be pregnant I do not want life-sustaining treatment withheld or withdrawn if it is possible that the embryo/fetus will develop to the point of live birth with the continued application of life-sustaining treatment" (Arizona Attorney General, n.d.).

The other four states (Idaho, Maryland, Minnesota, and Virginia) allow patients to decide if the AD shall be honored in the case of pregnancy.

Anatomical Gifts

The majority of ADs (24 of 34, or 70.59%) contain a section for patients to indicate whether or not they wish to make an anatomical gift upon their deaths. However, the states vary widely in terms of specificity. The simplest forms, such as the one from Arizona, ask if the patient wishes to make the following anatomical gift: "Any organ/tissue/My entire body/ Only the following organs/tissues" (Arizona Attorney General, n.d.).

More detailed anatomical gift sections, such as those seen on the Ohio and Oklahoma ADs, have several choices for what organs, tissues, or parts the patient wants to donate and the purpose(s) for which they want to donate. For instance, in addition to the more frequently seen anatomical gifts of eyes and organs, Ohio also allows the patient to select such items as fascia, tendons, skin, nerves, ligaments, bones, and veins (Department of Rehabilitation and Correction, 2021).

Disposal of Body and Autopsy

Some ADs allow patients to indicate how they want the medical facility to handle their bodies upon their deaths, but most ADs do not. Only six of the 34 (17.65%) ADs allow patients to indicate their wishes for the disposal of their bodies. The Arizona form contains a section titled "Funeral and Burial Disposition" that allows patients to select whether their bodies should be buried or cremated or whether a patient's appointed agent should make this decision (Arizona Attorney General, n.d.). Similarly, Georgia allows patients to indicate (under "Final Disposition of Body") whether they want their healthcare agent to decide what happens to their bodies after death, whether they want someone else to make this decision, or whether they want to include instructions of their own (in an open text box) on the form (Georgia Department of Health and Human Services, 2010). Meanwhile, Minnesota contains an open text box accompanying the phrase, "My wishes about what happens to my body when I die (cremation, burial)" (Office of Minnesota Attorney General, n.d.).

Questions regarding autopsies are even less prevalent. Only three states (Arizona, Georgia, and Missouri) ask the patient about their requests for an autopsy. Arizona explains in its AD that under Arizona law, an autopsy is not required unless the county medical examiner, the county attorney, or a superior court judge orders it to be performed.

Discussion

Our case study was guided by the overarching research question: How are ADs meeting criteria for comprehension and how might this complicate end-of-life decisions made by terminally ill patients? In this section, we discuss the implications of our findings.

READABILITY

Prior research on medical forms such as consent forms shows that readability scores for these forms tend to be low, in that they are written in language that is difficult to understand. For example, Choudhry et al. (2016) found that only 24% of patients had the reading skills to adequately comprehend the discharge summaries they were given.

Similarly, our quantitative analysis shows that the grade levels vary widely, from 7.9 to 16.1, and at an average of eleventh grade, which is still much higher than the eighth-grade reading level that is recommended for medical forms that are intended for patients, including consent forms (U.S. Food and Drug Administration, 2014).

Furthermore, although the explanatory/ancillary materials are intended to help the user or patient better understand the forms, the scores show that they do not necessarily lower the FRE or FKGL levels and in some cases actually increase those scores for states such as Michigan, New Mexico, and Utah.

As stated by Bange et al. (2019), these public-facing documents should be written at the appropriate reading levels and include the "the reader's point of view and goal in reading" (p. 878) the document. They also suggested simplifying the words and writing in a succinct manner.

INSTRUCTION TYPES

As noted above, the ADs vary on what they instruct the patient to do. Of the four types of instructions, "choose one" and "check all that apply" are standard on medical forms. As such, patients would likely be familiar with these types of instructions. The "open text boxes" are also fairly straightforward. The most likely issue with open text boxes is that the patient may not be specific enough in what they write, or they may leave the box unfilled. The "cross out" is the least-used instruction type, used only by three states. The Maryland form is particularly problematic in its use of cross out because it combines what appears to be a "select one" option with a "cross out" option (State of Maryland Office of the Attorney General, 2022). As Figure 7.1 shows, the "cross out" option could easily be missed if the patient doesn't carefully read the form.

Figure 7.1 shows that under the "Preference in Case of Terminal Condition" section are three choices, which at first glance would indicate that one choice should be checked. However, the patient is actually instructed to initial one of the choices *or* to cross out the entire section. The exact directions read, "If you want to state what your preference is, initial one only. If you do not want to state a preference here, cross through the whole section" (State of Maryland Office of the Attorney General, 2022).

Many ADs have more than one type of instruction. The tactic of using more than one instruction for options could add to unnecessary confusion surrounding what the patient should do with various parts of the form.

B. Preference in Case of Terminal Condition
(If you want to state what your preference is, initial **one** only. If you do not want to state a preference here, cross through the whole section.)

If my doctors certify that my death from a terminal condition is imminent, even if life-sustaining procedures are used:

1. Keep me comfortable and allow natural death to occur. I do not want any medical interventions used to try to extend my life. I do not want to receive nutrition and fluids by tube or other medical means.

   >>OR<<

2. Keep me comfortable and allow natural death to occur. I do not want medical interventions used to try to extend my life. If I am unable to take enough nourishment by mouth, however, I want to receive nutrition and fluids by tube or other medical means.

   >>OR<<

3. Try to extend my life for as long as possible, using all available interventions that in reasonable medical judgment would prevent or delay my death. If I am unable to take enough nourishment by mouth, I want to receive nutrition and fluids by tube or other medical means.

Figure 7.1. Maryland's Form Preference in Case of Terminal Condition Section. *Source*: State of Maryland Office of the Attorney General, 2022.

FORMATTING

The FPLG recommends the use of "ragged right margins where possible, rather than fully justifying your text" as "good document design" (p. 88). However, more than one third of the ADs use force justified alignment. Although the justified text may not directly lower the reading comprehension level of a form, ragged-right format is still preferable, as left-justified text has been shown to negatively affect dyslexic readers, especially in the web environment (Chen et al., 2015).

Furthermore, the FPLG recommends using "bold and italics to make important concepts stand out" (p. 82). We found that several states (9 of 34, or 26.47%) use all capital letters either as headings or warnings, which as the FPLG state, is not a good emphasis technique.

KEY ISSUES

The way the form presents key issues could also affect comprehension. These key issues include definitions (or parsing) of terms, mental health and pregnancy concerns, anatomical gifts, disposal of body, and autopsy. Similar to Gunter-Hunt et al.'s (2002) analysis, we found "substantial variability across the country both in the types of documents used by states and in the content of documents" (p. 54).

Definitions of Terms

The majority of ADs ask patients to make decisions depending on how their current (or future) conditions are defined. A number of ADs, for instance, differentiate between "serious illness," "terminal illness," and "frailty." The definitions of these options differ from state to state. For instance, Florida defines a terminal condition as one "caused by injury, disease, or illness from which there is no reasonable medical probability of recovery and which, without treatment, can be expected to cause death" (Patient Forms, n.d.). However, Wisconsin's definition of a terminal condition is more specific: It considers a terminal condition to be "an incurable condition caused by injury or illness that reasonable medical judgment finds would cause death imminently" (State of Wisconsin Department of Health Services, n.d.). The resulting question is, "What is considered imminent?" Meanwhile, Delaware's form defines serious illness or frailty as "a condition based on which the health-care practitioner would not be surprised if the patient died within the next year" (Delaware Health and Social Services, n.d.). By some standards, 1 year would not be considered imminent. This point shows the double-edged sword of defining terms. Definitions are helpful, but when the definitions are too specific, they are limiting or otherwise problematic.

Because the definitions are specific to the ADs, and these definitions may differ from how the patient commonly regards the meaning of the term, patients have to be careful about reading through definitions they might otherwise be inclined to skip. Furthermore, as noted earlier, New Jersey is the only state that specifically states in its AD that braindead is the legal standard for declaration of death (New Jersey Commission, n.d.).

Anatomical Gifts

Although a majority (24 of 34; 70.59%) of the ADs contain a section for patients to indicate whether they wish to make an anatomical gift upon their deaths, this section is not prevalent.

Some of the states that may not contain an anatomical gift section and simply expect patients to indicate their wishes on this topic in an open text box (for other wishes). For example, neither Alabama nor Texas includes a section explicitly for anatomical gifts, but both ADs contain open text boxes (labeled "other directions" [Alabama Public Health, n.d.] and "additional requests" [Texas Health and Human Services, 2023] respectively), where patients can at least theoretically note their intentions for anatomical gifts. However, of the 10 ADs that do not include an anatomical gift section, three of them (North Carolina, South Carolina, and Wisconsin) are also missing open text boxes for the patient to enter this information.

When patients are faced with a multitude of questions concerning end-of-life considerations, they may not immediately think about anatomical gifts but would do so if prompted (by an actual question). Therefore, we see value in having an anatomical gift section on an AD and having that section spell out the specific organs and tissues that are commonly used. (See Figure 7.2 for an example of such a section.)

As shown in Figure 7.2, Ohio allows patients to select either all organs, tissues and eyes, or to specify the following body parts they want to donate upon their deaths (Department of Rehabilitation and Correction, 2021). What might be even more helpful on such a form (in addition to listing the specific body parts) would be the option to select "other." This would allow for the donation of any other body parts that may not commonly be used but that may become transplantable in the future.

This form also provides patients with more agency in that it permits them to indicate the purposes for which they wish to authorize these gifts.

ANATOMICAL GIFT (optional)

Upon my death, the following are my directions regarding donation of all or part of my body:
In the hope that I may help others upon my death, I hereby give the following body parts:
[Check all that apply.]

☐ All organs, tissue and eyes for any purposes authorized by law.

OR

☐ The following selected items:

☐ Heart	☐ Lungs	☐ Liver (and associated vessels)	☐ Pancreas/Islet Cells
☐ Small Bowel	☐ Intestines	☐ Kidneys (and associated vessels)	☐ Eyes/Corneas
☐ Heart Valves	☐ Bone	☐ Tendons	☐ Ligaments
☐ Veins	☐ Fascia	☐ Skin	☐ Nerves

For the following purposes authorized by law:
☐All purposes   ☐Transplantation   ☐Therapy   ☐Research   ☐Education

Figure 7.2. Ohio's Form Anatomical Gift Section. *Source*: Department of Rehabilitation and Correction, 2021.

They may indicate that their body parts may be used for the following: all purposes, transplant, therapy, research, or education.

Mental Health, Pregnancy, Disposal of Body, and Autopsy

Mental health treatments are not a priority in these ADs, as only three states include this topic. This is concerning due to the increased prevalence of mental health issues. As advocated by Gunter-Hunt et al. (2002), more states need to include "issues of admission to a long-term care facility and advanced illness/dementia" (p. 56).

Pregnancy is another topic that appears in only a few ADs. Questions regarding the patient's end-of-life wishes if they are pregnant can be problematic due to overriding state laws, particularly if those state laws define a fetus as a viable human being. In fact, some states even indicate that the ADs are invalid if the patient is pregnant. This is in line with DeMartino et al.'s (2019) analysis of state laws regarding pregnant women without decisional capacity, when they found that "a majority of [United States] states restrict the health care options available to decisionally incapacitated women during pregnancy and do not disclose these restrictions in advance directive forms" (p. 1631).

Similarly, the forms vary widely in terms of how the body is handled or disposed of. The majority of ADs do not include considerations for disposal of the body. However, because some patients who donate organs or body parts upon their death also donate their entire cadaver (for purposes such as medical research or education), an alternative for the ADs is to combine the patient's wishes regarding the disposal of their body with the anatomical gifts section.

The question of whether patients want to request or deny an autopsy upon their death is similar to the questions regarding pregnancy, in that state laws may conflict with a patient's personal wishes. For instance, people with certain religious affiliations do not permit autopsies (Alhawari et al., 2019), but in the case of suspected foul play or homicide, an autopsy could be required by state laws (Centers for Disease Control and Prevention, n.d.).

Conclusion and Recommendations

Our exploratory, mixed methods approach to analyzing 34 state ADs found that these forms largely do not meet the criteria for typical patient comprehension. To begin with, the majority of ADs are beyond the recommended eighth-grade reading level. Additionally, the forms consistently lack clarity of instructions, formatting, and language. Furthermore, the key issues covered on the forms vary widely. Certain issues, such as instructions about pregnancy, mental health, organ donation, and disposal of the body are absent on a number of the forms. In addition, definitions (when they exist) vary widely, with some being overly vague and others being unnecessarily specific. These deficiencies complicate end-of-life decisions made by terminally ill patients.

The possibility of creating one, standard, federal AD is unlikely, due to varying state laws around medical treatments. The next best alternative is to rework the existing ADs so that they establish advocacy for those who are facing potential end-of-life choices and care. Each state might review its existing AD and consider the possibility of instituting the following modifications:

- Make the AD document readily accessible through the state government or legislature website.

- Include, as a standard practice on all ADs, sections about pregnancy, mental health, organ donation, and disposal of the body.

- Provide definitions and differentiations of terms used (such as terminal illness, serious illness, imminent death, and brain dead).

- Include visible instructions on the form that direct the patient to a particular phone number, website, or person to contact for questions about filling out the form.

- Provide emotional wellness support resources (in addition to technical support help) in filling out the form for those facing life/death situations.

- Format the anatomical gift section with a detailed checklist, where the patient can select which (if any) organs or body parts they want to donate and for what purposes (e.g., transplant, medical research, or other).

- Remove any instructions that ask the patient to perform more than one action per question. For example, instead of asking the patient to initial and/or strikethrough options, ask them to simply select (checkbox) the treatments they do want.

- Remove double negatives from statements on the form.

Such changes would strengthen the content and potential usefulness of the state ADs. These changes could also clear up confusion on the part of the patient and ensure that patients' responses on the form accurately represent their healthcare wishes. The result of this would be execution of the patient's actual wishes (not a misunderstanding on the part of the patient or a healthcare worker). In addition, these findings align with what Gunter-Hunt et al. (2002) found 20 years ago regarding ADs: that detailed ADs also improve the understanding of the wishes by a potential proxy (or healthcare power of attorney).

Limitations and Future Studies

This case study was limited in that we were not able to interview or survey the AD document creators/designers/writers to determine their rationale or decision-making processes, which could be impacted by each state's laws and regulations. Future studies could further explore other aspects of the FPLG, such as analyzing forms' sentence and paragraph structures to determine how well each state's form meets those guidelines. In addition, more insights could be gained by conducting usability testing of these forms to determine how patients (users) from various age groups, socio-economic backgrounds, and education levels comprehend these forms and use them to make critical life or death decisions. A contextual task analysis might also be useful to employ in future studies because the context of use is key to helping technical communicators identify where the forms fail to meet users' informational and emotional needs that are often external to technical information.

References

Adult Development and Aging. (n.d.). *Durable power of attorney for healthcare decisions*. Kansas State University. https://www.aging.k-state.edu/programs/adv-healthcare-planning/Advance%20Health%20Care%20Directives%20and%20Wallet%20Cards_Form%20Fillable.pdf

Alabama Public Health. (n.d.). *Advance directive for health care.* https://www. alabamapublichealth.gov/cancer/assets/advdirective.pdf

Alhawari, Y., Verhoff, M. A., Ackermann, H., & Parzeller, M. (2019). Religious denomination influencing attitudes towards brain death, organ transplantation and autopsy—A survey among people of different religions. *International Journal of Legal Medicine, 134*(3), 1203–1212. https://doi.org/10.1007/s00414-019-02130-0

Arizona Attorney General. (n.d.). *Life care planning packet.* https://www.azag. gov/sites/default/files/docs/seniors/life-care/2018/Life-Care-Planning-Packet-Complete.pdf

Bange, M., Huh, E., Novin, S. A., Hui, F. K., & Yi, P. H. (2019). Readability of patient education materials from radiologyinfo.org: Has there been progress over the past 5 years? *American Journal of Roentgenology, 213,* 875–879. https://doi.org/10.2214/AJR.18.21047

Bothun, L. S., Feeder, S. E., & Poland, G. A. (2021). Readability of participant informed consent forms and informational documents: From phase 3 COVID-19 vaccine clinical trials in the United States. *Mayo Clinic Proceedings, 96*(8), 2095–2101. https://doi.org/10.1016/j.mayocp.2021.05.025

Callahan, D. (1988). Vital distinctions, mortal questions: Debating euthanasia and health-care costs. *Commonwealth, 115*(13), 397–404.

Centers for Disease Control and Prevention. (n.d.). *Investigations and autopsies.* https://www.cdc.gov/phlp/publications/coroner/investigations.html

Chen, C. J., Keong, M. Y. W., Teh, C. S., & Chuah, K. M. (2015). Learners with Dyslexia: Exploring their experiences with different online reading affordances. *Themes in Science & Technology Education, 8*(1), 63–79.

Cherla, D. V., Sanghvi, S., Choudhry, O. J., Liu, J. K., & Eloy, J. A. (2012). Readability assessment of internet-based patient education materials related to endoscopic sinus surgery. *The Laryngoscope, 122*(8), 1649–1654. https://doi. org/10.1002/lary.23309

Choudhry, A. J., Baghdadi, Y. M. K., Wagie, A. E., Habermann, E. B., Heller, S. F., Jenkins, D. H., Cullinane, D. C., & Zielinski, M. D. (2016). Readability of discharge summaries: With what level of information are we dismissing our patients? *American Journal of Surgery, 211*(3), 631–636. https://doi. org/10.1016/j.amjsurg.2015.12.005

Committee on Law and the Elderly of the Delaware Bar Association. (n.d.). *Delaware's advance health care directive form (English).* https://www.dhss. delaware.gov/dsaapd/files/advancedirective.pdf

Delaware Health and Social Services. (n.d.). *Advance directive.* https://www.dhss. delaware.gov/dsaapd/files/advancedirective.pdf

DeMartino, E. S., Sperry, B. P., Doyle, C. K., Chor, J., Kramer, D. B., Dudzinski, D. M., & Mueller, P. S. (2019). U.S. state regulation of decisions for pregnant women without decisional capacity. *JAMA: Journal of the American Medical Association, 321*(16), 1629–1631. https://doi.org/10.1001/jama.2019.2587

Department of Rehabilitation and Correction. (2021). *Advance directives for health-care*. https://drc.ohio.gov/Portals/0/Policies/DRC%20Policies/69-OCH-03%20 (June%202021).pdf?ver=7SzaC94ua2w0sUogXtAAAQ%3D%3D

Dresser, R., & Astrow, A. B. (1998). An alert and incompetent self: The irrelevance of advance directives. *Hastings Center Report, 28*(1), 28–30. https://doi.org/10.2307/3527971

Federal Plain Language Guidelines. (2011). Plainlanguage.gov. https://www.plain-language.gov/media/FederalPLGuidelines.pdf

Frost, E., & Eble, M. F. (2015). Technical rhetorics: Making specialized persuasion apparent to public audiences. *Present Tense: A Journal of Rhetoric in Society, 4*(2): 1–5.

Georgia Department of Health and Human Services. (2010). *Georgia advance directive for health care*. https://aging.georgia.gov/document/publication/directivespdf/download

Gunter-Hunt, G., Mahoney, J. E., & Sieger, C. E. (2002). A comparison of state advance directive documents. *Gerontologist, 42*(1), 51–60. https://doi.org/10.1093/geront/42.1.51

Hadden, K. B., Prince, L. Y., Moore, T. D., James, L. P., Holland, J. R., & Trudeau, C. R. (2017). Improving readability of informed consents for research at an academic medical institution. *Journal of Clinical and Translational Science, 1*(6), 361–365. https://doi.org/10.1017/cts.2017.312

Idaho Department of Health and Welfare. (n.d.). *Advance directives and registry services*. https://healthandwelfare.idaho.gov/services-programs/birth-marriage-death-records/advance-directives-and-registry-services

Idaho Healthcare Directive Registry. (n.d.). *Advance directive registration form*. Idaho Department of Health and Welfare: Division of Public Health. https://publicdocuments.dhw.idaho.gov/WebLink/DocView.aspx?id=21779& dbid=0&repo=PUBLIC-DOCUMENTS&cr=1

Jindal, P., & MacDermid, J. C. (2017). Assessing reading levels of health information: Uses and limitations of Flesch formula. *Education for Health (Abingdon, England), 30*(1), 84–88. https://doi.org/10.4103/1357-6283.210517

Keränen, L. (2007). "Cause someday we all die": Rhetoric, agency, and the case of the "patient" preferences worksheet. *The Quarterly Journal of Speech, 93*(2), 179–210. https://doi.org/10.1080/00335630701425100

Kher, A., Johnson, S., & Griffith, R. (2017). Readability assessment of online patient education material on congestive heart failure. *Advances in Preventive Medicine, 2017*, 9780317–9780318. https://doi.org/10.1155/2017/9780317

Kopelson, K. (2019). Dying virtues: Medical doctors' epideictic rhetoric of how to die. *Rhetoric of Health & Medicine, 2*(3), 259–290. https://doi.org/10.5744/rhm.2019.1013

Levi, B. H., & Green, M. J. (2010). Too soon to give up: Re-examining the value of advance directives. *American Journal of Bioethics, 10*(4), 3–22.

Louisiana Elder Law Task Force. (2018). *Planning for incapacity: A self-help guide. Advanced directive forms for Louisiana.* https://web.archive.org/web/20220521164109/http://goea.louisiana.gov/assets/Legalservicesfiles/Planningforincapacity.pdf

Massachusetts law about health care proxies and living wills. (n.d.). Mass.gov. https://www.mass.gov/info-details/massachusetts-law-about-health-care-proxies-and-living-wills

Melonçon, L. (2017). Patient experience design: Expanding usability methodologies for healthcare. *Communication Design Quarterly, 5*(2), 19–28.

Microsoft. (n.d.). *About Flesch Kincaid readability and level statistics.* https://support.microsoft.com/en-us/office/about-flesch-kincaid-readability-and-level-statistics-8fbb787e-4e1d-4dcc-933f-c7d2f06e76ac

Miller, B. (2017, September 6). Nurses in the know: The history and future of advance directives. *Online Journal of Issues in Nursing, 22*(3).

National Institute on Aging. (n.d.). *Advance care planning: Health care directives.* U.S. Department of Health and Human Services, National Institutes of Health. https://www.nia.nih.gov/health/advance-care-planning-health-care-directives

New Jersey Commission on Legal and Ethical Problems in the Delivery of Health Care. (n.d.) *Instruction directive.* https://www.nj.gov/health/advancedirective/documents/instruction_directive.pdf

New York State Department of Health. (n.d.). *Advance care planning and advance directives FAQ.* https://www.health.ny.gov/community/advance_care_planning/faq.htm

Newman Giger, J., Davidhizar, R. E., & Fordham, P. (2006). Multi-cultural and multi-ethnic considerations and advanced directives: Developing cultural competency. *Journal of Cultural Diversity, 13*(1), 3–9.

Office of Minnesota Attorney General. (n.d.). *Minnesota Statute § 145C.* https://www.ag.state.mn.us/consumer/handbooks/probate/HealtCareDir.pdf

Office of the Attorney General. (n.d.). *Advance health care directive form.* State of California Department of Justice. https://oag.ca.gov/system/files/media/ProbateCodeAdvanceHealthCareDirectiveForm-fillable.pdf

Office of the State Public Health Director. (n.d.). *Oregon advance directive for health care.* https://sharedsystems.dhsoha.state.or.us/DHSForms/Served/le3905.pdf

O' Sullivan, L., Sukumar, P., Crowley, R., McAuliffe, E., & Doran, P. (2020). Readability and understandability of clinical research patient information leaflets and consent forms in Ireland and the UK: A retrospective quantitative analysis. *BMJ Open, 10*(9), e037994–e037994. https://doi.org/10.1136/bmjopen-2020-037994

Patient Forms. (n.d.). *Health care advance directives.* University of South Florida Student Health Services. https://www.usf.edu/student-affairs/student-health-services/documents/advanced_directive_info.pdf

Patient Self Determination Act of 1990, H.R. 4449, 101st Cong. (1990). https://www.congress.gov/bill/101st-congress/house-bill/4449

Power of attorney and advance directive resources. (n.d.). Michigan.gov. https://www.michigan.gov/orsschools/after-retirement/power-of-attorney-and-advance-directive-resources

Public Health, Subpart I Advanced Directives. C.F.R. § 489.100-104 (1992). https://www.ecfr.gov/current/title-42/chapter-IV/subchapter-G/part-489/subpart-I

Roberts, H., Zhang, D., & Dyer, G. S. (2016). The readability of AAOS patient education materials: Evaluating the progress since 2008. *Journal of Bone and Joint Surgery. American Volume, 98*(17), e70–e70. https://doi.org/10.2106/JBJS.15.00658

Shalowitz, D. I., Garrett-Mayer, E., & Wendler, D. (2006). The accuracy of surrogate decision makers: A systematic review. *Archives of Internal Medicine 166*(5):493–97.

Spalding, R. (2021). Accuracy in surrogate end-of-life medical decision-making: A critical review. *Applied Psychology: Health and Well-Being, 13*(1), 3–33. https://doi.org/10.1111/aphw.12221

St.Amant, K. (2017). The cultural context of care in international communication design: A heuristic for addressing usability in international health and medical communication. *Communication Design Quarterly, 5*(2), 62–70.

St.Amant, K. (2021). Cognition, care, and usability: Applying cognitive concepts to user experience design in health and medical contexts. *Journal of Technical Writing and Communication, 51*(4), 407–428. https://doi.org/10.1177/0047281620981567

State of Maryland Office of the Attorney General. (2022). *Maryland advance directive: Planning for future health care decisions.* https://www.marylandattorneygeneral.gov/health%20policy%20documents/adirective.pdf

State of Wisconsin Department of Health Services. (n.d.). *Declaration to health care professionals form.* https://dhs.wisconsin.gov/forms/advdirectives/f00060.pdf

Tennessee State Government. (n.d.). *Advance directive for health care.* https://www.tn.gov/content/dam/tn/health/documents/Advance_Directive_for_Health_Care.pdf

Texas Health and Human Services. (2023). *Directive to physicians and family or surrogates (Living will).* https://www.hhs.texas.gov/sites/default/files/documents/laws-regulations/forms/LivingWill/LivingWill.pdf

U.S. Department of Education. (2019, July). *Adult literacy in the United States.* National Center for Education Statistics. https://nces.ed.gov/pubs2019/2019179/index.asp

U.S. Food and Drug Administration. (2014, July). *Informed consent: Draft guidance for IRBs, clinical investigators, and sponsors.* https://www.fda.gov/regulatory-information/search-fda-guidance-documents/informed-consent#_ftnref39

Yadav, K. N., Gabler, N. B., Cooney, E., Kent, S., Kim, J., Herbst, N., Mante, A., Halpern, S. D., & Courtright, K. R. (2017). Approximately one in three US adults completes any type of advance directive for end-of-life care. *Health Affairs, 36*(7), 1244–1251. https://doi.org/10.1377/hlthaff.2017.0175

Chapter 8

Localizing Mental Health Resources within Collegiate Athletics

Extending User Advocacy and Patient-Experience Design

Mallory Henderson

Collegiate athletics has experienced a rise in mental health issues over the past several years (Lentz et al., 2018; NCAA Research, 2021, 2022).[1] Specifically, eating-disorder (ED) prevalence rates are of concern (Beals, 2004; Greenleaf et al., 2009; Sundgot-Borgen & Torstveit, 2010). Early intervention of EDs has led to higher success rates or chances to recover (Ambwani et al., 2020; Flynn et al., 2021; Treasure et al., 2015), suggesting that athletic personnel involved in intervention should have resources to help them detect signs and symptoms and take appropriate actions. For student-athletes, resources should appropriately encourage or guide them to seek help. Both student-athletes and coaches have called for better resources (Hainline, 2014; NCAA Research, 2021, 2022). Technical and professional communication (TPC) as well as rhetoric of health and medicine (RHM) scholars can answer this call.

The field of TPC has actively worked toward creating design methods and practices to improve user experience (UX) and to address users' specific needs. TPC scholars have decided that it is no longer adequate to solely consider users when creating materials. Instead, we must recognize and design for worlds around users (Simmons & Zoetewey, 2012,

p. 9) and balance, making design more specific, customizable, and local (Lancaster & King, 2024a) to the ecology of users. Localization practices are shifting to reconsider how materials are translated and designed to fit specific cultural contexts (Acharya, 2018; Agboka, 2014; Sun, 2002), while medical settings are employing patient-centered design and are progressively involving patients in the design process (Kessler et al., 2021; Melonçon, 2017b; Rose et al., 2017).

As a subset of health, mental health and wellness concerns within athletic environments pose a unique community for TPC/RHM work and scholarship. As the governing body of collegiate athletics, the National Collegiate Athletic Association (NCAA) has partnered with the Sports Science Institute (SSI) and other medical organizations to make significant efforts in providing educational resources to support the health and well-being of student-athletes. However, design research for such educational resources in TPC is largely unaddressed in existing literature.

This chapter aims to address this void. As a prior collegiate athlete now recovered from an ED, I argue for using collaborative efforts of patient experience design (PXD) and localized design to show how each methodology can better meet the contextualized needs of athletes and coaches as patient-users. This chapter illustrates a multimodal analysis to examine two NCAA resources using Shoemaker et al.'s (2013a, 2013b, 2013c) Patient Education Materials Assessment Tool (PEMAT): a tool meant to ensure that materials advocate for patients and are patient centered. Applying PXD and localized design as a recovered collegiate athlete, I show how each framework may better account for coaches' and athletes' contextualized needs. Specifically, I argue that participatory methods for resource design should engage individuals who are experienced in and have recovered from an ED: thereby creating, *experienced* patient design (EPD) or *recovered* patient experience design (R-PXD). Limitations of this study exist, but such research for resource design within collegiate athletics and extensions of PXD within the field of TPC/RHM show promise for future studies.

Research Objective

This study had four main objectives: (1) to use the disciplines of TPC, RHM, and literacy studies to collaboratively identify the need for design methods rooted in user advocacy; (2) to show how TPC can work in nonacademic spaces to better advocate for, and support patients' (users') mental health concerns; (3) to drive the momentum of participatory methods that involve users as their own voices of advocacy, rather than being imagined or spoken for by designers; and (4) to question and improve our own methodologies within the field of TPC. This study does not intend to identify strengths and shortcomings of the NCAA and SSI. The core of this research methodology stems from human-centered design (HCD) principles but then branches into participatory methods of PXD and localized UX design. These objectives can be broken down into the overarching goals:

1. Consider TPC, RHM, and literacy studies frameworks to highlight why ED resources for collegiate athletics require localized and PXD design methodology.

2. Apply localized and PXD methodology to show what this design offers that nonparticipatory or patient-centered design (PCD)/HCD principles cannot.

3. Align athletic contexts and EDs to reconsider and disturb our own methodologies within TPC and RHM.

4. Use this research to bring awareness to the issue of EDs and create motivation for TPC scholars to become involved in collegiate athletics.

By achieving these objectives, this research positions TPC both to challenge our practices and to seek balance in reaching audiences not widely recognized in the field, creating accessible materials for any individual involved in ED mental health concerns (while still accounting for individualized needs), hearing the voices of designers and patient-users, and recognizing the available means to do so (termed "equilibriUX" by Lancaster & King [2024a]). Ultimately, this work proposes design practices rooted in advocacy rather than traditional or convenient methods. In return, the hope is that this research will expand our fields and do the larger work of supporting mental health issues within collegiate athletics.

Literature Review

This research calls for and involves four main domains of existing scholarship: contextualized research on EDs within collegiate athletics, localization practices and UX within TPC, PXD, and literacy studies frameworks. The main scholarship within each create the exigency for reconsidering resource design practices or informing on how to implement new methods of ED resource design within collegiate athletics.

EATING DISORDERS AND COLLEGIATE ATHLETICS

EDs are psychiatric disorders that are clinically recognized on the Diagnostic and Statistical Manual of Mental Disorders (American Psychiatric Association, 2022), with the highest mortality rates of all mental illnesses (Arcelus et al., 2011; Treasure et al., 2022). Individuals of all ages, sexes, genders, ethnicities, and sexual orientations can develop an ED (Pike et al., 2013); however, female athletes have shown the highest prevalence rates across all population groups (Sundgot-Borgen & Torstveit, 2004), including female collegiate athletes (Greenleaf et al., 2009; Ravi et al., 2021; Sanford-Martens et al., 2005). Although disordered eating (DE) is also a serious concern in collegiate athletics (Plateau et al., 2014; Wells et al., 2020) and can lead to an ED if untreated (Beals & Manore, 1994; Jagim et al., 2022; Kennedy et al., 2021), this research focuses on ED resources based on prevalence rates, diagnostic ability, and my positionality within the community (a college athlete who recovered from an ED).

Research shows that ED prevalence rates or positive ED screening results range from 20% to 33% in female-identified collegiate athletes (Bratland-Sanda & Sundgot-Borgen, 2013; Flatt et al., 2021; Greenleaf et al., 2009). Additional studies show that 20%–25.5% of female collegiate athletes are symptomatic for an ED, with 5% meeting diagnosable criteria (Greenleaf et al., 2009; Sanford-Martens et al., 2005). Athletes in these studies were not seeking help, nor was it clear that they were aware they had a problem: Thus, detection and intervention become crucial. When thinking of resource design, athletes are not the only users of resources to consider; coaches must be actively involved.

Though coaches are not medical professionals and their roles within ED intervention are debated, those closest to athletes often are best suited to identify and intervene in ED symptoms due to the difficulty of detecting it (Sherman et al., 2005; Torres-McGehee et al., 2011). At large, EDs are difficult to identify because of their nature as isolating and self-preserving disorders; identification becomes increasingly difficult in sport environments because ED behaviors can manifest as "normal" traits of an athlete. For example, symptoms such as "clean" eating, excessive training, weight fluctuation, or menstrual-cycle irregularities may be perceived as "normal" in sport environments (NEDA, 2022; Thompson, 2014). Likewise, personality traits similar to those of individuals with EDs, such as perfectionism or high self-discipline, may be misperceived as traits of a dedicated athlete (Thompson et al., 2014). Rhetorically speaking, athletes establish an embodied ethos or title as an athlete that may allow them to start or maintain disordered behavior (Shellenberger, 2020). Detection becomes tricky. However, based on coaches' proximity, time spent with athletes, mentorship-based roles, and approval seeking from athletes, coaches have a significant influence on athletes (Sherman et al., 2005; Torres-McGehee et al., 2011; Turk et al., 1999). Therefore, a coach's role, and how resources cater to their knowledge and context of use, matters.

TPC and RHM frameworks can offer a way to continue the efforts of providing resources but also do the critical work of questioning how resources are adequately meeting the needs of athletes and coaches as users.

Technical and Professional Communication Scholarship: Localization and User Experience

A fundamental shift in the field of TPC occurred when we started to consider our responsibilities to create *useful* rather than *usable* materials (Hart-Davidson, 2013; Mirel, 2004). Largely, this shift attributes from the work of user-centered design or HCD practices that focus on user advocacy and improving UX. However, as the field of TPC has expanded across communities and internationally, it has recognized that an ecological approach to design methods is necessary to identify complex problems and usefully meet users' specific needs (Caruso & Frankel, 2010; Matheson, 2006; Rose, 2016; Spinuzzi & Zachry, 2000; St.Amant, 2013): that is, understanding and accounting for the world around users. One way that TPCs have done this is through localization or the process of creating products for use in a specific target country or community (Hoft, 1995).

In the past, localization has ensured that materials were accessible cross-culturally, based on conventional factors such as linguistic features, punctuation, and format (Ho et al., 2015; Sun, 2002). Though important, scholars have noted this does not account for the dimensions of culture that critically affect UX: Conventions are only the first layer, whereas factors such as social structures, ideologies, ableness, and access are the more complex or knotted layers that affect UX (Agboka, 2014; Rose, 2016; Sun, 2002). In this way, limiting design to conventions undermines the lived realities of users. Acharya (2018) and Sun (2002) have emphasized further, however, that users' layers of culture are not static but are a dynamic and living process that changes depending on the context. This, followed by the proliferation of information and advancement of technologies (Rose et al., 2017; Tham, 2021), has only enhanced the constant change of UX and the way users' contexts affect use. Given these factors, localizing design to specific user needs quickly becomes complicated yet necessary work.

The concept of usefulness based on context and timeliness have also been discussed in health and medical settings. St.Amant (2017) noted, "Health and medical communicators need to expand the scope of the design process to include the contexts—or settings/locations—in which patients use an item" (p. 63). To envision users' contexts and experiences when using materials, TPCs have relied on personas to imagine their users and the dynamic ways that their contexts affect use (Cooper, 1999; Getto & St.Amant, 2014; Goodwin, 2009; Melonçon, 2017a). Whereas personas center UX and aim to advocate for users, TPC social-justice scholars have emphasized that this creates issues of *who* designers might imagine and *who* they might exclude (Jones et al., 2016; Rose, 2016). Though usability testing has allowed for direct user feedback, both TPC and RHM scholars have critiqued that it still holds individuals as end-users of materials and static moments of use (Gouge, 2017; Johnson et al., 2007; Kessler et al., 2021; Sun, 2002). This, paired with the similar issue of localization practices that develop materials onsite without user feedback, has created the exigency for alternative design methods. For example, Acharya's (2018) *localized UX design* positions users as co-designers with developers "who offer 'valuable user input' that designers can employ to modify product design according to the use context" (p. 3). Agboka's (2014) *participatory localization* shows another alternative by placing designers in users' environments during the development process so that they can better understand the ecologies and context of use. These challenges of UX and design methods relate to collegiate athletics: how to advocate and account for athletes' and coaches' specific needs as resource users as well as how to account for the varying contexts that surround each individual.

From an athletic standpoint, localized and participatory methods can be useful, given the variance of mental health illnesses; divisions of universities (DI, DII, and DIII); and funding allocated in each that goes toward education and resources. Similarly, scholars have discussed the dilemma of localized versus globalized design (Agboka, 2014; Rose, 2016; Sun & Getto, 2017). With economic, technological, and political advancements, globalization demands that materials can be used across multiple contexts. However, regarding mental health concerns, universalizing resource design and content for ease of access or to minimally meet policy requirements are problematic. If resources are designed to address all mental health concerns, for example, coaches may miss signs or inappropriately approach an athlete with ED symptoms. Localized design hones linguistic features of specific disorders but incorporates those certified in specific disorders in the design process. However, as Melonçon (2017b) noted, localized framework places medical context aside: here, that is, athletes as patient-users with mild to severe mental health concerns and coaches who are using materials to better support their athletes' health. TPCs have noted that UX and localizing design in healthcare settings requires a shift from *user* to *patient* experience (Angeli & Norwood, 2016; Melonçon, 2017b; Rose et al., 2017). Melonçon's (2017b) PXD emphasizes this shift.

PATIENT-EXPERIENCE DESIGN

Positioning coaches and athletes as patients may seem strange. However, when they use resources to improve, support, change, and make informed decisions about health, they fall directly within the ecology involved in medical interactions (Greenhalgh, 2017; Melonçon 2017b). Greenhalgh (2017) explained that it is not just patients themselves but also those who help patients make decisions or support those individuals. In this case, athletes who are struggling are direct patients, but coaches are a subset of patients who use resources. Therefore, the ecology of users includes athletes as patients, coaches, and other athletic personnel.

Providers have used PCD to advocate for patient usability and to create materials that are accessible and understandable (e.g., using plain language, making sure material is readable; Melonçon, 2017b). For example, scholars and designers have used resources and assessment tools to meet readability and understandability metrics (Center for Disease and Control Prevention, 2019; Shoemaker et al., 2013c). Although the tools are patient centered, designers of these tools have acknowledged their inability to account for timeliness, comprehension, or context of use. As mentioned, conventional usability or imagined personas miss the complexities of patients' realities and contextualized needs that influence their experience with materials. Melonçon's (2017b) PXD has helped to fill these gaps. Simply, Melonçon (2017b) defined PXD as a "participatory methodological approach centered on contextual inquiry to understand the relationship between information (or technology) and human activities in health care" (p. 20). PXD uses participatory methods to center patients' perceptions in the design process to better meet patient's needs and ultimately improve quality of life. PXD offers an opportunity to extend user-advocacy by designing materials specifically for *and* with the lived experiences of patients: It captures patients *experiencing* rather than imagined or static experience (Figure 8.1).

Other scholars have extended Melonçon's (2017b) original PXD model (Bivens, 2019; Kessler et al., 2021; St.Amant, 2017). For example, St.Amant coined international-PXD (I-PXD) to consider patients' experiences cross culturally. He emphasized using patients' voices to identify variables that affect discrepancies of use with materials. Although most PXD research has involved traditional patients and diseases, Bivens's work places PXD in the context of mental health concerns such as within opioid users. Though PXD has not specifically extended to ED patients, usability testing has. For example, Langlet et al. (2021) studied ED clinical and technology personnel's experiences with a virtual reality app of ED treatment. Although an effective research method, the study excluded ED patients from participatory feedback, fearing negative impact on patients could occur. Both Bivens's (2019) and Langlet et al.'s (2021) research complicates the original proposal of PXD by highlighting that patients' conditions (and

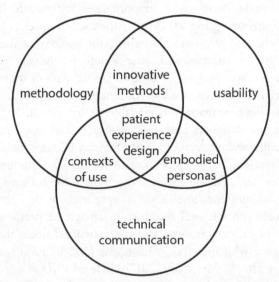

Figure 8.1. Melonçon's Reconfiguration of the Domains of Knowledge in PXD Methodology. *Source*: Melonçon, 2017b. Reprinted with permission.

indivisible conditions) may not afford them to be involved in the design process. This same issue arises in athletes who are struggling or coaches who may not know whether resources are actually useful. One solution may be to involve recovered, or *experienced*, patients as active voices in the design process: EPD or R-PXD.

Melonçon (2017b) called for research to "apply PXD as a methodology in other settings and sites to either prove, disprove, hone, or alter the idea" (p. 25). To apply PXD in other settings and contexts, as Bivens (2019) did, scholars can reconfigure PXD methodology to include individuals who have recovered from an ED in resource design. Melonçon's (2017b) PXD "unapologetically keeps the focus on the embodied user and the way the patient needs to use the object" (p. 23). Though it may seem counterintuitive to encourage patient-users' empowerment by using someone else's voice, athletes in denial or in early stages of developing an ED may not realize if resources are catering to their needs. Likewise, coaches who are not fully trained in ED may not know whether resources are actually advocating for them and their context of use. Therefore, rather than involving vulnerable athletes or coaches to advocate for UX, recovered individuals can embody their experiences, without risk of inducing harm in return. For coaches as patient-users, recovered individuals can help to identify content and design that better account for their roles of intervention or support within the patient ecology. For athletes, recovered individuals can voice lived experiences that conventional design misses.

ED LITERACY DEVELOPMENT: PATIENT-USER LITERACY

As mentioned within the considerations of TPC and healthcare literature, a multitude of factors exist that influence UX with health-related materials. For example, these factors may include, but are not limited to, setting, health literacy, patient(s) condition(s), and embodied experience (Kessler, 2021; Melonçon, 2017a). Although each are prominent factors that influence UX with ED resources, a focus on literacy demonstrates why these resources requires careful design.

From literacy studies, Brandt's (2001) notion of literacy sponsors and Prior and Shipka's (2003) concept of lamination help uncover how mental health literacy (MHL[2]) influences and complicates coaches' and athletes' use of ED resources. Brandt (2001) defined literacy sponsors as "any agents, local or distant, concrete or abstract, who enable, support, teach, and model, as well as recruit, regulate, suppress, or withhold, literacy—and gain advantage by it in some way" (p. 19). In collegiate athletics, there are multiple sponsors of ED MHL,[3] both helpful and harmful. Although formal education and training act as sponsors, literate activity happens in both official and nonofficial spaces (Bellwoar, 2012; Cushman, 2013). For example, as with other mental illnesses, the role of media as a source of knowledge is well documented. Social media has worked to destigmatize mental health concerns but also contributes to misinformation and reinforcement of mental health and ED stereotypes (Treasure et al., 2022; Zhu & Smith, 2021). Like DE and EDs in sport culture, diet and wellness culture is pervasive in media, making disordered behavior more unrecognizable. This representation ultimately continues to normalize, justify, or reinforce disordered behavior in sport cultures. Aside from social media, health-communication scholars have noted that the web and interpersonal communication are sites of conversations centered on health (Foss, 2021; Head et al., 2021). Here, agents act as sponsors with the potential to enable, model, diminish, or recruit, disordered behavior. Figure 8.2 conceptualizes Brandt's (2001) notion of literacy sponsors and its place in influencing patient-user experience.

Considering literary sponsorship on EDs highlights concerns for PXD for coaches and athletes. Further examining Prior and Shipka's (2003) concept of lamination shows another complexity: rather than one instance, literate experience is a lifelong accumulation of literate activities (Durst, 2019; Prior & Shipka, 2003; Roozen & Erickson, 2017). Therefore, when coaches and athletes use resources, they are not only drawing on one official source of ED literacy but a multitude of embodied experiences comprised of literacy agents. Even per Brandt's (2001) theory of sponsorship, EDs complicate the idea of who sponsors are. New literacy scholars have pushed against Brandt's (2001) implication that sponsorship is a hierarchical retaliation: that is, that "sponsees" are without agency (Cushman, 2013; Mahiri, 2004). In collegiate athletics, coaches traditionally are placed in an "expert," or sponsor, position. However, questions arise about who is sponsoring coaches' knowledge: educational training, media, interpersonal communication, or a combination. Furthermore, athletes themselves may be reinforcing disordered norms among each other, acting as their *own* sponsors. Papathomas and Lavalle (2012) noted that one athlete struggling with an ED or DE can significantly affect other teammates. In this sense, design methods that focus on readability and plain language quickly become insufficient for the ecologies that surround individuals during resource use: Literacy is just *one* piece in the ecology. Therefore, the "worlds" (St. Amant, 2017) surrounding coaches and athletes are crucial to consider

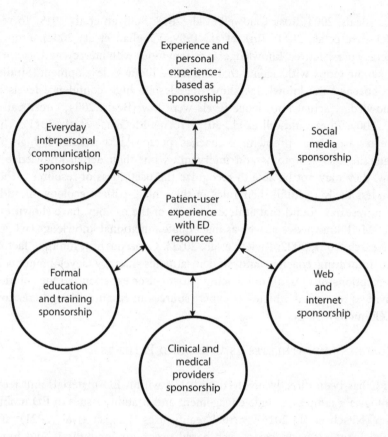

Figure 8.2. Representation of Possible Literacy Sponsors Involved in Patient-User Interactions with ED Related Issues. *Source*: Adapted from health-communication framework (Foss, 2021; Head et al., 2021; Tian & Robinson, 2021) and Brandt's (2001) theory of literacy sponsors, with personal experience with EDs and collegiate athletics.

and change the way resources should be localized and designed to meet the magnitude of mental health concerns.

Collegiate coaches' MHL and knowledge of EDs have been a common area of inquiry within sports medicine. Though no studies are specific to ED-MHL and collegiate coaches from a literacy studies standpoint, research exists on MHL education available for sport coaches (Hebard et al., 2021; Moreland et al., 2018) and factors affecting knowledge, such as resources and training procedures within Division I, II, and III universities (Beals, 2004; Rosa-Caldwell et al., 2018; Sullivan et al., 2019; Torres-McGehee et al., 2011; Turk et al., 1999; Vaughan et al., 2004). From a literacy perspective, "knowledge" is synonymous with literacy, and "factors" is synonymous with agents that influence literacy development. Studies on coaches' ED knowledge show that despite high confidence levels in knowledge, actual knowledge levels were low (Beals, 2004; Greenleaf et al., 2009; Rosa-Caldwell et al., 2018; Torres-McGehee et al., 2011). This creates a complex problem: If coaches' perceived knowledge is high for detecting ED symptoms and intervening, but their actual knowledge is low, they may not be able to recognize the usefulness of resources. They are less likely to fulfill their roles in the athlete-patient ecology. In addition, research found that athletes, as part of the ecology, have shown low ED-MHL knowledge as well as inadequate nutritional knowledge (NCAA Research, 2021, 2022; Riviere et al., 2021). Consequently, multiple factors are informing coaches' and student-athletes' literacy development and perceptions of EDs, all influencing patient-user experience. One solution involves recovered athletes as expert sources in resource design creation: PXD methodology.

COACHES AND ATHLETES AS USERS AND PATIENTS

TPC has been directly involved with UX within ED materials and technologies: examples include engagement and usability issues of ED mobile app (Nitsch et al., 2016); virtual reality uses (Langlet et al., 2021); and participants' experiences of web-based programs for bulimia and binge EDs (Yim et al., 2020). Few studies involving collegiate athletics exist (Hornbuckle, 2002; Parker, 2020); Kroshus et al. (2019) conducted a study on knowledge base before and after using the training module examined in this chapter. However, usability testing has not touched collegiate athletics nor has it affected coaches and athletes' perceptions of resources. Using PXD and localization practices that involve recovered ED athletes in the design process can give us a head start in understanding *if* and *why* resources are useful or not. Asking larger questions rooted in the user-advocacy framework helps to uncover the nuances of what is useful versus what is usable in mental health resources. Evaluating two NCAA resources using a TPC and RHM framework aims to demonstrate how involving recovered athletes in PXD processes might become a new model for future mental health PXD studies.

Methods

This study used an exploratory PXD approach for researching design and usefulness of ED resources for collegiate athletes, using multimodal analysis and EPD or R-PXD in the form of expert review.

The multimodal analysis examined two NCAA mental health resources published on the NCAA's official website (NCAA, 2022b). First, I examined the localized efforts of each resource in terms of development processes by locating publishing information on the document and researching involved stakeholders listed on the organizations' website. Second, I examined the resources' abilities as patient-centered resources using Shoemaker et al's (2013c) PEMAT assessment tool that evaluates *understandability* and *actionability*.[4] Understandability measures features such as *content, word choice, organization, layout and design*, and *visual aids*. Actionability measures features that collectively pertain to *action, audience, manageability, numeric calculations*, and *tables and graphics*. The purpose of evaluating each resource using PEMAT was to highlight what the resources already offered patient-users without PXD methodology.

The expert review included one user examining the resources from the perspective of a recovered athlete. Expert review, as a UX method, is defined as "a usability-inspection method in which (usually) one reviewer examines a design to identify usability problems" (Harley, 2018). Expert review has been a top UX method (Vredenburg et al., 2002) and was among "the methods most commonly employed by practitioners" in an extensive examination of UX industry methods (Robinson et al., 2018, p. 11). As a design expert and a recovered collegiate athlete who developed an ED, I applied localized and PXD methodologies to the resources for comparison. The distinction between the PEMAT evaluation and a localized, PXD informed evaluation was to highlight how the resources' designs are distinguished when using PCD, versus participatory methods that intentionally advocate for patient-user experience.

As an exploratory PXD approach, the methods employed in this study serve as one possible model for future studies to use and extend. The PXD study in this chapter does not claim to be rigorous, statistically strong, or generalizable. Rather, it is a first attempt to model how EPD or R-PXD might be used in design processes for ED/DE resources. Demonstrating a combined analysis with expert review methods, I offer an alternative for what resource design may look like with localized, PXD methods, as well as two tables of alternative resource content. Overall, this analysis includes five phases:

- Phase 1. Examine each resource's localization practices based on how it was developed.

- Phase 2. Examine *Resource 1* using the PEMAT-P (print) and provide results.

- Phase 3. Evaluate *Resource 2* using PEMAT-A/V (audio/visual) for the training module and provide the results.

- Phase 4. Discuss the PEMAT results using localized, PXD methodologies.

- Phase 5. Offer alternatives for developing resources and their current content.

THE NCAA AND ED MENTAL HEALTH RESOURCES

Both resources can be found on NCAA.org's website under the "Health and Safety/Sports Science Institute" tab (NCAA, 2022c). Each is listed on the "Mental Health and Education Resources" page under separate categories (NCAA, 2022b).

Resource 1. This resource is a mental health fact sheet specifically made for ED awareness titled "Eating Disorders" (NCAA Sports Science Institute, n.d.a.). The document's two purposes are (1) to inform readers of signs and symptoms of EDs and (2) to create actionable steps for athletes and athletic department staff to either seek help, intervene, treat, or increase education on EDs. The document is not gender, division, or sport specific. Rather, the resource is for all NCAA website users. This resource was selected because it is the only ED educational resource listed under the mental health fact sheets (NCAA, 2022b).

Resource 2. This resource is a web-based educational training module (NCAA Sports Science Institute, n.d.b.). The training module comprises three videos for student-athletes, coaches, and faculty athletic representatives. This resource was selected because it emphasizes coaches' roles in prevention and intervention. Overall, the module addresses signs and symptoms of mental illnesses, coaches' stigma about mental health, creation of help seeking environment, coaches' roles in treatment, and scenario-based questions on responding to nonemergency and emergency mental health situations (Kroshus et al., 2019). As mentioned, this resource

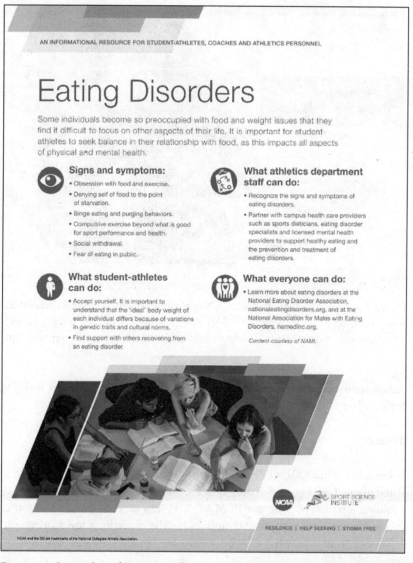

Figure 8.3. Screenshot of Resource 1 "Eating Disorders" Fact Sheet. *Source*: (NCAA Sports Science Institute, n.d.a.); https://ncaaorg.s3.amazonaws.com/ssi/mental/SSI_EatingDisordersFactSheet.pdf

is universalized to support all mental health concerns and, therefore, was not evaluated based on its applicability to EDs. Rather, I examined the module because it allowed for discussion on the positives and affordances of universalized design versus localized content for patient materials.

Figure 8.4. Screenshot of Resource 2 "Supporting Student-Athlete Mental Wellness" Training Module. *Source*: (NCAA Sports Science Institute, n.d.b.); https://www.ncaa.org/sports/2016/11/3/supporting-student-athlete-mental-wellness.aspx

Additionally, it opened the discussion for current mental health training for coaches and what is at stake when designing resources.

TOOLS

Shoemaker et al.'s (2013c) PEMAT form was used to evaluate both resources listed above (see Figure 8.5 on pages 239 and 240). PEMAT-P was used to evaluate *Resource 1*, whereas PEMAT-A/V was used to evaluate *Resource 2* (see Figure 8.6 on pages 241 and 242). The directions for scoring and how each tool was used can be found in Appendixes 8.A and 8.B on pages 256–262.

Results

Resource 1 is described in Table 8.1. (The sections below summarize the findings of Table 8.1 in terms of localized efforts and the PEMAT-P evaluation of understandability, and actionability.)

Title of Material:

Name of Reviewer: Review Date:

Read the PEMAT User's Guide (available at: http://www.ahrq.gov/professionals/prevention-chronic-care/improve/self-mgmt/pemat/) before rating materials.

UNDERSTANDABILITY

Item #	Item	Response Options	Rating
Topic: Content			
1	The material makes its purpose completely evident.	Disagree=0, Agree=1	
2	The material does not include information or content that distracts from its purpose.	Disagree=0, Agree=1	
Topic: Word Choice & Style			
3	The material uses common, everyday language.	Disagree=0, Agree=1	
4	Medical terms are used only to familiarize audience with the terms. When used, medical terms are defined.	Disagree=0, Agree=1	
5	The material uses the active voice.	Disagree=0, Agree=1	
Topic: Use of Numbers			
6	Numbers appearing in the material are clear and easy to understand.	Disagree=0, Agree=1, No numbers=N/A	
7	The material does not expect the user to perform calculations.	Disagree=0, Agree=1	
Topic: Organization			
8	The material breaks or "chunks" information into short sections.	Disagree=0, Agree=1, Very short material* N/A	
9	The material's sections have informative headers.	Disagree=0, Agree=1, Very short material* N/A	
10	The material presents information in a logical sequence.	Disagree=0, Agree=1	
11	The material provides a summary.	Disagree=0, Agree=1, Very short material* N/A	
Topic: Layout & Design			
12	The material uses visual cues (e.g., arrows, boxes, bullets, bold, larger font, highlighting) to draw attention to key points.	Disagree=0, Agree=1, Video=N/A	

---

* A very short print material is defined as a material with two or fewer paragraphs and no more than 1 page in length.

Figure 8.5. Patient Education Materials Assessment Tool for Printed Materials (PEMAT-P). *Source*: Agency for Health and Quality Research, U.S. Department of Health and Human Services. Screenshot from Patient Education Materials Assessment Tool (PEMAT) and User's Guide (2020) on public domain at https://www.ahrq.gov/health-literacy/patient-education/pemat.html

Item #	Item	Response Options	Rating
Topic: Use of Visual Aids			
15	The material uses visual aids whenever they could make content more easily understood (e.g., illustration of healthy portion size).	Disagree=0, Agree=1	
16	The material's visual aids reinforce rather than distract from the content.	Disagree=0, Agree=1, No visual aids=N/A	
17	The material's visual aids have clear titles or captions.	Disagree=0, Agree=1, No visual aids=N/A	
18	The material uses illustrations and photographs that are clear and uncluttered.	Disagree=0, Agree=1, No visual aids=N/A	
19	The material uses simple tables with short and clear row and column headings.	Disagree=0, Agree=1, No tables=N/A	

Total Points: _____

Total Possible Points: _____

Understandability Score (%): _____

(Total Points / Total Possible Points) × 100

ACTIONABILITY

Item #	Item	Response Options	Rating
20	The material clearly identifies at least one action the user can take.	Disagree=0, Agree=1	
21	The material addresses the user directly when describing actions.	Disagree=0, Agree=1	
22	The material breaks down any action into manageable, explicit steps.	Disagree=0, Agree=1	
23	The material provides a tangible tool (e.g., menu planners, checklists) whenever it could help the user take action.	Disagree=0, Agree=1	
24	The material provides simple instructions or examples of how to perform calculations.	Disagree=0, Agree=1, No calculations=NA	
25	The material explains how to use the charts, graphs, tables, or diagrams to take actions.	Disagree=0, Agree=1, No charts, graphs, tables, or diagrams=N/A	
26	The material uses visual aids whenever they could make it easier to act on the instructions.	Disagree=0, Agree=1	

Total Points: _____

Total Possible Points: _____

Actionability Score (%): _____

(Total Points / Total Possible Points) × 100

Figure 8.5. Continued.

Title of Material:

Name of Reviewer: Review Date:

Read the PEMAT User's Guide (available at: http://www.ahrq.gov/professionals/prevention-chronic-care/improve/self-mgmt/pemat/) before rating materials.

UNDERSTANDABILITY

Item #	Item	Response Options	Rating
Topic: Content			
1	The material makes its purpose completely evident.	Disagree=0, Agree=1	
2	The material does not include information or content that distracts from its purpose.	Disagree=0, Agree=1	
Topic: Word Choice & Style			
3	The material uses common, everyday language.	Disagree=0, Agree=1	
4	Medical terms are used only to familiarize audience with the terms. When used, medical terms are defined.	Disagree=0, Agree=1	
5	The material uses the active voice.	Disagree=0, Agree=1	
Topic: Use of Numbers			
6	Numbers appearing in the material are clear and easy to understand.	Disagree=0, Agree=1, No numbers=N/A	
7	The material does not expect the user to perform calculations.	Disagree 0, Agree 1	
Topic: Organization			
8	The material breaks or "chunks" information into short sections.	Disagree=0, Agree=1, Very short material* N/A	
9	The material's sections have informative headers.	Disagree 0, Agree 1, Very short material* N/A	
10	The material presents information in a logical sequence.	Disagree 0, Agree 1	
11	The material provides a summary.	Disagree=0, Agree=1, Very short material* N/A	
Topic: Layout & Design			
12	The material uses visual cues (e.g., arrows, boxes, bullets, bold, larger font, highlighting) to draw attention to key points.	Disagree 0, Agree 1, Video=N/A	

* A very short print material is defined as a material with two or fewer paragraphs and no more than 1 page in length.

Patient Education Materials Assessment Tool for Printable Materials (PEMAT-P)   3

Figure 8.6. Patient Education Materials Assessment Tool for Audio Visuals (PEMAT-AV). *Source:* Agency for Health and Quality Research, U.S. Department of Health and Human Services. Screenshot from Patient Education Materials Assessment Tool (PEMAT) and User's Guide (2020) on public domain at https://www.ahrq.gov/health-literacy/patient-education/pemat.html

Item #	Item	Response Options	Rating
Topic: Use of Visual Aids			
15	The material uses visual aids whenever they could make content more easily understood (e.g., illustration of healthy portion size).	Disagree=0, Agree=1	
16	The material's visual aids reinforce rather than distract from the content.	Disagree=0, Agree=1, No visual aids=N/A	
17	The material's visual aids have clear titles or captions.	Disagree=0, Agree=1, No visual aids=N/A	
18	The material uses illustrations and photographs that are clear and uncluttered.	Disagree=0, Agree=1, No visual aids=N/A	
19	The material uses simple tables with short and clear row and column headings.	Disagree=0, Agree=1, No tables=N/A	

Total Points: _____

Total Possible Points: _____

Understandability Score (%): _____

(Total Points / Total Possible Points) × 100

ACTIONABILITY

Item #	Item	Response Options	Rating
20	The material clearly identifies at least one action the user can take.	Disagree=0, Agree=1	
21	The material addresses the user directly when describing actions.	Disagree=0, Agree=1	
22	The material breaks down any action into manageable, explicit steps.	Disagree=0, Agree=1	
23	The material provides a tangible tool (e.g., menu planners, checklists) whenever it could help the user take action.	Disagree=0, Agree=1	
24	The material provides simple instructions or examples of how to perform calculations.	Disagree=0, Agree=1, No calculations=NA	
25	The material explains how to use the charts, graphs, tables, or diagrams to take actions.	Disagree=0, Agree=1, No charts, graphs, tables, or diagrams=N/A	
26	The material uses visual aids whenever they could make it easier to act on the instructions.	Disagree=0, Agree=1	

Total Points: _____

Total Possible Points: _____

Actionability Score (%): _____

(Total Points / Total Possible Points) × 100

Patient Education Materials Assessment Tool for Printable Materials (PEMAT-P)

Figure 8.6. Continued.

Table 8.1. PEMAT-P Examination: Resource 1 "Eating Disorder" Fact Sheet

Resource 1.	Understandability	Actionability
# of Questions	17	7
# of Relevant Questions	14	5
Totals	14/14	4/5
Score	**100%**	**80%**
Localized Efforts	Publisher: NCAA and SSI. Content attributed to the NAMI (NCAA Sports Science Institute, n.d.a.)	

Source: Created by the author.

RESOURCE 1: LOCALIZED EFFORTS

This resource was created by the NCAA and the SSI; however, the content itself was attributed to NAMI. PEMAT measures of *understandability* and *actionability* scored 100% and 80%, respectively. For *understandability*, 14 of the 17 questions applied and thus were scored out of 14 points. Questions that did not apply regarded *use of numbers* or *visual aids* such as tables and graphs. For *actionability*, two of the seven questions received N/A and were, therefore, scored out of 5 points. Questions that did not apply pertained to *numeric calculations* and *tables and graphs*.

PEMAT-P Evaluation: Understandability

PEMAT's purpose was to create patient materials that were patient-centered: Materials were understandable and could be acted on. *Understandability* included features such as *content, word choice, organization, layout and design*, and *visual aids. Content* made its purpose evident through the heading as an "informational resource for student athletes, coaches, and athletic personnel." No miscellaneous information distracted users from its purpose. *Word choice* avoided using medical jargon. However debatable, not every user understood the terminology. For example, "purging and binging behaviors" were familiar words, but identifying what these behaviors were or how they appeared in athletic contexts were points to consider. *Organization* included separate "chunks" of information using bullet points and sectional headings. *Layout and design* followed a consistent pattern of font, bolded headings, and color scheme. *Visual aids* included headings accompanied by an icon that represented the audience or action item. No images or figures were unclear or cluttered.

PEMAT-P Evaluation: Actionability

Actionability was more difficult to access because the document was not meant to be an extensive resource. Its purpose was brevity[5] rather than thoroughness. Actionability collectively pertained to *audience and action, manageability, numeric calculations,* and *tables and graphics.* The resource effectively created *calls to action* for student-athletes and athletic personnel by making an intentional effort to locate specific patient-users rather than generalizing each. For example, the resource created a clear *audience* divide into "what athletes can do" and "what athletic department staff can do." Further, the document used personalized language when it addressed athletes as "yourself." Each of these details encourages patient-user centeredness. However, at the top, the resource addressed patient-users as "student athletes, coaches, and athletic personnel." In the section heading dedicated to "athletic personnel," no distinction between which personnel should be involved was made: This was left ambiguous. For example, it was unclear whether coaches, athletic trainers, or sports dietitians were responsible for recognizing the student-athletes' issues. This ambiguity makes *manageability* difficult to evaluate.

Conventionally, *manageability* received a positive score. The resource states two clear action items for each audience member to complete, listed as two separate bullets. Each step also had clear action words and phrases such as "partner with" and "recognize signs." However, when considering who should do what (i.e., coaches versus athletic trainers), and if actions are separated into manageable steps for the intended audience, manageability was questionable. For example, if coaches are encouraged to "partner with campus health care providers such as dietitians," DII or DIII coaches may not have access to or funding for registered dieticians like DI programs do. For "recognizing signs and symptoms," an assumption existed that coaches know the signs and symptoms *and* can apply them in a real-life context. For the athletes, the resource called for two primary actions: "accept yourself" and "find support with others recovering from an eating disorder." Though actions were separated into two steps and appear manageable based on conventional features, how to complete these steps was also not evident. For example, for those struggling with mental illness, no information was offered on self-acceptance. Though *manageability* overall received a positive score conventionally, completing those actions or steps was ambiguous when considering audience or context of use.

Lastly, *Resource 1* did not contain *numeric* values or calculations and was disregarded. In terms of *visual aids,* as discussed in *understandability,* each heading was accompanied by an icon that represented the audience or action item. No images or figures were unclear or cluttered or created distraction from the content.

Resource 2 is described in Table 8.2. (The sections below summarize the findings of Table 8.2 in terms of localized efforts and the PEMAT-A/V evaluation of understandability and actionability.)

RESOURCE 2: LOCALIZED EFFORTS

Like *Resource 1*, the training module was created by the SSI and NCAA. However, it was unclear if any other partners were involved in the development of the resource. The resource scored a 100% *understandability* score and a 100% *actionability* score. For *understandability,* 12 of the 13 questions applied to the resource and thus scored 12 points total. The question pertaining to *visual aids* of tables and figures was disregarded. For actionability, three of four questions applied to the resource and scored 3 points total. The question pertaining to tables and figures was disregarded.

Table 8.2. Results Summary of PEMAT-P Evaluation for Resource 2 "Supporting Student-Athlete Mental Wellness"

Resource 1.	Understandability	Actionability
# of Questions	13	4
# of Relevant Questions	12	3
Totals	12/12	3/3
Score	**100%**	**100%**
Localized Efforts	Known Publisher: The SSI and NCAA (NCAA Sports Science Institute, n.d.b.)	

Source: Created by the author.

PEMAT-A/V Evaluation: Understandability

Content made this module's purpose evident by stating the collective objective was guiding coaches to "support student-athlete mental wellness." For *word-choice and style*, the resource used plain language and refrained from overusing medical jargon. Instances of potentially unfamiliar terminology included vernacular language or provided examples to clarify meaning. For instance, part 2 of the module listed "severe functional impairment" to describe a stage on a continuum of wellness (Figure 8.7). To illustrate what severe or functional impairment may look like, the module provided follow-up examples such as "aggressive, withdrawn, depressed, or anxious," "significant difficulty with emotions and thinking," and "constant fatigue" (Figure 8.7).

Organization was seen in the module's systemically familiarizing coaches by addressing stigma, signs and symptoms first, and then shifted to taking action and assessing scenarios. The order appeared logical. *Layout and design* are apparent in sections accompanied by pictures or visual cues to represent the content. For example, in the "fighting stigma" segment,

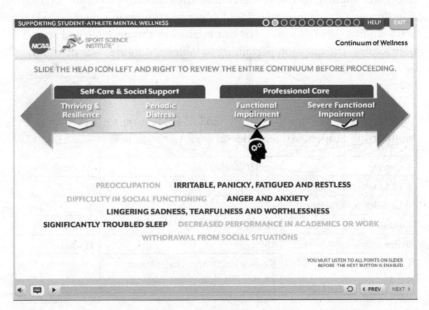

Figure 8.7. Screenshot of Interactive Component of MH Related Content and Terminology in Resource 2. *Source*: NCAA Sport Science Institute.

the module suggested that coaches create an environment that fosters conversation and is open to discussing mental health. Fostering conversation was represented by a simple icon of two human figures conversing with a speech cloud between them. Other aspects of *layout and design* were also clear. For example, the audio and text on the screen were clear, with closed captioning available. Lastly, tables and figures were not evaluated in *visual aids*; each image and icon were clear and uncluttered.

PEMAT A/V Evaluation: Actionability

Audience and action were presented in the module's identifying coaches as its target audience, and thus, directly addressed them as patient-users from the beginning. The module also directly used personalized language by stating "you as a coach can . . ." or "it is crucial that you . . ." (Sports Science Institute, n.d.). *Action* was seen in the module's identifying multiple action items for coaches to complete and separate chunks of information as interactive icons preventing patient-users to skip to the next section. Explicit steps were presented in the module informing coaches on an aspect of mental health or creating a call to action, followed by interactive icons that state what coaches can do. For example, the module's fourth segment was dedicated to information on how coaches may unknowingly feed the stigma attached to mental illness. The module then shifted to four icons that provide directives for actions coaches could implement or change. For instance, the module instructed coaches to "watch their language" and then provided words to avoid using, such as "crazy" or "psycho" (Sports Science Institute, n.d.).

Manageability was visible in the module's step-by-step action items. For example, during the scenario-based quiz section, the module asked coaches, "Would you know what to do?" The module engaged them to choose the appropriate action step either calling 911/The National Suicide Hotline or following routine protocol. However, going past conventional measures, the module even appeared at times to consider the context of use that affects patient-user experience. When selecting "follow the routine protocol," the module stated that "you [coaches] are a key component of your department's referral protocol and should follow the specific referral process outlined by your institution's . . ." (see Figure 8.8). However, the module also commented that "if one [a plan or protocol] has not yet been established, you should call 911" (Sports Science Institute, n.d.). Therefore, the resource acknowledged that each institutional procedure is different and recognized variances across division and or department levels. This instance also demonstrated why localized design is necessary for ecological variances that affect patient-user experience within collegiate athletics. *Numeric calculations* and *tables and graphics* were not present in this module.

Discussion

In considering the outcomes of this analysis, I considered localized design and participatory design methods in applying PXD.

THE APPLICATION OF PXD AND LOCALIZED DESIGN METHODS

A PEMAT examination shows that the fact sheet and educational module are effective as informational resources. Each resource scored 100% in *understandability* and 80% or higher in *actionability*. The PEMAT criteria

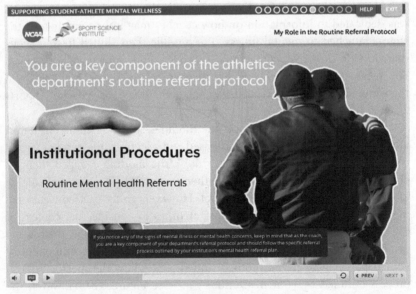

Figure 8.8. Screenshot of Department Specific Protocol Referral from Resource 2. *Source*: NCAA Sport Science Institute.

and features align with PCD principles that encourage ease of use and emphasize the importance of plain language, accessibility, and information processing (Melonçon, 2017b). However, content and comprehension concerns discussed in *word choice* and *manageability* of actionable steps may explain why the resource designs may not meet UX needs and advocate for coaches and athletes as patient-users. PXD methodology (Melonçon, 2017b) calls for emphasizing patient embodiment and contexts of use in patient materials. To embody patient-athletes, however, requires an innovative approach to protect this vulnerable population: reasons that involving recovered athletes may be justified.

RECONSIDERING LOCALIZATION AND PXD PARTICIPATORY METHODS

As stated on the resources themselves, the stakeholders involved in the resource design were the NCAA, SSI, and NAMI. Stakeholders within the SSI include MDs, certified and licensed athletic trainers, and other communication specialists (NCAA, 2022a). Though NAMI's listed partnerships include multiple organizations involved with other mental health concerns such as the American Foundation for Suicide Prevention, the American Psychiatric Organization and International Crisis Intervention teams, ED-specific partnerships are not apparent. In terms of MH concerns, resource developers must partner with MH organizations.

However, complexities of ED patient ecologies require further steps to localize and include those who specialize in EDs. Sun and Getto (2017) and Lancaster and King (2024b) called for moving past localized design. However, they also acknowledge the importance and challenge of balancing design. In the context of health communication, we must move past group-based approaches to contextualize culture (Hsieh, 2021). "When interventions are designed for specific population-based labels, we may overlook populations that are not named" (p. 451). Although destigmatizing and supporting mental health in collegiate athletics is a large "population label" that must be addressed, we are in danger of generalizing or inadequately treating signs and symptoms of specific disorders if we treat them "globally" or as a group—pointing to a truly complex design issue for generating ED resources.

As has been discussed, the severity, complex sponsorship, and normalization of EDs call for resource design that is localized and intentional

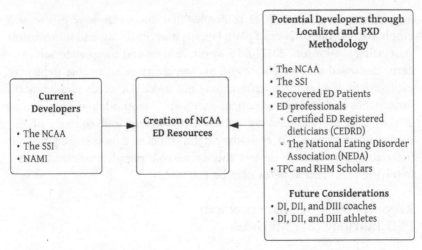

Figure 8.9. Alternative and Additional Developers in NCAA Resource Design using Experienced Patient Design (EPD) and Localized Methodologies. *Source*: Created by the author.

to EDs. Figure 8.9 offers alternatives of who could be involved in resource design development.

In Figure 8.9, "Current Developers" shows those currently involved in the design process of *Resource 1* and *2*, and "Potential Developers" shows what the development process may look like with localized and PXD participatory involvement. This figure shows the involvement of recovered patients and other alternative options to embody patient experience and advocate for patient-users by emphasizing contexts of use. For example, instead of mental health organizations at large, NEDA could be a developing partner. In addition to the medical staff on the SSI, certified ED registered dieticians (CEDRD) (rather than dieticians at large) could help to highlight nuances in signs and symptoms. For context of use, and while potentially not versed in collegiate athletic environments, recovered patients can identify variables that make resources useful rather than usable for coaches and athletes. For example, a recovered patient can recognize the difficulty in the action on *Resource 1* of "recognizing signs and symptoms" because they may have experienced the symptoms or have even hidden them themselves.

The methodology must involve "identifying, understanding, and addressing the variables affecting usability in different contexts of care—or settings where care-related activities take place" (St.Amant, 2017, p. 64). Engaging recovered patients as valuable input, designers of such resources can begin to identify variables that make resources useful or capture the complexities of EDs in sport environments. To demonstrate, Table 8.3 and 8.4 present PXD methodology of a recovered individual. The alternative column aims to show what a shift in content may look like with a focus on embodied and the contextualized needs.

Table 8.3. Alternative Content for Resource 1 Based on EPD and Localized Methodology

Current Language for Athletes	Alternative Suggestions	Need for Change
"Accept yourself."	*Accept that you may need help outside of yourself. Accept that you may need support to heal your relationship with....*	Literacy sponsors and history Stigma Embodiment
"It is important to consider that the ideal body weight does not exist"	*It is important to consider that your habits around food and exercise may be harmful to your....*	Misconceptions Stigma Literacy sponsors and history
"Find support with others recovering from an ED."	*Hear stories from other athletes... Explore these resources that offer support...*	Denial Stigma Traits of an athlete
Current Language for Coaches	**Alternative Suggestions**	**Need for Change**
"Recognize the signs and symptoms of eating disorders." "Partner with campus healthcare providers such as dieticians...."	*Check whether you can recognize the signs....*	ED-MHL literacy Divisional variances, access, funding

Source: Created by the author.

Table 8.4. Alternative Content for Resource 2 Based on EPD and Localized Methodology

Module Topic	Current Content: Universal MH Concerns	Transition Content: Specific to EDs in Athletic Contexts	Alternative Content Examples
Signs and symptoms	Withdrawal from peers Legal issues Confusion Suicidal thoughts Lack of motivation Worried Tired or weak	ED signs and symptoms	*Increased isolation from teammates, social withdrawal* *Bringing separate food to team functions* *Memory loss* *Difficulty concentrating* *Sensitivity to cold* *Increased fatigue or perceived exertion* *Training outside of/ beyond requirements*
Stigma and Creating Supportive Environment	Watch Your Language "Crazy, psycho, nuts, lunatic"	Watch language that may be triggering or reinforcing behaviors	*"You're so disciplined in your training routine! Keep pushing yourself."* *"You really earned this meal or __ food today."* *"It's great how clean you eat! How can we get your teammates to be more like you?"* *"Let's be sure we're watching our intakes of __. It's bad for us."*
Intervention or Referral	Follow department plan, call 911, or call the Suicide Hotline	Provide action items specific to ED intervention	*It is better to do or say something rather than nothing.* *Inform and talk to your athletic trainer (all DI-DIII schools have).* *Approach athlete with ease and patience.*

Source: Created by the author.

ETHICAL CONSIDERATIONS AND LIMITATIONS

Significant considerations and limitations within this analysis should not be ignored. The first is that the examination of localization practices must be taken with care. Though the NCAA, SSI, and NAMI are listed as co-developers of the resources, the exact design process of these materials is unknown. Therefore, this analysis is not to suggest or assume anything about developer identities or about the creation process that is not apparent. It is also not to diminish the work that the NCAA or SSI have done to support athletes. Rather, this analysis aims to capture what localized and PXD informed design may look like in collegiate athletics to advocate for patient-users.

The second is PEMAT as a form of evaluation. PEMAT's function was to show why these resources may appear useful on the surface but become less so when evaluated from the perspective of a recovered individual (i.e., using PXD methodology) and to open discussion for alternative design approaches. However, this comparison may have been better informed with an assessment tool specifically meant to evaluate design process and/or content, but no known tools exist to evaluate content based on context. There is also no way to examine the design process without speaking to the developers themselves. This brings up the conversation of assessment tools at large, as well as directions this research could take, which are discussed in the next section.

Lastly, this evaluation and research study was limited to one perspective (an expert reviewer who is also a recovered athlete). However, my voice represents athletes who are struggling, or have struggled. My voice also acts as an ally for coaches who deserve and require resources that support their role in the patient ecology: for now. Involving experienced and recovered athletes in PXD practices may address ethical considerations for recruiting vulnerable populations, such as student-athletes who have (or may develop) an ED. It does not address, however, complications in recruiting such individuals. Questions about how stakeholders would oversee recruiting or funding this process would need to be addressed in future research and during collaborative efforts of resource creators.

Future Implications and Conclusion

Implications from this study for future research include using innovative design methods to advocate for and improve patient-user experience. TPC and RHM scholars may find needs for unconventional features of design, redefining usability, advocating for participatory design methods, and reconsidering localization practices. Scholars are accounting for more than patient-user experience, including lived experiences and the worlds (ecologies) surrounding patient-users.

Melonçon's (2017b) PXD is a promising method geared toward doing this work in TPC/RMH. However, mental health and EDs specifically complicate the notions of who we consider a patient. When a current patient cannot advocate for themselves (i.e., if a coach does not realize resources are not catering to their contexts); when an individual does not consider themselves a patient (i.e., an athlete is reluctant to receive treatment); and when involving current patients in design causes harm; are three specific cases that call for alternative PXD approaches. This research presents future implications for PXD methodology by considering the idea of an EPD or R-PXD. By using experienced and recovered "patients," designers can glean useful insights about *if* and *why* a resource is useful and usable.

Melonçon's (2017b) call to innovate PXD methods and the ecology of patient experience noted by Greenhalgh (2017) also questions preserving the patient's centrality or where centrality should be placed within PXD methodology. For example, as mentioned, student-athletes struggling with an ED are not the only patients involved in decision making or the treatment process. Coaches, sports dieticians, and athletic trainers also fall into this ecology. The participatory method proposed in this study involves recovered patients as valuable sources of input in the design process of ED resources. However, involving others such as CEDRD, organizations such as NEDA, or collaborating with coaches could be future directions design research could take. Though student-athletes are the central patients, TPC/RHM scholars may consider extending the patient focus in PXD methodology further within the ecology. Involving individuals for participatory methods may be based on specific patient condition and context, for example. For EDs and collegiate athletics, this again could mean recovered individuals but also voices that could support athletes and coaches when they may not be able to support themselves.

CONVENIENCE OF MATERIALS

As a side, this research poses an additional conversation for the concept of ease. Though *understandability* and *actionability* scores for each resource were high, a significant shift from *Resource 1* to *Resource 2* exists for involvement, content, and manageable steps. *Resource 2*'s multimodality allowed for thorough explanation of concepts through examples, demonstration through interactive steps, application questions, and retained attention by disallowing individuals to skip through segments without completion. However, *Resource 2* was a multimodal resource with the purpose of universalizing and informing coaches of *all* mental health concerns. Circling back to balancing localized design, this approach was both beneficial and harmful. Though *Resource 1* was "convenient," it missed crucial content that reinforced comprehension. Though *Resource 2* was extensive, it did not specify ED symptoms because of its purpose to cover general mental health and wellness concerns rather than specific ED related content.

Moving forward, TPC and RHM as collective fields might embrace Dilger's (2000) original lead in questioning the concept of ease, as well as educational or informational genres. TPC and RHM is concerned with making easy-to-use materials. However, there may be times when we simplify material too much. For example, in what ways do the genre of fact sheets complete their purpose of convenience? In what situations are topics such as mental health too complex to be reduced to a fact sheet? As the evaluation through PEMAT has shown, design risks exist when assessing conventional usefulness, or even what *we* as designers imagine as useful. We share a responsibility to consider complex design questions but also to involve users of our products in the design process. Participatory methods of PXD and localized design moves us in this direction.

This research creates a promising space for TPC and RHM scholars to enter into and engage in critical work. It is my hope that others come to this research study to improve on it: by continuing to answer the call for creating innovative yet balanced design research methods and by humbly answering the call from athletes and coaches who have asked for support in mental health and wellness.

Appendix 8.A

Patient Education Materials Assessment Tool for Audiovisual Materials (PEMAT-A/V)

How To Use the PEMAT To Assess a Material

There are seven steps to using the PEMAT to assess a patient education material. The instructions below assume that you will score the PEMAT using paper and pen. If you use the **PEMAT Auto-Scoring Form**, a form that will automatically calculate PEMAT scores once you enter your ratings, you can skip Step 5. The form is available at: http://www.ahrq.gov/professionals/prevention-chronic-care/improve/self-mgmt/pemat/index.html.

Step 1: Read through the PEMAT and User's Guide. Before using the PEMAT, read through the entire User's Guide and instrument to familiarize yourself with all the items. In the User's Guide a (P) and (A/V) are listed after an item to indicate whether it is relevant for print and audiovisual materials, respectively.

Step 2: Read or view patient education material. Read through or view the patient education material that you are rating in its entirety.

Step 3: Decide which PEMAT to use. Choose the PEMAT-P for printable materials or the PEMAT-A/V for audiovisual materials.

Step 4: Go through each PEMAT item one by one. All items will have the answer options "Disagree" or "Agree." Some—but not all—items will also have a "Not Applicable" answer option. Go one by one through each of the items, 24 for printable materials and 17 for audiovisual materials, and indicate if you agree or disagree that the material is meeting a specific criterion. Or, when appropriate, select the "Not Applicable" option.

You may refer to the material at any time while you complete the form. You don't have to rely on your memory. Consider each item from a patient perspective. For example, for "Item 1: The material makes its purpose completely evident," ask yourself, "If I were a patient unfamiliar with the subject, would I readily know what the purpose of the material was?"

Step 5: Rate the material on each item as you go. After you determine the rating you would give the material on a specific item, enter the number (or N/A) that corresponds with your answer in the "Rating" column of the PEMAT. Do not score an item as "Not Applicable" unless there is a "Not Applicable" option. Score the material on each item as follows:

If Disagree	Enter 0
If Agree	Enter 1
If Not Applicable	Enter NA

Suggested Citation:

Shoemaker SJ, Wolf MS, Brach C. Patient Education Materials Assessment Tool for Audiovisual Materials (PEMAT-A/V). (Prepared by Abt Associates, under Contract No. HHSA290200900012I, TO 4). Rockville, MD: Agency for Healthcare Research and Quality; October 2013. AHRQ Publication No. 14-0002-EF.

Figure 8.10. Patient Education Assessment Tool (PEMAT-P) for Print Materials. Instructions on how to access material and completed evaluation score sheet of Resource 1 "Eating Disorders. *Source*: Agency for Health and Quality Research, U.S. Department of Health and Human Services.

Additional Guidance for Rating the Material on Each Item (Step 5):

- Rate an item "Agree" when a characteristic occurs throughout a material, that is, nearly all of the time (80% to 100%). Your guiding principle is that if there are obvious examples or times when a characteristic could have been met or could have been better met, then the item should be rated "Disagree." The User's Guide provides additional guidance for rating each item.
- Do not skip any items. If there is no "Not Applicable" option, you must score the item 0 (Disagree) or 1 (Agree).
- Do not use any knowledge you have about the subject before you read or view the patient education material. Base your ratings ONLY on what is in the material that you are rating.
- Do not let your rating of one item influence your rating of other items. Be careful to rate each item separately and distinctly from how you rated other items.
- If you are rating more than one material, focus only on the material that you are reviewing and do not try to compare it to the previous material that you looked at.

Step 6: Calculate the material's scores. The PEMAT provides two scores for each material— one for understandability and a separate score for actionability. Make sure you have rated the material on every item, including indicating which items are Not Applicable (N/A). Except for Not Applicable (N/A) items, you will have given each item either 1 point (Agree), or 0 points (Disagree). To score the material, do the following:

- *Sum the total points* for the material on the understandability items only.
- *Divide the sum by the total possible points,* that is, the number of items on which the material was rated, excluding the items that were scored Not Applicable (N/A).
- *Multiply the result by 100* and you will get a percentage (%). This percentage score is the understandability score on the PEMAT.

  - **Example:** If a print material was rated Agree (1 point) on 12 understandability items, Disagree (0 points) on 3 understandability items, and N/A on one understandability item (N/A), the sum would be 12 points out of 15 total possible points (12 + 3, excluding the N/A item). The PEMAT understandability score is 0.8 (12 divided by 15) multiplied by 100 = 80%.

To score the material on actionability, repeat Step 6 for the actionability items.

Step 7: Interpret the PEMAT scores. The higher the score, the more understandable or actionable the material. For example, a material that receives an understandability score of 90% is more understandable than a material that receives an understandability score of 60%, and the same goes for actionability. If you use the PEMAT to rate the understandability and actionability of many materials, you may get a sense of what score indicates exceptionally good or exceptionally poor materials.

Title of Material: Supporting Student-Athlete Mental Wellness

Name of Reviewer: *Mallory Henderson*   Review Date: 3/30/22

Read the PEMAT User's Guide (available at: http://www.ahrq.gov/professionals/prevention-chronic-care/improve/self-mgmt/pemat/) before rating materials.

UNDERSTANDABILITY

Item #	Item	Response Options	Rating
Topic: Content			
1	The material makes its purpose completely evident.	Disagree=0, Agree=1	1
Topic: Word Choice & Style			
3	The material uses common, everyday language.	Disagree=0, Agree=1	1
4	Medical terms are used only to familiarize audience with the terms. When used, medical terms are defined.	Disagree=0, Agree=1	1
5	The material uses the active voice.	Disagree=0, Agree=1	1
Topic: Organization			
8	The material breaks or "chunks" information into short sections.	Disagree=0, Agree=1, Very short material[1]=N/A	1
9	The material's sections have informative headers.	Disagree=0, Agree=1, Very short material[*]=N/A	1
10	The material presents information in a logical sequence.	Disagree=0, Agree=1	1
11	The material provides a summary.	Disagree=0, Agree=1, Very short material[*]=N/A	1
Topic: Layout & Design			
12	The material uses visual cues (e.g., arrows, boxes, bullets, bold, larger font, highlighting) to draw attention to key points.	Disagree=0, Agree=1,, Video=N/A	1
13	Text on the screen is easy to read.	Disagree=0, Agree=1, No text or all text is narrated=N/A	1
14	The material allows the user to hear the words clearly (e.g., not too fast, not garbled).	Disagree=0, Agree=1, No narration=N/A	1

---

[1] A very short audiovisual material is defined as a video or multimedia presentation that is under 1 minute, or a multimedia material that has 6 or fewer slides or screenshots.

Patient Education Materials Assessment Tool for Audiovisual Materials (PEMAT-AV)   3

Figure 8.10. Continued.

Appendix 8.B

Patient Education Materials Assessment Tool for Printable Materials (PEMAT-P)

How To Use the PEMAT To Assess a Material

There are seven steps to using the PEMAT to assess a patient education material. The instructions below assume that you will score the PEMAT using paper and pen. If you use the **PEMAT Auto-Scoring Form**, a form that will automatically calculate PEMAT scores once you enter your ratings, you can skip Step 5. The form is available at:
http://www.ahrq.gov/professionals/prevention-chronic-care/improve/self-mgmt/pemat/index.html.

Step 1: Read through the PEMAT and User's Guide. Before using the PEMAT, read through the entire User's Guide and instrument to familiarize yourself with all the items. In the User's Guide a (P) and (A/V) are listed after an item to indicate whether it is relevant to print and audiovisual materials, respectively.

Step 2: Read or view patient education material. Read through or view the patient education material that you are rating in its entirety.

Step 3: Decide which PEMAT to use. Choose the PEMAT-P for printable materials or the PEMAT-A/V for audiovisual materials.

Step 4: Go through each PEMAT item one by one. All items will have the answer options "Disagree" or "Agree." Some—but not all—items will also have a "Not Applicable" answer option. Go one by one through each of the items, 24 for printable materials and 17 for audiovisual materials, and indicate if you agree or disagree that the material meets a specific criterion. Or, when appropriate, select the "Not Applicable" option.

You may refer to the material at any time while you complete the form. You don't have to rely on your memory. Consider each item from a patient perspective. For example, for "Item 1: The material makes its purpose completely evident," ask yourself, "If I were a patient unfamiliar with the subject, would I readily know what the purpose of the material was?"

Step 5: Rate the material on each item as you go. After you determine the rating you would give the material on a specific item, enter the number (or N/A) that corresponds with your answer in the "Rating" column of the PEMAT. Do not score an item as "Not Applicable" unless there is a "Not Applicable" option. Score the material on each item as follows:

If Disagree	Enter 0
If Agree	Enter 1
If Not Applicable	Enter N/A

Suggested Citation:
Shoemaker SJ, Wolf MS, Brach C. Patient Education Materials Assessment Tool for Printable Materials (PEMAT-P). (Prepared by Abt Associates under Contract No. HHSA290200900012I, TO 4). Rockville, MD: Agency for Healthcare Research and Quality; October 2013. AHRQ Publication No. 14-0002-EF.

Figure 8.11. Patient Education Assessment Tool (PEMAT-P) for Print Materials. Instructions on how to access material and completed evaluation score sheet of Resource 2 "Supporting Student-Athlete Mental Wellness." *Source*: Agency for Health and Quality Research, U.S. Department of Health and Human Services.

Additional Guidance for Rating the Material on Each Item (Step 5)

- Rate an item "Agree" when a characteristic occurs throughout a material, that is, nearly all of the time (80% to 100%). Your guiding principle is that if there are obvious examples or times when a characteristic could have been met or could have been better met, then the item should be rated "Disagree." The User's Guide provides additional guidance for rating each item.
- Do not skip any items. If there is no "Not Applicable" option, you must score the item 0 (Disagree) or 1 (Agree).
- Do not use any knowledge you have about the subject before you read or view the patient education material. Base your ratings ONLY on what is in the material that you are rating.
- Do not let your rating of one item influence your rating of other items. Be careful to rate each item separately and distinctly from how you rated other items.
- If you are rating more than one material, focus only on the material that you are reviewing and do not try to compare it to the previous material that you looked at.

Step 6: Calculate the material's scores. The PEMAT provides two scores for each material—one for understandability and a separate score for actionability. Make sure you have rated the material on every item, including indicating which items are Not Applicable (N/A). Except for Not Applicable (N/A) items, you will have given each item either 1 point (Agree) or 0 points (Disagree). To score the material, do the following:

- *Sum the total points* for the material on the understandability items only.
- *Divide the sum by the total possible points,* that is, the number of items on which the material was rated, excluding the items that were scored Not Applicable (N/A).
- *Multiply the result by 100* and you will get a percentage (%). This percentage score is the understandability score on the PEMAT.

  o **Example:** If a print material was rated Agree (1 point) on 12 understandability items, Disagree (0 points) on 3 understandability items, and N/A on one understandability item (N/A), the sum would be 12 points out of 15 total possible points (12 + 3, excluding the N/A item). The PEMAT understandability score is 0.8 (12 divided by 15) multiplied by 100 = 80%.

To score the material on actionability, repeat Step 6 for the actionability items.

Step 7: Interpret the PEMAT scores. The higher the score, the more understandable or actionable the material. For example, a material that receives an understandability score of 90% is more understandable than a material that receives an understandability score of 60%, and the same goes for actionability. If you use the PEMAT to rate the understandability and actionability of many materials, you may get a sense of what score indicates exceptionally good or exceptionally poor materials.

Figure 8.11. Continued.

Title of Material: Eating Disorders Awareness

Name of Reviewer: *Mallory Henderson*   Review Date: 3/28/22

Read the PEMAT User's Guide (available at: http://www.ahrq.gov/professionals/prevention-chronic-care/improve/self-mgmt/pemat/) before rating materials.

UNDERSTANDABILITY

Item #	Item	Response Options	Rating
Topic: Content			
1	The material makes its purpose completely evident.	Disagree=0, Agree=1	1
2	The material does not include information or content that distracts from its purpose.	Disagree=0, Agree=1	1
Topic: Word Choice & Style			
3	The material uses common, everyday language.	Disagree=0, Agree=1	1
4	Medical terms are used only to familiarize audience with the terms. When used, medical terms are defined.	Disagree=0, Agree=1	1
5	The material uses the active voice.	Disagree=0, Agree=1	1
Topic: Use of Numbers			
6	Numbers appearing in the material are clear and easy to understand.	Disagree=0, Agree=1, No numbers=N/A	N/A
7	The material does not expect the user to perform calculations.	Disagree=0, Agree=1	1
Topic: Organization			
8	The material breaks or "chunks" information into short sections.	Disagree=0, Agree=1, Very short material*=N/A	1
9	The material's sections have informative headers.	Disagree=0, Agree=1, Very short material*=N/A	1
10	The material presents information in a logical sequence.	Disagree=0, Agree=1	1
11	The material provides a summary.	Disagree=0, Agree=1, Very short material*=N/A	1
Topic: Layout & Design			
12	The material uses visual cues (e.g., arrows, boxes, bullets, bold, larger font, highlighting) to draw attention to key points.	Disagree=0, Agree=1 Video=N/A	1

---

* A very short print material is defined as a material with two or fewer paragraphs and no more than 1 page in length.

Patient Education Materials Assessment Tool for Printable Materials (PEMAT-P)   3

Figure 8.11. Continued.

Item #	Item	Response Options	Rating
Topic: Use of Visual Aids			
15	The material uses visual aids whenever they could make content more easily understood (e.g., illustration of healthy portion size).	Disagree=0, Agree=1	1
16	The material's visual aids reinforce rather than distract from the content.	Disagree=0, Agree=1, No visual aids=N/A	1
17	The material's visual aids have clear titles or captions.	Disagree=0, Agree=1, No visual aids=N/A	N/A
18	The material uses illustrations and photographs that are clear and uncluttered.	Disagree=0, Agree=1, No visual aids=N/A	1
19	The material uses simple tables with short and clear row and column headings.	Disagree=0, Agree=1, No tables=N/A	N/A

Total Points: 14

Total Possible Points: 14

Understandability Score (%): 100%

(Total Points / Total Possible Points) × 100

ACTIONABILITY

Item #	Item	Response Options	Rating
20	The material clearly identifies at least one action the user can take.	Disagree=0, Agree=1	1
21	The material addresses the user directly when describing actions.	Disagree=0, Agree=1	1
22	The material breaks down any action into manageable, explicit steps.	Disagree=0, Agree=1	1
23	The material provides a tangible tool (e.g., menu planners, checklists) whenever it could help the user take action.	Disagree=0, Agree=1	0
24	The material provides simple instructions or examples of how to perform calculations.	Disagree=0, Agree=1, No calculations=NA	N/A
25	The material explains how to use the charts, graphs, tables, or diagrams to take actions.	Disagree=0, Agree=1, No charts, graphs, tables, or diagrams=N/A	N/A
26	The material uses visual aids whenever they could make it easier to act on the instructions.	Disagree=0, Agree=1	1

Total Points: 4

Total Possible Points: 5

Actionability Score (%): 80%

(Total Points / Total Possible Points) × 100

Figure 8.11. Continued.

Notes

1. NCAA Research is responsible for conducting a research study called the "Student-Athlete Well-Being Study." The 2021 survey examined the impact of the experiences and well-being of student-athletes during the COVID-19 pandemic. The 2022 study was a follow-up and continuation of the Spring 2020 and Fall 2020 surveys. The survey was designed by NCAA Research, the Sport Science Institute, and the NCAA's Division I-III Student-Athlete Advisory Committees (SAACs).

2. Regarding this subset of health literacy, Kutcher et al. (2016) has defined mental health literacy (MHL) as "understanding how to obtain and maintain positive mental health; understanding mental disorders and their treatments; decreasing stigma related to mental disorders; and, enhancing help-seeking efficacy" (p. 155).

3. Mond (2014) coined the term "eating disorders mental health literacy" (ED-MHL) as a derivative of MHL. He acknowledged ED-MHL as an understudied area of research in health literacy.

4. Shoemaker et al.'s (2013c) definition of understandability and actionality are as follows: Understandability: Patient education materials are understandable when consumers of diverse backgrounds and varying levels of health literacy can process and explain key messages. Actionability: Patient education materials are actionable when consumers of diverse backgrounds and varying levels of health literacy can identify what they can do based on the information presented.

5. Brevity or convenience and concern this may raise will be mentioned further in the discussion section.

References

Agboka, G. Y. (2014). Decolonial methodologies: Social justice perspectives in intercultural technical communication research. *Journal of Technical Writing and Communication, 44*(3), 297–327. https://doi.org/10.2190/TW.44.3.e

Ambwani, S., Cardi, V., Albano, G., Cao, L., Crosby, R. D., Macdonald, P., Schmidt, U., & Treasure, J. (2020). A multicenter audit of outpatient care for adult anorexia nervosa: Symptom trajectory, service use, and evidence in support of "early stage" versus "severe and enduring" classification. *The International Journal of Eating Disorders, 53*(8), 1337–1348. https://doi.org/10.1002/eat.23246

American Psychiatric Association. (2022). *Diagnostic and Statistical Manual of Mental Disorders: Text Revision (DSM-5-TR)* (5th ed.). American Psychiatric Association Publishing.

Angeli, E., & Norwood, C. (2016). Responding to public health crises: collective mindfulness, high reliability, and user experience. *Communication Design Quarterly, 5*(2), 29–39. https://doi.org/10.1145/3131201.3131204

Arcelus, J., Mitchell, AJ., Wales, J., & Nielsen S. (2011). Mortality rates in patients with anorexia nervosa and other eating disorders. A meta-analysis of 36 studies. *Arch Gen Psychiatry. 68*(7), 724–731. https://doi.org/10.1001/archgenpsychiatry.2011.74

Acharya, K. (2018). Usability for user empowerment: Promoting social justice and human rights through localized UX design. In *Proceedings of The 36th ACM International Conference on the Design of Communication,* 1–7. https://doi.org/10.1145/3233756.3233960

Beals, K. (2004). *Disordered eating among athletes: A comprehensive guide for health professionals.* Human Kinetics.

Beals, K. & Manore, M. M. (1994). The prevalence and consequences of subclinical eating disorders in female athletes. *International Journal of Sport Nutrition, 4*(2), 175–195. https://journals.humankinetics.com/view/journals/ijsnem/4/2/article-p175.xml

Bellwoar, H. (2012). Everyday matters: Reception and use as productive design of health-related texts. *Technical Communication Quarterly, 21*(4), 325–345. https://doi.org/10.1080/10572252.2012.702533

Bivens, K. (2019). Reducing harm by designing discourse and digital tools for opioid users' contexts: The Chicago Recovery Alliance's community-based context of use and PwrdBy's technology-based context of use. *Communication Design Quarterly Review, 7*(2), 17–27. https://doi.org/10.1145/3358931.3358935

Brandt, D. (2001). *Literacy in American Lives.* Cambridge University Press.

Bratland-Sanda, S. & Sundgot-Borgen, J. (2013). Eating disorders in athletes: Overview of prevalence, risk factors and recommendations for prevention and treatment. *European Journal of Sport Science, 13*(5), 499–508. https://doi.org/10.1080/17461391.2012.740504

Center for Disease and Control Prevention (CDC). (2019). *CDC clear communication index user guide: A tool for developing and assessing CDC Public Communication Products.* CDC. gov. https://www.cdc.gov/ccindex/pdf/clear-communication-user-guide.pdf

Caruso, C., & Frankel, L. (2010). Everyday people: Enabling user expertise in socially responsible design. In D. Durling, R. Bousbaci, L. Chen, P. Gauthier, T. Poldma, S. Roworth-Stokes & E. Stolterman (Eds.), *Design and complexity-DRS International Conference 2010,* 7–9 July, Montreal, Canada. https://dl.designresearchsociety.org/drs-conference-papers/drs2010/researchpapers/24

Cooper, A. (1999). *The inmates are running the asylum: Why high-tech products drive us crazy and how to restore the sanity.* Sams Publisher.

Cushman, E. (2013). Elias Boudinot and the Cherokee Phoenix: The sponsors of literacy they were and were not. In J. Duffy, J. N. Christoph, E. Goldblatt, N. Graff, R. S. Nowacek, & B. Trabold (Eds.), *Literacy, economy, power:*

Writing and research after literacy in American lives (pp. 13–29). Southern Illinois University Press.

Dilger, B. (2000). The ideology of ease. *Journal of Electronic Publishing, 6*(1). https://doi.org/10.3998/3336451.0006.104

Durst, S. (2019). Disciplinarity and literate activity in civil and environmental engineering: A lifeworld perspective. *Written Communication, 36*(4), 471–502. https://doi.org/10.1177/0741088319864897

Flatt, R. E., Thornton, L. M., Fitzsimmons-Craft, E. E., Balantekin, K. N., Smolar, L., Mysko, C., Wilfley, D. E., Taylor, C. B., DeFreese, J. D., Bardone-Cone, A. M., & Bulik, C. M. (2021). Comparing eating disorder characteristics and treatment in self-identified competitive athletes and non-athletes from the National Eating Disorders Association online screening tool. *International Journal of Eating Disorders, 54*(3), 365–375. https://doi.org/10.1002/eat.23415

Flynn, M., Austin, A., Lang, K., Allen, K., Bassi, R., Brady, G., Brown, A., Connan, F., Franklin-Smith, M., Glennon, D., Grant, N., Jones, W. R., Kali, K., Koskina, A., Mahony, K., Mountford, V., Nunes, N., Schelhase, M., Serpell, L., & Schmidt, U. (2021). Assessing the impact of first episode rapid early intervention for eating disorders on duration of untreated eating disorder: A multi-centre quasi-experimental study. *European Eating Disorders Review: Journal of the Eating Disorders Association, 29*(3), 458–471. https://doi.org/10.1002/erv.2797

Foss, K. (2021). Health and the media. In Y. Tian & J. D. Robinson (Eds.), *The Routledge handbook of health communication* (3rd ed., pp. 277–289). Routledge. https://doi.org/10.4324/9781003043379-26

Getto, G., & St.Amant, K. (2014). Designing globally, working locally: Using personas to develop online communication products for international users. *Communication Design Quarterly, 3*(1), 24–46. https://doi.org/10.1145/2721882.2721886

Greenhalgh, T. (2017). *How to implement evidence-based healthcare.* Wiley Blackwell.

Greenleaf, C., Petrie, T. A., Carter, J., & Reel, J. J. (2009). Female collegiate athletes: Prevalence of eating disorders and disordered eating behaviors. *Journal of American College Health, 57*(5), 489–496. https://doi.org/10.3200/JACH.57.5.489-496

Goodwin, K. (2009). *Designing for the digital age: How to create human-centered products and services.* Wiley.

Gouge, C. (2017). Improving patient discharge communication. *Journal of Technical Writing and Communication, 47*(4), 419–439.

Hainline, B. (2014). Introduction. *Mind, body, and sport: Understanding and supporting student-athlete mental wellness.* National Collegiate Athletic Association.

Harley, A. (2018, February 25). *UX expert reviews.* Nielsen Norman Group. https://www.nngroup.com/articles/ux-expert-reviews

Hart-Davidson, W. (2013). What are the work patterns of technical communication? In J. Johnson-Eilola & S. Selber (Eds.), *Solving problems in technical communication* (pp. 50–73). University of Chicago Press.

Head, K., Bute, J., & Ridley-Merriweather, K. (2021). Everyday interpersonal communication about health and illness. In Y. Tian & J. D. Robinson (Eds.), *The Routledge handbook of health communication* (3rd ed., pp. 149–162). Routledge. https://doi.org/10.4324/9781003043379-13

Hebard, S. P., Bissett, J. E., Kroshus, E., Beamon, E. R., & Reich, A. (2021). A content analysis of mental health literacy education for sport coaches. *Journal of Clinical Sport Psychology*. Advanced online publication. https://journals. humankinetics.com/view/journals/jcsp/aop/article-10.1123-jcsp.2021-0011/ article-10.1123-jcsp.2021-0011.xml

Ho, E. Y., Tran, H., & Chesla, C. A. (2015). Assessing the cultural in culturally sensitive printed patient-education materials for Chinese Americans with type 2 diabetes. *Health Communication, 30*(1), 39–49. https://doi.org/10.1 080/10410236.2013.835216

Hornbuckle, V. L. (2002). *An analysis of usability of women's collegiate basketball websites based on measurements of effectiveness, efficiency, and appeal* (Publication No. 3059984). [Doctoral dissertation, University of Northern Colorado]. ProQuest Dissertations & Theses Global. https://www.proquest. com/dissertations-theses/analysis-usability-womens-collegiate-basketball/ docview/305517188/se-2

Hoft, N. L. (1995). *International technical communication: How to export information about high technology.* John Wiley & Sons.

Hsieh, E. (2021). Intercultural health communication: Rethinking culture in health communication. In Y. Tian & J. D. Robinson (Eds.), *The Routledge handbook of health communication* (3rd ed., pp. 441–455). Routledge. https:// doi.org/10.4324/9781003043379-37

Jagim, A. R., Fields, J., Magee, M. K., Kerksick, C. M., & Jones, M. T. (2022). Contributing factors to low energy availability in female athletes: A narrative review of energy availability, training demands, nutrition barriers, body image, and disordered eating. *Nutrients, 14*(5), 986. https://doi.org/10.3390/ nu14050986

Johnson, R., Salvo, M., & Zoetewey, M. (2007). User-centered technology in participatory culture: Two decades "beyond a narrow conception of usability testing." *IEEE Transactions on Professional Communication, 50*(4), 320–332. https://doi.org/10.1109/TPC.2007.908730

Jones, N., Moore, K., & Walton, R. (2016). Disrupting the past to disrupt the future: An antenarrative of technical communication. *Technical Communication Quarterly, 25*(4), 211–229. https://doi.org/10.1080/10572252.2016.1224655

Kennedy, S. F., Kovan, J., Werner, E., Mancine, R., Gusfa, D., & Kleiman, H. (2021). Initial validation of a screening tool for disordered eating in adolescent athletes. *Journal of Eating Disorders, 9*(21), 1–11. https://doi.org/10.1186/s40337-020-00364-7

Kessler, M. M., Breuch, L-A. K., Stambler, D. M., Campeau, K. L., Riggins, O. J., Feedema, E., & Misono, S. (2021). User experience in health and medicine: Building methods for patient experience design in multidisciplinary collaborations. *Journal of Technical Writing and Communication, 51*(4), 380–406. https://doi.org/10.1177/00472816211044498

Kroshus, E., Wagner, J., Wyrick, D. L., & Hainline, B. (2019). Pre-post evaluation of the "Supporting Student-Athlete Mental Wellness" module for college coaches. *Journal of Clinical Sport Psychology, 13*(4), 668–685. https://journals.humankinetics.com/view/journals/jcsp/13/4/article-p668.xml

Kutcher, S., Wei, Y., & Coniglio, C. (2016). Mental health literacy: Past, present, and future. *The Canadian Journal of Psychiatry, 61*(3),154–158. https://doi.org/10.1177/0706743715616609

Lancaster, A., & King, C. S. T. (2024a). Introduction to *EquilibriUX*: Designing for Balance and User Experience. In A. Lancaster & C. S. T. King (Eds.), *Amplifying voices in UX: Balancing design and user needs in technical communication*. State University of New York Press.

Lancaster, A., & King, C. S. T. (Eds.). (2024b). *Amplifying voices in UX: Balancing design and user needs in technical communication*. State University of New York Press.

Langlet, B. S., Odegi, D., Zandian, M., Nolstam, J., Södersten, P., & Bergh, C. (2021). Virtual reality app for treating eating behavior in eating disorders: Development and usability study. *JMIR Serious Games, 9*(2), e24998. https://doi.org/10.2196/24998

Lentz, B., Kerins, M. L., & Smith, J. (2018). Stress, mental health, and the coach–athlete relationship: A literature review. *The Applied Research in Coaching and Athletics Annual, 33,* 214–238.

Mahiri, J. (2004). New literacies in a new century. In J. Mahiri (Ed.), *What they don't learn in school: Literacy in the lives of urban youth* (pp. 1–17). Peter Lang.

Matheson, F. (2006). Designing for a moving target. In *Proceedings of the 4th Nordic conference on Human-computer interaction: Changing roles* (pp. 495–496).

Melonçon, L. (2017a). Embodied personas for a mobile world. *Technical Communication, 64*(1), 50–65.

Melonçon, L. (2017b). Patient experience design: Expanding usability methodologies for healthcare. *Communication Design Quarterly, 5*(2), 19–28. https://doi.org/10.1145/3131201.3131203

Mirel, B. (2004). *Interaction design for complex problem solving: Developing useful and usable software*. Morgan Kaufmann.

Mond, J. (2014). Eating disorders "mental health literacy": An introduction. *Journal of Mental Health, 23*(2), 51–54. https://doi.org/10.3109/09638237.2014.889286

Moreland, J. J., Coxe, K. A., & Yang, J. (2018). Collegiate athletes' mental health services utilization: A systematic review of conceptualizations, operationalizations, facilitators, and barriers. *Journal of Sport and Health Science, 7*(1), 58–69. https://doi.org/10.1016/j.jshs.2017.04.009

Nitsch, M., Dimopoulos, C. N., Flaschberger, E., Saffran, K., Kruger, J. F., Garlock, L., Wilfley, D. E., Taylor, C. B., & Jones, M. (2016). A guided online and mobile self-help program for individuals with eating disorders: An iterative engagement and usability study. *Journal of Medical Internet Research, 18*(1), e7. https://doi.org/10.2196/jmir.4972

NCAA. (2022a). *About the SSI.* https://www.ncaa.org/sports/2016/8/23/about-the-ssi.aspx?id=116

NCAA. (2022b). *Mental health educational resources.* https://www.ncaa.org/sports/2016/8/4/mental-health-educational-resources.aspx

NCAA. (2022c). *Sports Science Institute.* https://www.ncaa.org/sports/2021/5/24/sport-science-institute.aspx

NCAA Research. (2021). Student-athlete well-being study: Survey results. https://www.ncaa.org/sports/2020/5/22/ncaa-student-athlete-well-being-study.aspx

NCAA Research. (2022). Student-athlete well-being study: Survey results. https://ncaaorg.s3.amazonaws.com/research/other/2020/2022RES_NCAA-SA-Well-BeingSurveyPPT.pdf

NCAA Sports Science Institute. (n.d.a.). *Eating disorders.* NCAA.org https://ncaaorg.s3.amazonaws.com/ssi/mental/SSI_EatingDisordersFactSheet.pdf

NCAA Sports Science Institute. (n.d.b.). *Supporting student-athlete mental wellness* [online module]. NCAA.org. https://www.ncaa.org/sports/2016/11/3/supporting-student-athlete-mental-wellness.aspx

NEDA. (2022). *Statistics and research on eating disorders.* https://www.nationaleatingdisorders.org/statistics-research-eating-disorders

Papathomas, A., & Lavalle, D. (2012). Eating disorders in sport: A call for methodological diversity. *Revista de Psicología del Deporte, 21*(2), 387–392.

Plateau, C. R., McDermott, H. J., Arcelus, J., & Meyer, C. (2014). Identifying and preventing disordered eating among athletes: Perceptions of track and field coaches. *Psychology of Sport and Exercise, 15*(6), 721–728. https://doi.org/10.1016/j.psychsport.2013.11.004

Pike, K. M., Dunne, P. E., & Addai, E. (2013). Expanding the boundaries: Reconfiguring the demographics of the "typical" eating disordered patient. *Current Psychiatry Reports, 15*(11), 411. https://doi.org/10.1007/s11920-013-0411-2

Parker, P. (2022). *Competition and athlete engagement to improve nutrition: A usability study of the mobile application BSUathlEATS* (Publication No. 5510). [Master's thesis, Ball State University]. DuraSpace. http://cardinalscholar.bsu.edu/handle/123456789/201819

Prior, P., & Shipka, J. (2003). Chronotopic lamination: Tracing the contours of literate activity. In C. Bazerman & D. Russell (Eds.), *Writing selves,* writing societies: Research from activity perspectives (pp. 180–238). WAC Clearinghouse.

Ravi, S., Ihalainen, J. K., Taipale-Mikkonen, R. S., Kujala, U. M., Waller, B., Mierlahti, L., Lehto, J., & Valtonen, M. (2021). Self-reported restrictive eating, eating disorders, menstrual dysfunction, and injuries in athletes competing at different levels and sports. Nutrients, 13(9), 3275. https://doi.org/10.3390/nu13093275

Riviere, A. J., Leach, R., Mann, H., Robinson, S., Burnett, D. O., Babu, J. R., & Frugé, A. D. (2021). Nutrition knowledge of collegiate athletes in the United States and the impact of sports dietitians on related outcomes: A narrative review. Nutrients, 13(6), 1772. https://doi.org/10.3390/nu13061772

Robinson, J., Lanius, C., & Weber, R. (2018). The past, present, and future of UX empirical research. Communication Design Quarterly Review, 5(3), 10–23. https://doi.org/10.1145/3188173.3188175

Roozen, K., & Erickson, J. (2017). Expanding literate landscapes: Persons, practices, and sociohistoric perspectives of disciplinary development. Computers and Composition Digital Press/Utah State University Press. http://ccdigitalpress.org/expanding

Rosa-Caldwell, M. E., Todden, C., Caldwell, A. R., & Breithaupt, L. E. (2018). Confidence in eating disorder knowledge does not predict actual knowledge in collegiate female athletes. PeerJ, 6, e5868. https://doi.org/10.7717/peerj.5868

Rose, E. (2016). Design as advocacy: Using a human-centered approach to investigate the needs of vulnerable populations. Journal of Technical Writing and Communication, 46(4), 427–445. https://doi.org/10.1177/0047281616653494

Rose, E., Racadio, R., Wong, K., Nguyen, S., Kim, J., & Zahler, A. (2017). Community-based user experience: Evaluating the usability of health insurance information with immigrant patients. IEEE Transactions on Professional Communication, 60(2), 214–231. https://doi.org/10.1109/TPC.2017.2656698

Sanford-Martens, T. C., Davidson, M. M., Yakushko, O. F., Martens, M. P., & Hinton, P. (2005). Clinical and subclinical eating disorders: An examination of collegiate athletes. Journal of Applied Sport Psychology, 17(1), 79–86. https://doi.org/10.1080/10413200590907586

Shellenberger, L. (2020). Imagining an embodied ethos: Serena Williams' "defiant" black ethe. Peitho, 22(2). https://cfshrc.org/article/imagining-an-embodied-ethos-serena-williams-defiant-black-ethe

Shoemaker, S., Wolf, M., & Brach, C. (2013a). Patient education materials assessment tool for audiovisual materials (PEMAT-A/V). Prepared by Abt Associates, Inc. under contract no. HHSA290200900012I, TO 4. Rockville, MD: Agency for Healthcare Research and Quality, AHRQ Publication No. 14-0002-EF.

Shoemaker, S., Wolf, M., & Brach, C. (2013b). Patient education materials assessment tool for printable materials (PEMAT-P). Prepared by Abt Associates,

Inc. under contract no. HHSA290200900012I, TO 4. Rockville, MD: Agency for Healthcare Research and Quality, AHRQ Publication No. 14-0002-EF.

Shoemaker, S., Wolf, M., & Brach, C. (2013c). Patient education materials assessment tool (PEMAT) and user's guide. Prepared by Abt Associates, Inc. under Contract No. HHSA290200900012I, TO 4). Rockville, MD: Agency for Healthcare Research and Quality, AHRQ Publication No. 14-0002-EF.

Simmons, M. & Zoetewey, M. (2012). Productive usability: Fostering civic engagement and creating more useful online spaces for public deliberation. *Technical Communication Quarterly, 21*(3), 251–276. https://doi.org/10.108 0/10572252.2012.673953

Spinuzzi, C. & Zachry, M. (2000). Genre ecologies: an open-system approach to understanding and constructing documentation. *ACM Journal of Computer Documentation, 24*(3), 169–181. https://doi.org/10.1145/344599.344646

St.Amant, K. (2013). What do technical communicators need to know about international environments? In J. Johnson-Eilola & S. A. Selber (Eds.) *Solving problems in technical communication* (pp. 479–499). University of Chicago Press.

St.Amant, K. (2017). The cultural context of care in international communication design: A heuristic for addressing usability in international health and medical communication. *Communication Design Quarterly, 5*(2), 62–70. https://doi.org/10.1145/3131201.3131207

Sun, H. (2002). Why cultural contexts are missing: A rhetorical critique of localization practices. In *Proceeding for the 2002 Annual Conference, Society for Technical Communication, 49,* 164–168.

Sun, H., & Getto, G. (2017). Localizing user experience: Strategies, practices, and techniques for culturally sensitive design. *Technical Communication, 64*(2), 89–94.

Sundgot-Borgen, J., & Torstveit, M. K. (2004). Prevalence of eating disorders in elite athletes is higher than in the general population. *Clinical Journal of Sport Medicine, 14*(1), 25–32. https://doi.org/10.1097/00042752-200401000-00005

Sundgot-Borgen, J., & Torstveit, M. K. (2010). Aspects of disordered eating continuum in elite high-intensity sports. *Scandinavian Journal of Medicine & Science in Sports, 20*(s2), 112–121. https://doi.org/10.1111/j.1600-0838.2010.01190.x

Sullivan, P., Murphey, J., & Blacker, M. (2019). The level of mental health literacy among athletic staff in intercollegiate sport. *Journal of Clinical Sport Psychology, 13*(3), 440–450. https://doi.org/10.1123/jcsp.2018-0052

The Patient Education Materials Assessment Tool (PEMAT) and User's Guide for Audio Visuals (PEMAT-A/V). (2020a). Agency for Healthcare Research and Quality, Rockville, MD. https://www.ahrq.gov/health-literacy/patient-education/pemat.html

The Patient Education Materials Assessment Tool (PEMAT) and User's Guide for Printed Materials (PEMAT-P). (2020b). Agency for Healthcare Research and

Quality, Rockville, MD. https://www.ahrq.gov/health-literacy/patient-education/pemat.html

Thompson, R. (2014). *Eating disorders. Mind, body, and sport: Understanding and supporting student-athlete mental wellness* (pp. 25–28). National Collegiate Athletic Association. https://www.naspa.org/images/uploads/events/Mind_Body_and_Sport.pdf

Tian, Y., & Robinson, J. (2021). Social media and health. In Y. Tian & J. D. Robinson (Eds.), *The Routledge handbook of health communication* (3rd ed., pp. 304–317). Routledge. https://doi.org/10.4324/9781003043379-26

Torres-McGehee, T. M., Leaver-Dunn, D., Green, J. M., Bishop, P. A., Leeper, J. D., & Richardson, M. T. (2011). Knowledge of eating disorders among collegiate administrators, coaches, and auxiliary dancers. *Perceptual and Motor Skills, 112*(3), 951–958. https://doi.org/10.2466/02.13.PMS.112.3.951-958

Treasure, J., Stein, D., & Maguire, S. (2015). Has the time come for a staging model to map the course of eating disorder from high risk to severe enduring illness? An examination of the evidence. *Early Intervention in Psychiatry, 9*(3), 173–184. https://doi.org/10.1111/eip.12170

Treasure, J., Hübel, C., & Himmerich, H. (2022). The evolving epidemiology and differential etiopathogenesis of eating disorders: Implications for prevention and treatment. *World Psychiatry, 21*(1), 147–148. https://doi.org/10.1002/wps.20935

Turk, J. C., Prentice, W. E., Chappell, S., & Shields, E. W. (1999). Collegiate coaches' knowledge of eating disorders. *Journal of Athletic Training, 34*(1), 19–24.

Vredenburg, K., Mao, J. Y., Smith, P. W., & Carey, T. (2002). A survey of user-centered design practice. In *Proceedings of the SIGCHI Conference on Human Factors in Computing System* (pp. 471–478). ACM Digital Library. https://doi.org/10.1145/503376.503460

Wells, K. R., Jeacocke, N. A., Appaneal, R., Smith, H. D., Vlahovich, N., Burke, L. M., Hughes, D. (2020). The Australian Institute of Sport (AIS) and National Eating Disorders Collaboration (NEDC) position statement on disordered eating in high performance sport. *British Journal of Sports Medicine, 54*(21), 1247–1258.

Yim, S. H., Bailey, E., Gordon, G., Grant, N., Musiat, P., & Schmidt, U. (2020). Exploring participants' experiences of a web-based program for bulimia and binge eating disorder: Qualitative study. *Journal of Medical Internet Research, 22*(9), e17880–e17880. https://doi.org/10.2196/17880

Zhu, X., & Smith, R. (2021). Stigma, communication, and health. In Y. Tian & J. D. Robinson (Eds.), *The Routledge handbook of health communication* (3rd ed., pp. 77–90). https://doi.org/10.4324/9781003043379-8

Chapter 9

Inclusive Measures

Establishing Audio Description Tactics
that Impact Social Inclusion

Brett Oppegaard and Michael K. Rabby

Visual media has not dominated the world—like it does today—for long. For most of humanity, in fact, only the most privileged people even had access to it. But during the democratizing movements of the Renaissance, ocularcentrism began to bloom as a part of our global media ecology, bringing to the masses vision-oriented technologies such as linear perspective, eyeglasses, the microscope, the telescope, and the printing press (Dolmage, 2014; Meadows, 2002). With such sudden availability of these eyeball extensions, everyone suddenly wanted to look, regardless of who was left out in that process, and a frenzied demand for visual media started to spread (Comolli, 1980; Mirzoeff, 2016).

All media systems tend to favor some types of people over others through the choices made by the designers. At this point in time, those who can see and see well—across demographics, mediums, interfaces, cultures—tend to benefit most from just about any media-design choice. In the latest round of social separation between the sighted and the visually impaired, digital channels of all kinds have sprouted in the past two decades and now are pumping out unprecedented amounts of visual media, including the posting *every minute* of about 240,000 photos on Facebook;

273

65,000 photos on Instagram; and 500 hours of video on YouTube (Statista, 2020, 2021). Most of that visual information circulating throughout the global community is inaccessible to people who cannot see it.

Based on a tally by the World Health Organization (2022), though, billions of people around the world are now blind, Deafblind, or with low vision. Instead of receiving due recognition as a massive and important subset of the global population, these people have been transformed into pariahs unable to function in many social situations. Their offense? They have poor vision. They are not adept at navigating the complexities of a visual environment that requires not only sensory data through the eyes but also the ability to use that data to make sense of a complicated set of intellectual and physical relations, combining information, imagination, and insights into a material and psychic space (Mirzoeff, 2011). Modern media designers, with a goal of social justice, can do something significant about this massive disconnection and do it without disrupting the global media ecology (Agboka, 2013; Getto & Sun, 2017; Gonzales & Zantjer, 2015; Shivers-McNair, 2017). There's a common media-enriching approach called Audio Description, for example, and this chapter will present research resulting in five novel, experimental, and straightforward best practices for designers to use to try to connect with non-visual audiences through it. This approach could serve as a model for ways in which accessibility can be explored through intersections with technical communication research, especially in cases where researchers take an active part in the development of production processes and public products (Melonçon, 2013; Moore, 2017; Oswal, 2013).

Like alt-text, only more-detailed and evocative, Audio Description typically is conceptualized as an accommodation, a service, or an add-on feature, aimed at helping blind people know what can be seen in their vicinity. As a remediation process, it involves a symbiotic relationship between multiple people, including at least one sighted person willing to share what is being seen. That person will provide an audible description that can be heard, or a textual description that can be audibilized via a screen reader, or a textual description that can be made tactile by a Braille e-reader. Empirical research about Audio Description, though, exists in only a nascent state, especially outside of journals and books that specialize in disability studies. This chapter demonstrates its potential for wider appeal.

While health communication is a common research area across disciplines, for example, few research studies to date link Audio Description to potential impacts on public health, even in tangential ways. One possible path of inquiry to connect these ideas therefore is through the concept of social exclusion/inclusion. Anyone can be socially excluded (or included) in any specific context, but people with disabilities, in general, tend to be excluded far more often than most people in all sorts of situations. They also typically have poorer health because of such widespread exclusion, including a higher likelihood of having secondary conditions, broadly defined as "medical, social, emotional, family, or community problems that a person with a primary disabling condition likely experiences" (Castro et al., 2018, p. 2).

People who cannot see or see well, as a subset of people with disabilities, have complex and compounded issues around social exclusion/inclusion due to the visual indications of a disability (such as a white cane or a guide dog) combined with a built environment hostile and unwelcoming to people without strong vision. Aside from the physical barriers, people who are blind, Deafblind, or who have low vision also perpetually are at a disadvantage in mediated social situations. The increased difficulties of gathering visual information around them, from subtle interpersonal social cues to visually mediated messages can be disenfranchising, regardless of any individual's levels of intelligence, wit, and charm.

In that vein, this chapter examines the potential for links between accessible media, such as audio-described media, and public health effects caused by social exclusion or inclusion. Designing for broad inclusion, including creating an appropriate balance between the needs of the few and the many, requires accessibility as its core. But such an approach also demands a state of equilibrium within the mediated context that feels welcoming to all, including participatory steps not too steep or foreboding for anyone. Accessibility, from that perspective, can be considered a basic and binary requirement, like an open door. Is the door open, yes or no? But media design also has an enormous service component to it that can vary greatly in quality but still not violate any state or federal laws. Even if allowed in the accessible door, in other words, what is inside waiting for you? Do people feel like they belong? Or does it feel like a hostile environment, continually trying to prod that person back outside? Such a topic could be approached from multiple angles and at various levels of abstraction to triangulate larger phenomena. From our scholar-practitioner paradigm, though, we wanted to keep our research grounded in everyday practices. To this end, we analyzed compositional approaches to descriptions. In other words, when sighted people described a piece of visual media for a person who cannot see it, what did they say, and how did they say it, and how does that approach affect feelings of social inclusion/exclusion?

We did not seek a thick case-study examination of just a single description process, or even a project-level process; instead, we wanted to identify broader compositional approaches that could impact feelings of social inclusion/exclusion through an identification of their general characteristics and a conversion of those ideas into best practices. We think of that position in the process as the bridging moment, in which a describer has chosen to describe a specific piece of visual media, has picked the strategic perspective from which to describe the image and determined its aim. These tactics we have identified will allow the describer to carry out the work in ways that measurably improve social inclusion.

To identify these core compositional tactics, the authors hosted a hackathon-like "Descriptathon" that brought together more than 100 people from throughout the US and Canada in a three-day intensive workshop in February 2021 designed to audio describe print brochures of communally important public places (16 national parks). Those descriptions then were evaluated by people who are blind, Deafblind, or who have low vision using validated scales measuring aspects of Social Inclusion (Jason et al., 2015; McColl et al., 2001; Peterson et al., 2008). This chapter will present the results of that study and the identification and development of the Social Inclusion tactics and will convert these characteristics of socially inclusive description into best practices that can be used and tested in an effort to improve feelings of inclusion. (We also address our research in Oppegaard and Rabby, 2022.)

For our foundational research questions, we asked two questions:

RQ1: What characteristics of Audio Description do blind/ visually impaired people identify as important?

RQ2: Does a relationship exist between effective Audio Description and social inclusion?

Audio Description as an Antidote to Overarching Ocularcentrism

Audio Description, as a research field and as a practice, primarily has been studied in relation to dynamic media, such as television, theater, and opera, all forms that put severe time constraints on what can or cannot be done, in terms of the description. Mostly, description in dynamic media fits in the narrow gaps between dialogue, leaving little room for introspection, inclusiveness, or rich public dialogue. In that context, listeners usually are happy with whatever visual information they can get audibly. But in the area of Audio Description we are most concerned about—"static" media, including photographs, illustrations, collages, charts, timelines, maps, etc.—much still needs to be studied and learned about description that can extend to virtually unlimited lengths, with digital media and distribution systems, unrestrained by time.

Audio Description is a rich interpretive and descriptive activity, not a straightforward transcription service. It also is a learned skill, which needs improved educational materials and more training programs about it, through which people can practice describing and get better at this skill, while researchers iteratively circle around the processes and products. Access to these evolving educational materials, refined by associated empirical studies of them, creates a realistic path to improving Audio Description in a holistic manner. As a technique for translating visual media into audible media and as a way to offer equivalent experiential access, Audio Description already has been widely accepted by this underrepresented community as a primary solution, including by several national associations of people who are blind or Deafblind or who have low vision (e.g., Accessible Media Inc. and The Canadian Association of Broadcasters, 2015; American Council of the Blind, 2022; Hutchinson et al., 2020; Rai et al., 2010). It clearly needs more attention, broader public support, and use, though, to reach its potential.

Even though Audio Description has been studied by academics for roughly four decades, it has grown rapidly as an area of scholarly interest only in the past few years, including through the publishing of a first wave of book-length manuscripts and the creation of dedicated conferences, such as the biennial Advanced Research Seminar on Audio Description (Fryer, 2016; Koirala & Oppegaard, 2022; Maszerowska et al., 2014; Matamala & Orero, 2016). Meanwhile, the U.S. National Park Service has collected and developed research initiatives around "Healthy Parks, Healthy People," which attempts to quantify and qualify the public-health impacts of its heritage sites as cornerstones of American health and wellbeing (National Park Service, 2020). Some of that research includes evidence that spending time in nature is associated with good health (White et al., 2019); lowers levels of stress, depression, and anxiety (Bratman et al., 2015; Cox et al., 2017; Haluza et al., 2014); and socially equalizes health disparities (Lachowycz & Jones, 2014; Mitchell & Popham, 2008; Wolch et al., 2014).

Meanwhile, an extensive and complicated debate about disability rhetoric intersects with multiple areas of technical communication, which is beyond the scope of this chapter. In summary, though, people who are blind typically are socialized in ways that establish and reinforce their exclusion from society, regardless of the will or skill of the person, and assumptions that people who are blind are "helpless, docile, dependent, or incapacitated" tend to lead those who even subconsciously hold those views, including designers, to create exclusionary and perpetuating media systems that deny a person with low vision, Deafblindness, or blindness any opportunity to attain independence or societal achievements (Scott, 1969, p. 17). Social exclusion is a contested term as well, with many facets, and settling that terminology debate is beyond the scope of this chapter. For our purposes here, we define social exclusion along the lines of van Bergen et al. (2017, p. 257), as socioeconomic inequalities associated with multiple dimensions, such as "limited social participation," "material deprivation," "inadequate access to basic social rights," and "lack of normative integration." Such social exclusion causes significant secondary health risks that include anxiety, depression, and chronic pain (Kim et al., 2016; Wilber et al., 2002).

Media Design Picks Winners and Losers

Consider the time, energy, and resources involved in simply designing a corporate logo. For example, Google (one of our research project's supporters) initiated intense international debates about typefaces, colors, and graphics (Zjawinski, 2008). But to anyone with a screen reader as their interface, the Google logo is almost invisible. In the alt-text, picked up by screen readers, the logo is represented by just a single word: "Google." As of this writing, that alt-text says nothing more to the billions of people who use the page every day, not what typeface was chosen, what colors the letters are, what size the font is, not even "Google *logo*," and that is the embedded information directly available on the company's ubiquitous web-search webpage about this carefully crafted paragon of graphic design for people who cannot see it.

As an extended thought experiment, recollect the media designs in your life that do not privilege sight above all and instead might equally emphasize sound, touch, taste, or smell. That probably did not take you long to list, because taste or smell interfaces are almost non-existent, and touch typically is a complementary interface, not a primary one. Visual interfaces clearly are the dominant apex interfaces, so that leaves sound-oriented interfaces in a secondary or niche role. But how far below sight is sound in media design? Even radio or podcast interfaces typically require users to navigate visual radio or podcast screens before an audience member can listen. Even Alexa or Google Home devices, the possible harbingers of the audible interfaces of the future, have visual features and buttons that provide key controls, and they can be difficult to use anyway for any complex commands beyond getting them to play songs or answer trivia questions.

Where do the media designs that offer a balanced and redundant sensory mix exist? Again, you might have trouble coming up with many examples. This is not building an argument to regulate or diminish or restrict visual media in any way. This argument provides a chance to reflect, and a challenge to imagine media designs that include everyone, regardless of their abilities to gather data through any particular sensory organ. By not favoring one sense over any other and giving more senses a chance to collect data at equivalent scales, more people inherently would feel included by the diversity of options to gather the information. Yet old sight-privileging design habits are hard to break. They pervade new systems that simultaneously increase media use and exclusionary designs in exponential fashion without concern or addressing those left further behind by these choices. Just as not all media should be fully audible or fully tactile, to flip the situation over, not all media should be processed predominately through the eyes without understanding how such design choices favor some people and disenfranchise others. Although designers might have been able to make the case, at least in the analog world, that production and dissemination options were too limited to make such accessible media, digital-media designers, producers, and creators have plenty of options now. Accessible design might require more time in the media workshop, additional planning and resources, and potentially compromises for sighted audiences, which are the dominant consumers. But it could happen if the will exists to include everyone.

This chapter's primary provocation, therefore, is not about restricting what can be seen but instead challenges how our media combined with our many senses conjure ideas and images in our minds and could include more people in those processes. If you have ever listened to a great radio program or podcast, you can imagine how powerful audible media can be and that it can rival the information richness provided by visual media. Pragmatically, though, it is easier for media designers to package and sell a photo or a video than to capitalize on a piece of audio, with no visuals, which aligns capitalism with visual media production and consumption and helps to explain its dominance in the global media economy and ecology.

Yet even if the only focus of media design is on pure capitalistic domination—and all legal, ethical, and moral obligations are disregarded—major economic incentives still exist for creating more-accessible media en masse. In just raw numbers, and focused only on visual acuity, about 27 million, or roughly 10% of, Americans have a visual impairment (American Foundation for the Blind, 2020). During the next three decades, the population of adults with vision impairment and age-related eye diseases is expected to double, because of the rapidly aging U.S. population (Centers for Disease Control, 2020). In addition to impediments that visual impairments cause for many everyday activities—such as reading, watching television, driving, cooking, cleaning, paying bills—blindness and low vision also affect many important public interactions and have been medically linked to increases in stress and illnesses, plus higher levels of premature deaths (Centers for Disease Control and Prevention, 2020).

In other words, people who are blind or Deafblind or who have low vision want to experience a full life, too. They want to feel included. They have families and friends, associates and colleagues, rivals and competitors, etc., who are sighted and who have access to all of this rapidly circulating information. Such an accessibility divide—based only on visual acuity—creates a fractured society as well as multiple health issues for people who are blind or Deafblind or who have low vision. Media designers can address this issue directly, if they just make more-accessible media a priority.

From a National Legacy of Exclusion to Inclusion

As a country founded during the rise of visual media, the US has a long history of excluding and mistreating people with disabilities, including people who are blind or Deafblind or who have low vision (Mitra et al., 2019). Questions about such treatment, related to the public health of people with disabilities, have been gaining scholarly interest across disciplines, but disciplinary-bounded approaches to the topic so far often have been at abstract and theoretical levels rather than at the ground level (Krahn et al., 2015). As a growing chorus of unified voices, transcending disciplines, some scholars have chosen to focus on specific contexts and people with specific physical impairments as a way to cut through the noise. This precision-medicine approach—which aims to adapt generalized programs, such as treating all people with varying "disabilities" as a homogenous group—has been compartmentalized into customized therapeutic options that account for lifestyle commonalities (Sabatello, 2018). In the Social Model of Disability (Shakespeare, 2006), which stresses the built nature of most contemporary environments, Audio Description is a method of recognizing those inequitable structures and adapting the environment to make it more inclusive. Health, meanwhile, is affected by the degree to which persons with disabilities enjoy full rights and inclusion in a society (Mitra et al., 2019). Following such scholarly flows, this chapter gets grounded in its places, national parks representing prominent public resources, and its research participants, people who are blind or Deafblind or who have low vision, to impact social inclusion in those contexts with those people in ways that can increase precision in media accessibility in general.

As a primary example of how widespread opportunities for multimodal or even audio-oriented media could radically reshape society, consider the prominent public resources that are national parks and what it would mean to the country's psyche if all citizens could enjoy them in roughly equal ways. Much of the allure of a national park, meanwhile, is its pure visual spectacle, from the Statue of Liberty to Yellowstone National Park to the Grand Canyon. Visual media about these places carries rich content, too, but without the accompanying audible equivalents, much can be lost by those who cannot see them.

American taxpayers contribute almost $3 billion annually for the operations of these publicly funded places that help to communally protect more than 400 culturally significant sites throughout the country, and that stewardship is philosophically intended for everyone's benefit (Repanshek, 2022). U.S. National Park Service sites address conservation of wildlife and preservation of landscapes and also what these places mean to Americans, via interpretive storytelling, complemented by artifacts, performances, and other aspects of the built environment meant to convey the story of the place. These sites are considered of at least national interest but sometimes also of irreplaceable international importance. The National Park Service has become a worldwide leader in accessibility initiatives of all sorts.

When family members and friends want to do something enjoyable together at a national park, a person who is blind still may feel excluded because of the lack of visual access to much of the cultural media about the place, including brochures, wall texts, trail markers, exhibits, maps, signs, and any other number of visual cues, as well as any viewpoints that typically are not audio described. These parks, even as some of the most accessible public places in the world, can compound feelings of exclusion. Such feelings can be projected as being spread exponentially by the visually oriented media at all of the world's other gathering places, such as national wildlife refuges and forests, botanical gardens, museums, zoos and aquariums, plus nonprofit and for-profit attractions that are aimed at attracting the general public but are functionally inaccessible to a person who cannot see or see well. Expand that sphere of exclusion to include regional, state, and local points of interest—if those places are fully or partially inaccessible for people who are blind or Deafblind or who have low vision, the individual and their family and friends have to make tough decisions about their time together that often offer no great alternatives, like the coffee shop analogy earlier: Do they go together to these places anyway and risk that the person who cannot see or see well will have a frustrating time; do they go somewhere else that might be more accessible, if such a place exists; or do they leave their family member or friend out altogether?

Such social-exclusion problems, though, mostly are self-evident. Thus, our focus here is not on belaboring the idea that there are social-exclusion problems for people who are blind or Deafblind or who have low vision. Instead, we take this pragmatic approach: *A dearth of descriptions in public places exists, resulting in a woeful need for more.*

In addition to the need for quantity, Audio Description has important qualitative aspects and under-researched mental and physical health impacts. Audio Description offers a systemic approach to improving social inclusivity for people who are blind or Deafblind or who have low vision in ways similar to what widespread captioning has achieved in the past few decades for people who are deaf or hard of hearing. Its unique usefulness to people who are blind creates a latent need for research into its potential for health benefits. We propose that Audio Description can affect perceptions of social inclusion and that this social inclusion perception can affect mental and physical health. But we also want to know how: in what ways and to what effects? Social exclusion, for example, has been linked to damaging secondary medical conditions, such as anxiety, depression, and chronic pain (Castro et al., 2018; Kim et al., 2016; Wilber et al., 2002). As a start, we wondered if we could improve in any material ways inclusivity in public places and increase feelings of social inclusion and the health of people who are blind or visually impaired through Audio Description.

Method and Results

This study is part of a long-term research project aimed at improving inclusive content in national parks for people who are blind or Deafblind or who have low vision. The UniDescription Project (www.unidescription. org) began as a grant-funded initiative in fall 2014 with the objective of audio describing 40 U.S. National Park Service's Unigrid brochures (the distinctive black brochures featured at National Parks in the US). The Descriptathon evolved from these initial practical concerns as a way to increase motivation and engagement in participants as well as to make the experience more enjoyable for those involved. These data-collection processes were approved in advance by the lead author's Institutional Review Board as well as by the U.S. Office of Management and Budget.

A Descriptathon comprises a multi-day event when teams of site staff, volunteers, and people who are blind or Deafblind or who have low vision compete together in a March-Madness-style competition to write the best possible Audio Descriptions in the time allotted. Besides helping their teams write the descriptions, participants who are blind or Deafblind or who have low vision also serve as judges of the descriptions.

Winning teams advance in the friendly competition bracket, while losing teams have opportunities to participate in consolation rounds. Everyone gets to continue describing, and judges continue to provide feedback throughout the event, regardless of who wins or loses. More details about the Descriptathon and its development can be found in Oppegaard (2020) and on the Descriptathon website. Figure 9.1 shows the results of Descriptathon 7 as a visualization of the friendly competition.

Data from this study emanated from Descriptathon 7, held February 9–11, 2021. In terms of participants, 119 sighted and non-sighted people joined the workshop from throughout the US and Canada, grouped into 16 teams, which ranged from 5 to 10 people on each team. These teams created Audio Descriptions for multiple national parks brochures and presented them to be judged. These descriptions were then evaluated by 22 blind and visually impaired judges. This process generated 122

Descriptathon 7 Tourney Bracket
Feb. 9-11, 2021 (16 Teams)

Ulysses S. Grant NHS

Fort Scott NHS

Fort Scott NHS

Ozark NSR

Buffalo NR

Ozark NSR

Ozark NSR

Ozark NSR

Charles Young Buffalo Soldiers NM

Minuteman Missile NHS

Minuteman Missile NHS

Lincoln Boyhood NM

Lincoln Boyhood NM

Lincoln Boyhood NM

Agassiz NWR

Lincoln Boyhood NM / Charles Young Buffalo Soldiers NM / Fort Larned NHS / Ash Meadows NWR / HKNC Deaf-Blind Immersion / Hot Springs NP / Minuteman Missile NHS

Tourney CHAMPION
Sleeping Bear Dunes NL

Sleeping Bear Dunes NL

Sleeping Bear Dunes NL

www.unidescription.org

Sleeping Bear Dunes NL

Bonus Rounds

Dr. Denise Decker
Memorial
Second Chances
Round
Agassiz NWR

**Shiny Objects
Third Chances**
Minuteman Missile NHS /
Hot Springs NP

A Final Face to Face:
Round 4
Fourth Chances

Ash Meadows NWR

Cuyahoga Valley NP

Cuyahoga Valley NP

Hot Springs NP

Sleeping Bear Dunes NL

Sleeping Bear Dunes NL

Fort Rodd Hill NHS

Fort Rodd Hill NHS

Fort Larned NHS

Fort Rodd Hill NHS

HKNC Deaf-Blind Immersion

HKNC Deaf-Blind Immersion

Homestead NHP

Figure 9.1. Descriptathon 7 Final Results. *Source*: Created by the author.

qualitative responses. The judges decided among the pairings who wrote the description better, and, as a part of their process, they explained why they preferred one description more than the other. These descriptions provided the empirical data for clues of characteristics of effective audio description (and consequently promote social inclusion). This also helped to ensure their inclusion in this process.

Using this feedback from judges, we took a grounded-theory approach (Glaser & Strauss, 2017) to open-code these statements, clustered them into themes, and then analyzed the themes. Eventually, due to the thematic clusters that emerged, the data were parsed into five broader labels plus an "other" characteristic, reserved for statements that did not meet one of these themes.

Specifically, the first research question involved identifying the characteristics of effective Audio Description—in other words, what should people strive for when writing Audio Description? Using the remarks judges made about why they selected a particular description, a list of common themes was derived from the data and was recorded. The second author reviewed the statements multiple times, deriving key characteristics that emerged from the data (Creswell, 2015). These terms were then grouped into larger themes, reviewed, and recorded as the first draft of the typology. After several revisions and refinements, a second coder (also a blind person) unfamiliar with the typology was brought in and trained. Upon completion of this phase, modifications to the typology were made, and, in some cases, distinctions between the codes were clarified. Through this, the final codebook emerged. Having a blind person involved at this stage helped the inclusivity of our research and enabled an insider positionality to emerge as the categories evolved. When the two coders seemed to agree on what each category constituted, they proceeded to work independently for reliability.

A statement comprised one unit, as each one involved the judge invoking why they selected that description in that round. Given the varied and often detailed nature of some of the comments, several units received more than one code. Of the 121 statements coded (six winners did not share reasons and thus did not receive a code), 105 had one code, 12 had two codes, 3 had three codes, and 1 had four codes. If a statement had a clear preference, or one dominant reason, then that was the code. Statements received two or more code in cases for which multiple reasons were indicated: e.g., This was a difficult choice, as both descriptions were very similar. However, I preferred this description, because he had more details about the face of the person such as the nose and the lips. Furthermore, this description suggested regarding the person's sex/gender, that without putting an emphasis on it, which allows the listener to make his or her own judgment. A final pass of 30 of the statements by both coders resulted in 33 codes, with a reliability of 78.6% (Cohen's Kappa= 0.71). With strong reliability established, the second coder completed the remaining statements.

RQ1: Categorizing Characteristics of Audio Description that Blind/Visually Impaired People Identify as Important

The coding typology consisted of five major categories plus an *Other*. The *Other* comments noted something not encompassed by the five major categories but were worth noting. For example, this statement from the four round about Ash Meadows National Wildlife Refuge indicated specificity of labels: "It is difficult to try to describe a collage like this and convey meaning along with labels. I think Ash Meadows did a nice job of not only listing labels, but also giving me a sense of predators and prey, plants and reapers." On other occasions, a judge would indicate something non-descript: e.g., they thought it was "better." In total, 17 comments received an *Other* code.

Given that, to our knowledge, delineation of specific factors for what makes Audio Description effective has not been done before, these preferences merit further illustration before continuing. The five major categories, in order of most to least coded, were *Detailed* (coded 55 times), *Organization* (31), *Quality* (14), *Objective* (13), and *Concise* (12) (see Table 9.1).

Detailed (D)—For comments deemed as *Detailed*, the judges indicated that the description gave a sufficient sense of the image. The judges indicated this description gave the user the elements they wanted, without any further questions. The judges felt the descriptions gave them a sense of the contents of the image without overwhelming them. Although seemingly in contradiction to the category of concise, the two can work symbiotically—descriptions that give enough detail without wasting the user's time. Examples of comments coded as *Detailed* included, "I found the description slightly more detailed, and provided information regarding the individual's clothing and metals" and "Although they did not identify it as Death Valley their description was very good and detailed."

Organization (Org)—Descriptions with this code had a solid and recognizable writing structure, which followed a certain logic. It focused

Table 9.1. Audio Description Characteristics and their Definitions

Characteristic	Definition
Detailed	The description had sufficient detail
Highly Organized	The description had a logical, clear, and easy-to-follow organization, particularly crucial for maps
Quality	The description was well-written
Objective	The description was written from a detached and objective position
Concise	The description did not contain extraneous information

Source: Created by the author.

specifically on organization/layout description and could also refer to the organization of information within the app. Frequently, judges noted the logical layout, the ease that the description allowed them to picture the image, and the ability to logically visualize something without much effort. Judges indicated the helpfulness of properly orienting the user to comprehend the image easily. Sometimes the image involved a path, and the description used that to orient the user. Others used logical patterns, such as starting at the top left and moving clockwise. Examples of organized statements included, "The description of the person in the photo was featured more prominently in the beginning of the description. The other description had me waiting until the end for a description of the person," and "This description gave me very clearly all of the information about the map, what was there and what could be enjoyed."

Quality (Q)—Descriptions coded as *Quality* focused on the perceived caliber of the writing. Judges noted a description was written well with polish, skill, and flair. The execution and establishment of what constituted quality was perhaps largely based on the judge's opinion, but it came up frequently enough as a reason that a judge selected a statement. Examples of comments coded as *Quality* included, "I liked the wording better" and "I felt the description was clearer and more engaging."

Objective (Ob)—Descriptions coded as *Objective* based on the judge's comments were written from a detached, objective, and neutral perspective. Several times, the judges commented that the descriptions did not sound loaded or "politically correct." Essentially, the descriptions skirted politics and notations of division in the minds of the judges. Examples of objective comments included, "very informative, without going overboard or making assumptions" and "Good neutral description of gender features with strong explanation of accessories."

Concise (Con)—Items coded as *Concise* gave judges the information they felt they needed, with nothing extraneous. The descriptions did not use overly flowery language, nor did they get bogged down in overlong sentences and phrases. Given that in Audio Descriptions, the user cannot gloss over things in a way that one could in reading visually, conciseness seems particularly pertinent. Comments about descriptions that highlighted conciseness included, "Both descriptions were to the point and the words were chosen Wisely [sic] and well. I could really get a view of this portrait" and "Good description with enough details to allow me to picture it in my mind. Didn't get caught up in a lot of unnecessary details."

RQ2: PUTTING AUDIO DESCRIPTION CHARACTERISTICS TO THE
SOCIAL-INCLUSION TEST

The second research question examined the connection between social inclusion and the five identified characteristics. As noted earlier, one of the goals and potential benefits of high-quality Audio Description is to create higher levels of connectedness. The social connection variables consisted of three items derived from Peterson et al. (2008), rated on a 1–7 scale: "I feel connected to this place," "This place helps me fulfill my needs," and "I feel like I can get what I need in this place" (α=.89, M=4.84, SD=1.19).

When comparing the Audio Description "winners" the judges chose in contrast to the non-winners, the preferred Audio Descriptions rated higher on our social inclusion measure. An independent groups t test revealed judges rated the winners (M=5.32, SD=1.11) significantly higher than the non-winners: (M=4.61, SD=1.19), t(314)=4.51, p <.001, d=.51. These findings suggest the connection between quality Audio Description and feelings of social inclusion exist and are positively related (see Table 9.2).

Further examination of the data revealed which specific characteristics had higher social inclusion means. Table 9.2 displays the differences in the means, with *Concise* having the highest mean on social inclusion (5.56) and *Objective* having the lowest (4.98). When comparing the social inclusion means of an Audio Description with a particular characteristic against one without that characteristic (e.g., descriptions identified as objective vs. those that were not), a series of t tests revealed only one characteristic, *Concise*, with a significant difference: M=5.56, SD=0.81 vs. M=4.81, SD=1.20, t(314)=2.12, p<.05, d=.62.

Table 9.2. Audio Description Characteristics with Social Inclusion Means

Audio Description Characteristic	Social Inclusion Mean
Detailed (n=54)	5.12
Concise (n=12)	5.56
Organization (n=30)	5.23
Quality (n=14)	5.29
Objective (n=13)	4.98

Source: Created by the author.

Discussion

This research is novel in the area of Audio Description, not only because of its grounding in the Descriptathon context but also because this research includes social-inclusion values. It represents an important foundational step in improving descriptions and the experiences of the people who use them, as well as addressing the public health concerns associated with social exclusion. At least two key points from the present study merit further attention. We offer some tentative best practices for writing quality Audio Description based on this initial research, and the combination of these findings and analysis could be beneficial in many contexts, including within classrooms teaching common technical communication topics such as document design, media accessibility, and communicative aspects of social justice. In those ways, this chapter could serve as a primer for larger issues in this area or as a guide, in practical contexts, for creating better description overall.

At a theoretical level, we make a case here for a new approach to Audio Description research, one that aims at mid-level on the ladder of abstraction, to identify broad areas of common concern among targeted audiences and also to establish conceptual ground above the specifics of any individual description but below the line that argues that *all* description is good and helpful. We contend that there is more to Audio Description than quantity, and if quality is at stake, then how do we measure and improve it? At the most practical level, for example, designers could use this research to create concrete checks on the quality of their descriptions by simply asking Deafblind, blind, and low-vision listeners, "Is this description detailed enough for you?" "Is it well organized?" "Is it of high quality?" and so on, adding to the end of each yes-or-no question and "Why?" to generate a helpful heuristic measure of how the descriptions are performing for their targeted audiences.

The evidence indicates a connection to well-composed Audio Description and social inclusion. Access to opportunities is a key component of social inclusion (Oveido-Caceres et al., 2021), and creating effective Audio Descriptions provides one avenue for this. For all of the negatives of technology, perhaps the greatest positive potential involves its ability to connect people. The COVID-19 pandemic would have been much worse without the Internet to provide the vital means to keep people connected to their jobs, their families, and their community. By this same token, the use of technology to supplement and equalize experiences presents a fruitful vein to tap. It is important to design and supplement with intentionality to include people of all abilities (Chang et al., 2022).

The importance of intentional design for visual impairments as a means of inclusion has received support, such as Siu's (2011) case study covering public toilets. Baldarelli and Cardillo (2022) examined a tactile museum, designed to provide people with different abilities the chance to have a similar experience as a means of social inclusion. The effective use of technology (in the present case, the use of Audio Descriptions) can connect people to places unlike the tactile museum (which is specifically designed that way), such as national parks. And not all attempts to be inclusive are effective; interventions do not always have a clear and unitary impact. Candlin (2003), in a study concerning blind people's experiences in art museums, noted the difficulties of separate interventions to offer instruction and experiences specifically for blind people. She found a bifurcated set of opinions among her interviewees—either people felt included through a special event for visually impaired people, or they felt the experience "dreadful," as if they were singled out (made to feel separate from others).

The challenge exists to maintain social inclusion through effective Audio Descriptions without pulling people into a completely separate experience, which leads to the second key finding—the key characteristics of effective Audio Description. This is the first time in our knowledge of a typology of this sort that has been identified through empirical testing. If the goal is to improve Audio Description, a key step is to listen to the people who use them. The authors took the opinions of blind and visually impaired people to derive the typology and then involved another member of that community to construct its final version.

The findings now lead us to some age-old issues. For example, what is the balance between enough detail while staying concise? Both of these emerged in different ways in this study. For example, judges whose reactions were coded as *Concise* reported the highest level of social inclusion (.3 higher than the next closest mean). However, *Concise* was also identified the least number of times by the judges, emerging in only 12 of the comments (versus *Detailed*, which emerged in 54). This indicates that when it mattered, it really mattered.

Given the relative newness of this genre and its likely increasing importance, further refinement of what these characteristics may entail merits study. If we take the approach that specific mediums have their own unique standards (e.g., Hayles, 2004), it holds that developing effective Audio Description requires a unique set of rules and best practices. The present study represents an important first step in this boundary-defining process.

BEST PRACTICES

A primary goal of Audio Description is to equilibrate visual media and simultaneously transform it into something useful for people who cannot see it. At a strategic level, we are researching primarily at prominent public places because these sites, including national parks, are publicly supported and citizens invest in them for the good of all. Working on public places and making those fully accessible presents a logical place to start, and then we can use them as models for the rest of built society. Pragmatically, as a part of our results, we want our research to have a positive impact on the Audio Description community, including helping to create, share, and model accessible media products and processes for making those products.

The present study represents a step in this direction—a first, but important one. Using the goals identified in the judges' preferences, we look for opportunities in operational procedures where our research can create empirically sound arguments for novel best practices. In the case of this study, we recommend five best practices. Specifically, these are goals to keep in mind when writing Audio Descriptions.

Experimental Best Practice for Better Social Inclusion 1:
Description should be fact-based about what can be seen

The raucous debates over the term *Objectivity* have been well-chronicled in Journalism Studies, so there's no reason for Audio Description to use that term and repeat the mistakes made in another field. Scholars in art critique both what's inside the visual media frame and what's left out, but Audio Description is culturally considered a descriptive art, not a cultural-critical one. We recommend that describers generally focus on what can be seen in the visual media, including direct visual hints about what's outside the frame. But they should avoid tangents beyond that, historical or cultural, unless adding that context helps in the conceptualization of what is being shown to the viewer. More importantly, from this "Objective" perspective, describers should avoid any triggering words or phrases that can lead to inadvertent distractions based on the describer's political, philosophical, or moral positioning. (This specific concern appeared several times in the comments.)

Experimental Best Practice for Better Social Inclusion 2:
Description should be well-organized

Think about description organization this way: This description is the primary interface to the public place for people who are blind or Deafblind or who have low vision. When they perceive an organizational mess in the description about the place, the site might just as equivalently (for sighted visitors) leave an impression of dirt streaks on the front door and trash bags throughout the lobby. Disorganization can be interpreted as a sign of poor management or a lack of care for the visitor. Either way, disorganization will make the listener feel unimportant, unwanted, and socially outcast.

Experimental Best Practice for Better Social Inclusion 3:
Description should be detailed

Although some people on occasion want just a quick description of a sentence or two, to get the general idea of what the visual media shows, a much more common request for describers, in our experience, is to provide more details. This best practice generally contradicts the predominant alt-text culture of, say, providing a single line of text to describe any photograph. With its contradictory position to the dominant ideology, the idea needs more investigation to understand its nuances. But alt-text originally was a technological constraint, not an empirically tested best practice for description composition. This finding leads us to recommend detailed descriptions with a caveat, that they also are both well-organized and concise (see also Best Practices 2 and 4). Brevity for the sake of brevity, we have found in this research and in other studies, is not necessarily desirable in descriptions as an overriding characteristic. Neither is detail just for the sake of detail. We hypothesize that lack of detail could lead to perceptions of the listener feeling unimportant or not getting good service, although we did not test perceptions at those detailed levels. The aim here is that the describer provides enough details, so all major questions have been answered, but not too much, so the description feels bogged down by extraneous scraps of information. The challenge for now is that the right amount will need to be determined on a case-by-case basis through a localization process.

Experimental Best Practice for Better Social Inclusion 4:
Description should be concise

Concision seems interwoven with other best practices that relate to details and organization, but similar to Best Practice 3, Best Practice 4 likely refers to the "Goldilocks" principle of listeners wanting the amount of information that is "just right," instead of too little or too much. The comments in this coded cluster lead us to hypothesize that arbitrary lengths or word counts should not constrain descriptions, but a good editor should. Admittedly, this is one of the toughest dialectics in writing to manage in Audio Description and all types of writing, technical and otherwise.

Experimental Best Practice for Better Social Inclusion 5:
Description should be created with a writerly quality

Because Audio Description is generally scripted in advance and describers have ample time to consider what they are going to say and how they are going to say it, listeners expect a professional quality to the descriptions. They appreciate writerly flourishes, with active verbs and descriptive nouns and adjectives leading to poetic phrasing. When the listener feels the care and concern of the writer, then the listener is likely to feel more valuable and more included. Again, in opposition to most alt-text guidelines, thoughtful and considered writing matters here.

Conclusion

The field of Audio Description is new, but growing. Through identifying and describing five major characteristics for Audio Description—*Detailed, Highly Organized, Quality, Objective,* and *Concise*—and converting them into Trial Best Practices that connect them to social inclusion, this chapter opens fertile ground for further investigations into the efficacy of Audio Description to make people who cannot see or see well feel valued and important to public places, potentially increasing feelings of social inclusion, while also make the places more accessible for everyone.

References

Accessible Media Inc. and The Canadian Association of Broadcasters. (2015, June). *Post production described video best practices.* https://www.ami.ca/sites/default/files/2020-07/PP_Described_Video_Best_Practices_0.pdf

Agboka, G. Y. (2013). Participatory localization: A social justice approach to navigating unenfranchised/disenfranchised cultural sites. *Technical Communication Quarterly, 22*(1), 28–49.

American Council of the Blind. (2022, May 15). *The Audio Description project.* https://www.acb.org/adp

American Foundation for the Blind. (2022, May 15). *Statistical snapshots from the American Foundation for the Blind.* https://www.afb.org/research-and-initiatives/statistics

Baldarelli, M.-G., & Cardillo, E. (2022). Managerial paths, social inclusion, and NBS in tactile cultural products: Theory and practice. *Journal of Hospitality & Tourism Research, 46*(3), 544–582. https://doi.org/10.1177/1096348020944440

Bratman, G. N., Hamilton, J. P., Hahn, K. S., Daily, G. C., & Gross, J. J. (2015). Nature experience reduces rumination and subgenual prefrontal cortex activation. In *Proceedings of the National Academy of Sciences, 112*(28), 8567–8572. https://doi.org/10.1073/pnas.1510459112

Candlin, F. (2003). Blindness, art and exclusion in museums and galleries. *International Journal of Art & Design Education, 22*(1), 100–110.

Castro, O., Ng, K., Novoradovskaya, E., Bosselut, G., & Hassandra, M. (2018). A scoping review on interventions to promote physical activity among adults with disabilities. *Disability and Health Journal, 11*(2), 174–183.

Centers for Disease Control and Prevention. (2022, May 15). *The Burden of Vision Loss.* https://www.cdc.gov/visionhealth/basic_information/vision_loss_burden.htm

Chang, I., Castillo, J., & Montes, H. (2022). Technology-based social innovation: Smart city inclusive system for hearing impairment and visual disability citizens. *Sensors, 22*(3), 848. https://doi.org/10.3390/s22030848

Comolli, J. L. (1980). Machines of the visible. In T. de Lauretis & S. Heath (Eds.), *The cinematic apparatus* (pp. 121–142). Palgrave Macmillan.

Cox, D. T. C., Shanahan, D. F., Hudson, H. L., Fuller, R. A., Anderson, K., Hancock, S., & Gaston, K. J. (2017). Doses of nearby nature simultaneously associated with multiple health benefits. *International Journal of Environmental Research and Public Health, 14*(2). https://doi.org/10.3390/ijerph14020172

Creswell, J. W. (2015). *30 essential skills for the qualitative researcher.* Sage.

Dolmage, J. T. (2014). *Disability rhetoric.* Syracuse University Press.

Fryer, L. (2016). *An introduction to Audio Description: A practical guide.* Routledge.

Getto, G., & Sun, H. (2017). Localizing user experience: Strategies, practices, and techniques for culturally sensitive design. *Technical Communication, 64*(2), 89–94.

Gonzales, L., & Zantjer, R. (2015). Translation as a user-localization practice. *Technical Communication, 62(4)*, 271–284.

Haluza, D., Schönbauer, R., & Cervinka, R. (2014). Green perspectives for public health: A narrative review on the physiological effects of experiencing outdoor nature. *International Journal of Environmental Research and Public Health, 11 (5)*, 5445–5461. https://doi.org/10.3390/ijerph110505445

Hayles, K. H. (2004). Print is flat, code is deep: The importance of media-specific analysis. *Poetics Today, 25(1)*, 67–90.

Hutchinson, R., Thompson, H., & Cock, M. (2020). *Describing diversity.* VocalEyes. https://vocaleyes.co.uk/?request_file=14091

Jason, L. A., Stevens, E., & Ram, D. (2015). Development of a three-factor psychological sense of community scale. *Journal of Community Psychology, 43(8)*, 973–985. https://doi.org/10.1002/jcop.21726

Kim, K., Kim, D., Shin, Y., & Chul Yoo, D. (2016). Social exclusion of people with disabilities in Korea. *Social Indicators Research, 129(2)*, 761–773. https://doi.org/10.1007/s11205-015-1123-2

Koirala, S., & Oppegaard, B. (2022). The light bulb went on: A historiography-based approach to disentangling Audio Description's influential U.S. roots from its common practices. *Journal of Visual Impairment & Blindness, 116(4)*, 1–12. https://doi.org/10.1177/0145482X221116903

Krahn, G., Walker, D., & Correa-De-Araujo, R. (2015). Persons with disabilities as an unrecognized health disparity population. *American Journal of Public Health, 105(S2)*, S198–S206. https://doi.org/10.2105/AJPH.2014.302182

Lachowycz, K., & Jones, A. P. (2014). Does walking explain associations between access to greenspace and lower mortality? *Social Science & Medicine, 107*, 9–17.

Maszerowska, A., Matamala, A., & Orero, P. (Eds.). (2014). *Audio Description: New perspectives illustrated.* John Benjamins Publishing Company.

Matamala, A., & Orero, P. (2016). *Researching audio description.* Palgrave Macmillan.

Meadows, M. S. (2002). *Pause & effect: The art of interactive narrative.* New Riders.

Melonçon, L. (Ed.). (2013). *Rhetorical accessibility: At the intersection of technical communication and disability studies.* Routledge.

McColl, M. A., Davies, D., Carlson, P., Johnston, J., & Minnes, P. (2001). The community integration measure: development and preliminary validation. *Archives of Physical Medicine and Rehabilitation, 82(4)*, 429–434. https://doi.org/10.1053/apmr.2001.22195

Mitchell, R., & Popham, F. (2008). Effect of exposure to natural environment on health inequalities: An observational population study. *The Lancet, 372(9650)*, 1655–1660. https://doi.org/10.1016/S0140-6736(08)61689-X

Mitra, M., Long-Bellil, L., & Powell, R. (2019). Persons with disabilities and public health ethics. In A. C. Mastroianni, J. P. Kahn, & N. E. Kass (Eds.), *The Oxford handbook of public health ethics* (pp. 219–213). Oxford University Press.

Mirzoeff, N. (2011). *The right to look.* Duke University Press.

Mirzoeff, N. (2016). *How to see the world: An introduction to images, from self-portraits to selfies, maps to movies, and more.* Basic Books.

Moore, K. R. (2017). The technical communicator as participant, facilitator, and designer in public engagement projects. *Technical Communication, 64*(3), 237–253.

National Park Service. (2020). *Healthy parks, healthy people.* https://www.nps.gov/orgs/1078/index.htm

Oppegaard, B. (2020). Unseeing solutions: From failures to feats through increasingly inclusive design. In J. Majewski, R. Marquis, N. Proctor, & B. Ziebarth (Eds.), *Inclusive digital interactives: Best practices, innovative experiments, and questions for research* (pp. 219–242). Access Smithsonian, The Institute for Human Centered Design, & Museweb.

Oppegaard, B., & Rabby, M. K. (2022). Gamifying good deeds: User experience, agency, and values in play during a Descriptathon. *Technical Communication, 69*(4), 27–43.

Oswal, S. K. (2013). Exploring accessibility as a potential area of research for technical communication: A modest proposal. *Communication Design Quarterly Review, 1*(4), 50–60.

Peterson, N. A., Speer, P. W., & McMillan, D. W. (2008). Validation of a brief sense of community scale: Confirmation of the principal theory of sense of community. *Journal of Community Psychology, 36*(1), 61–73. https://doi.org/10.1002/jcop.20217

Rai, S., Greening, J., & Petré, L. (2010). *A comparative study of audio description guidelines prevalent in different countries.* London: Media and Culture Department, Royal National Institute of Blind People (RNIB). https://unidescription.org/storage/app/uploads/public/5f1/a3e/bb1/5f1a3ebb17896460620035.pdf

Repanshek, K. (2022, March 17). National Park Service got a small budget bump in FY22 funding package. *National Parks Traveler.* www.nationalparkstraveler.org/2022/03/national-park-service-got-small-budget-bump-fy22-funding-package

Scott, R. (1969). *The making of blind men: A study of adult socialization.* Russell Sage Foundation.

Sabatello, M. (2018). Precision medicine, health disparities, and ethics: The case for disability inclusion. *Genetics in Medicine, 20*(4), 397–399. https://doi.org/10.1038/gim.2017.120

Shakespeare, T. (2006). The social model of disability. *The Disability Studies Reader, 2,* 197–204.

Shivers-McNair, A. (2017). Localizing communities, goals, communication, and inclusion: A collaborative approach. *Technical Communication, 64*(2), 97–112.

Siu, K. W. M. (2011). Designing public toilets to enhance the well-being of the visually impaired. *International Journal of Health, Wellness & Society, 1*(3), 137–145. https://doi.org/10.18848/2156-8960/CGP/v01i03/41175

Statista. (2021, June 16). *Media usage in an internet minute as of August 2021*. https://www.statista.com/statistics/195140/new-user-generated-content-uploaded-by-users-per-minute

Statista. (2020, April 4). *Hours of video uploaded to YouTube every minute as of February 2020*. https://www.statista.com/statistics/259477/hours-of-video-uploaded-to-youtube-every-minute

Van Bergen, A. P., Hoff, S. J., Schreurs, H., van Loon, A., & van Hemert, A. M. (2017). Social exclusion index for health surveys (SEI-HS): A prospective nationwide study to extend and validate a multidimensional social exclusion questionnaire. *BMC Public Health, 17*(1), 253–266. https://doi.org/10.1186/s12889-017-4175-1

Web Accessibility Initiative. (2020, May 15). Audio description of visual information. https://www.w3.org/WAI/media/av/description/

White, M. P., Alcock, I., Grellier, J., Wheeler, B. W., Hartig, T., Warber, S. L., Bone, A., Depledge, M. H., & Fleming, L. E. (2019). Spending at least 120 minutes a week in nature is associated with good health and wellbeing. *Scientific Reports, 9*(1), 7730. https://doi.org/10.1038/s41598-019-44097-3

Wilber, N., Mitra, M., Walker, D. K., & Allen, D. (2002). Disability as a public health issue: findings and reflections from the Massachusetts survey of secondary conditions. *The Milbank Quarterly, 80*(2), 393–421. https://doi.org/10.1111/1468-0009.00009

Wolch, J. R., Byrne, J., & Newell, J. P. (2014). Urban green space, public health, and environmental justice: The challenge of making cities 'just green enough.' *Landscape and Urban Planning, 125*, 234–244. https://doi.org/10.1016/j.landurbplan.2014.01.017

Zjawinski, S. (2008, 12 March). How Google got its colorful logo. *Wired*. https://www.wired.com/2008/03/gallery-google-logos

Chapter 10

BabyTree App

Localizing Usability in the App Design to Accommodate China's Sociocultural and Healthcare Exigencies

Hua Wang

With the increasing popularity of pregnancy and mothering apps, engaging with these technologies is becoming a regular part of pregnant women's and new mothers' daily lives. Although pregnancy and motherhood are normal life phases, they are involved in social, cultural, material, and healthcare practices. Thus, apps that relate to these life phases need to provide a wide range of functions with various features to meet pregnant women's and new mothers' needs. They include resources such as knowledge of expectant mothers' body changes and fetus development, gestation, pregnancy tracking, social support during pregnancy and after childbirth, entertainment, shopping for pregnancy-/newborn-related products, and parenting (Johnson, 2014; Ledford et al., 2016; Lupton & Pedersen, 2016; Wang, H., 2021). As they are commercial apps, anyone can download and install them for free on a smartphone.

The popularity of these apps indicates that from the perspective of technical communication (TC), their design may implement some effective usability principles such as readability, learnability, credibility, efficiency, accessibility, and satisfaction. Usability determines "the extent to which a system, product or service can be used by specified users to achieve

299

specified goals with effectiveness, efficiency and satisfaction in a specified context of use" (International Organization for Standardization, 2018). This definition emphasizes that usability is a quality of the use context rather than a product or a system (Acharya, 2019; Sun, 2012).

In the fields of TC and usability studies, recent scholarship advocates that technology designers and technical communicators should pay attention to the specific use context, especially effective localization in design, as localizing usability can empower users (Acharya, 2019; Agboka, 2012; Sun, 2006, 2012).

Additionally, designers and technical communicators must consider all users, which for apps like BabyTree, include new and expecting mothers as well as physicians, small-business owners, and the Chinese government. This balance of diverse users creates an equilibrium in the design. Therefore, designers need to consider a variety of perspectives and voices to ensure that all users have agency.

In the case of pregnancy and mothering apps, their popularity may offer new ways to think of usability and the use context—social, cultural, material, healthcare, and other factors involved in pregnancy and motherhood, as well as the apps' dynamic interactions in a specific context. For example, the Chinese commercial app BabyTree enjoys an enormous popularity and market success in China. Why is the app so popular? To answer the question is significant because it relates to how the app is developed to fit into the Chinese context and meet local users' needs. This investigation will bridge the scholarly gap existing in recent scholarship on localization usability that mainly focuses on how TC and technology for internationalization design failed to fit into local contexts (Acharya, 2019; Agboka, 2012; Dorpenyo, 2020; Sun, 2012). Put another way, this case study will provide a nuanced perspective to theorize localization usability in a broader domain: how a local technology successfully accommodates the local needs.

This chapter investigates how the Chinese pregnancy app BabyTree makes localization usability efforts to amplify users' agency and accommodate China's social, cultural, material, and reproductive healthcare exigencies. To begin, it summarizes and critiques recent scholarship on localization usability and positions my research. After discussing the current reproductive discourse in China, the article briefly introduces the app BabyTree, its vision, and research questions. Then, it describes and justifies the cognitive walkthrough method, perspective-based inspection, and the digital rhetoric lens; walks through the app step-by-step to complete representative tasks; and examines its efficiency, flexibility, and consistency for task completion. Next, it analyzes how the interface design facilitates engagement with the app and amplifies users' agency through high-level interactivity and engaging users with various schemas of social connections. The article ends with implications for future research on localization usability.

Usability, Localization, and Social Justice

The term "usability" originated from Nielsen (1993), who uses an engineering approach and defines it with five metrics: learnability, efficiency, memorability, errors, and satisfaction. This approach of usability primarily focuses on the precise measurement of a product or system. In this sense, usability has nothing to do with "social affordances" (Sun, 2012). Because the usability research scope was expanded in the early 21st century, scholarship on usability studies has been paying particular attention to user-centered technology design, critically examining usability methods, practices, and the relationship between usability and sociocultural issues (Johnson et al., 2007; Schneider, 2005; Scott, 2008). Among the scholarship, one theoretical thread addresses localizing usability with a focus on the use context. For example, Sun (2006) examined how mobile messaging technology failed to address the local use in different cultural contexts and suggests a cultural usability method for technology design and localization studies. She pointed out the problematic aspect of technology localization that puts little effort in user situations in a localization process. Furthermore, she distinguished user localization and developer localization and put forward that user localization should happen at the user's site. Her study has been valid and extended by subsequent scholarship that pays more attention to localization usability in the use context of a system or product. For example, St.Amant (2017) emphasized accessibility of technology design that allows communicators of health and medicine to address different cultural contexts of care effectively; Roy (2013) suggested increasing user engagement by creating a community on the interface instead of the mere interaction between the system and the user; and Ladner (2015) argued that technology design should empower users.

With scholarship on technical and professional communication taking a social justice turn, more scholarship has been investigating social injustice and oppression of human rights and believes that localization usability of a system or product can promote social justice (Acharya, 2019; Agboka, 2013a, 2014; Haas & Eble, 2018; Jones, 2016; Walton, 2016). For example, Agboka (2013b) investigated the poorly localized documentations used by countries like China to market sexuopharmaceuticals in Ghana and pointed out the ineffectiveness of the localized TC at the developer's site. He advocated "a participatory approach that takes into account user linguacultural, political, economic, legal, and local knowledge systems in the localization process" (p. 28). He argued that through effective localization that emphasizes contexts, situatedness and locality, social justice can be accomplished in domains such as technology design, communication, and usability. Through examining the localized Global North biomedical equipment and devices in Nepal hospitals, Acharya (2019) analyzed how the localized biotechnology failed to meet the local medical practitioners' needs. He proposed that Global North technology designers should consider the usability expectations of local users in the Global South so that social justice can be promoted and the divide between the North and the South can be narrowed. Similarly, Dorpenyo (2020) illustrated another case to display the problematic aspect of technology localization: the biometric verification technology broke down in Ghana's 2012 general elections. He suggested that local users use their own local language, expertise, and experiences to solve the issues of localization and that they "redesign technologies or documents to fit local exigencies" to claim and assert rhetorical agency (p. 146). Nevertheless, Agbozo (2022) pointed out that "localization at users' contexts does not necessarily overcome localization challenges" (p. 8). Through analyzing the artifact of GhanaPostGPS, he argued that developers' neglect of users' geo-epistemology and the postcolonial African contexts that are like power reticulations constrain the localization at users' sites, and localization in users' contexts cannot completely bridge the gap between developer culture and user culture. His investigation provides nuanced insight into technology localization and further complicates the localization usability theory. To the researcher's knowledge, Agbozo's work is the only scholarship that focuses on a local technology in the Global South that fails to meet local users' needs. The present study continues the focus of a local technology in Global South; however, unlike the current scholarship mainly examining how technology localization fails to meet local users' needs and expectations, it investigates how a local technology successfully empowers local users, further complementing the previous research on the localization usability theory. Through the cognitive walkthrough and perspective-based inspection approaches and the lens of digital rhetorical analysis, the following sections analyze how a local commercial app BabyTree amplifies local users' agency and mitigates sociocultural, material, and reproductive healthcare exigencies in China.

Cultural Context: Medical Care and Pregnancy in China

In China, due to the hegemonic biomedical discourse, Chinese pregnant women's agency is constrained and their demand for high-quality medical knowledge cannot be met (Wang, H., 2021). Since 1978, the central government has launched several rounds of reform to replace a primary-health-care-orientated system with a market-oriented and intensive healthcare system (Gao et al., 2013). The healthcare reform policies primarily focus on the medical aspects of reproduction, normalizing medical checkups and screening of pregnancy and encouraging total hospitalization of births (Gao et al., 2013; Hellerstein et al., 2015). Consequently, the rates of cesarean delivery increased dramatically in China. In some urban areas, c-section rates even reached 100% (Huang et al., 2004). In the reformed healthcare system that promotes the extensive use of obstetricians, obstetric nurses, and doulas and encourages instrumental in-hospital births, women are subject to biomedical technologies with little agency.

Pregnant Chinese women and new mothers also suffer gender discrimination in employment. Many women have lost their jobs during pregnancy and after childbirth according to surveys (Wang, H., 2021). In 2016, China launched the two-child policy, which allows couples to have two children. The revised child policy makes the gender discrimination worse in employment. In the report *China's Two-Child Policy and Workplace Gender Discrimination*, Yaqiu Wang (2021) stated the following:

> After the two-child policy went into effect, a majority of women surveyed by various Chinese companies and women's groups reported they had been subjected to gender and pregnancy-based discrimination in pursuit of employment. Countless job ads specify a preference or requirement for men, or for women who have already had children. Numerous women have described, on social media, to the Chinese media, or in court documents, their experiences being asked about their childbearing status during job interviews, being forced to sign

contracts pledging not to get pregnant and being demoted or
fired for being pregnant. (para. 6)

Although China's constitution guarantees that women enjoy equal rights as
men do, the Chinese government fails to stop gender and pregnancy-based
discrimination in employment. The laws and regulations against gender
and pregnancy-based discrimination in employment do not clearly define
gender discrimination. They provide vague enforcement mechanisms,
leaving victims with limited legal avenues for redress. Due to unclear legal
standards or failing to meet evidentiary requirements, some women victims'
claims do not succeed. "Even when women win their cases, compensation
awarded to victims is too often too small to justify going through the legal
system and penalties imposed on companies too insignificant to serve as
a deterrent for future violators" (para. 9).

In addition, the stiff medical system in China causes advanced medical
equipment and best physicians to be concentrated in large, state-owned
public hospitals in big cities. The result is the uneven medical resources
distribution and imbalanced public health service (Zhou et al., 2017).
China's medical system is completely dominated by the government, and
best medical professionals and advanced equipment and technology are
gathered in the state-owned public first-class hospitals. Patients usually go
to the local rural hospitals or community hospitals for common diseases.
A large number of patients are coming to the public hospitals, which leads
to physicians' overtime work and results in lower standards of medical
service with which patients are dissatisfied. Meanwhile, there is a serious
shortage of physicians, especially pediatricians and gynecologists, across
the country. In 2015, the Chinese government announced the second-child
policy, and it is predicted that the population will increase by 1.5–2.5
million in the following 5 years (Liu et al., 2015). However, because of the
overload of work, lower pay, and the increasing medical disputes, many
physicians have resigned. In 2016, White Paper of Pediatric Resources
in China stated that 10.7% (14,310) of pediatricians across the country
resigned from 2011 to 2014, and the hospitals across China need 86,042
pediatricians so far according to the goal of 0.69 pediatricians per 1,000
persons in 2020 (Yao et al., 2020).

Apparently, the reproductive discourse in the Chinese context is
unjust, oppressive, problematic, and even exigent. Because of the above
factors, many pregnant Chinese women and new mothers turn to the
Internet and social media for high-quality healthcare, medical consultation, and assistance.

The App BabyTree and Research Questions

The app BabyTree was launched in 2007 (Wen, 2012). The Chinese Tencent app store shows that its rating is 4.8/5.0 points, with more than 60 million downloads. According to the 2018 Sullivan Report, the BabyTree company provides the largest and most active maternal and child-raising community platform in China in terms of monthly active users (Li, 2018). Its vision is to meet four core needs of the new generation of maternal users, including information and knowledge about pregnancy and parenting, medical consultation and social networking, healthy growth, and optimal shopping, so that Chinese mothers and babies worldwide can share a beautiful life journey (Ge, 2018; Wang, H., 2021; Wen, 2012).

In this research, the following questions were put forward to guide the investigation of how the app BabyTree is designed locally to facilitate users' rhetorical agency and social justice:

1. How is the technology design and usability principles of the app localized to fit into the Chinese context?

2. How are China's sociocultural, material, healthcare, and other contextual factors articulated in the app design to meet Chinese users' needs?

3. How might the app amplify its users' agency and promote social justice?

Methods and Data Collection

This research included two stages of assessment. The first stage of assessment was conducted through usability testing. It utilizes the cognitive walkthrough and the perspective-based usability inspection to complete representative tasks. The task-oriented cognitive method afforded the researcher to assess the app's technological mechanism and usability as a new user while the perspective-based usability inspection focused inspection on the particular perspective of efficiency, consistency, and flexibility in facilitating completion of user tasks on the app. The second stage of assessment adopted digital rhetorical analysis to examine discourse, agency, and community practices. Each of the methods is explained below.

Cognitive Walkthrough and Perspective-based Usability Inspection

"The cognitive walkthrough was first presented by Clayton Lewis and his colleagues in 1990. It was developed for evaluating walk-up-and-use interfaces such as kiosks and ATM machines, where the users' ability to understand and use the interface with no prior knowledge or formal training is critical" (Salazar, 2022). This method involves a team of reviewers with expertise to walk through early conceptual prototypes and answer prescribed questions to uncover the design problems of the interface from a new user's point of view. This evaluative method is used during the development of a new technological product or system to identify design problems that challenge its learnability for a new user. As it does not involve real users, this method is relatively cheap to conduct compared with other usability evaluation methods. Over the years, it has been widely used to assess all kinds of interfaces and websites.

This research adopted the cognitive walkthrough method to walk through the representative tasks that the app promotes as its vision and to evaluate the app's usability from the perspective of a new user. This tasked-based method facilitates direct engagement with the interfaces of the app and provides step-by-step observation of each task to examine the app's technological mechanisms, "slowing down the mundane actions and interactions that form part of normal app use in order to make them salient and therefore available for critical analysis" (Light et al., 2018, p. 882). Due to not orienting toward improving the app's usability and the quality of user experience, this research does not assess the app's learnability; instead, it is focused on the app's vision that tells users what they are expected to do to illuminate technological mechanisms of the app and "critically examine the workings of the apps as a social and cultural artefact" (p. 886). Therefore, the cognitive walkthrough method in this research was created to merely walk through the representative tasks that the app promotes while not considering "how easy it is for users to accomplish a task the first time they encounter the interface and how many repetitions it takes for them to become efficient at that task" (Joyce, 2019). Because the cognitive walkthrough method merely illustrated technological features of the app, not examining the icons, the colors, the fonts and the organization of the interfaces, a perspective-based usability inspection technique was introduced to check the visual design and organization of the app's interfaces in facilitating task completion.

Perspective-based usability inspection "divides the large variety of usability issues along different perspectives and focuses each inspection session on one perspective" (Zhang et al., 1999, p. 43). Because the combination of different perspectives covers the scope of usability inspection as much as possible, perspective-based usability inspection is believed to be more effective in detecting usability issues. For the sake of research, the expert use perspective was chosen to examine whether the visual appearance and the organization of the app's interfaces facilitate each task completion in an efficient and flexible way.

The usability testing in the research was simplified as the research did not consider the app's usability improvement; rather, it focused on whether an ideal user could achieve the expected use and what the ideal user would encounter when completing the expected tasks through a cognitive walkthrough method and a perspective-based inspection technique. As BabyTree aims to engage users with learning, sharing, and shopping, which includes information and knowledge about pregnancy and parenting, medical consultation and social networking and children's healthy growth, these adapted inspection techniques, which merely examine the app's vision or the expected use, allow researchers to understand what the app is supposed to do. This could provide a baseline for identifying usage subversion and appropriation.

However, both the cognitive walkthrough and the perspective-based usability inspections are engineering techniques and do not involve a user's interactions with or responses to the app. They do not enable a full understanding of the app's technical affordance, embedded rhetorical and cultural references, the social affordances it extends, and empowerment of a user; therefore, the research was also framed by the lens of digital rhetoric to add an important layer to the analysis. Such scholarly attempts have been conducted in computer and composition studies. For example, Carnegie (2009) argued that an interface functions rhetorically to enable empowerment and enact control via three modes of interactivity. Bjork (2018) promoted a combination of usability testing with digital rhetoric theories in online writing instruction. He argued that usability theory tends to see interfaces as static objects manipulated by users, while digital rhetoric views interfaces as dynamic, real-time interactions (p. 4). Similarly, this research focused more on what symbolic representation and social and cultural implications the app conveys when a potential user engages with it.

Digital Rhetoric: Agency and Social Communities

Digital rhetoric moves the cognitive walkthrough and the perspective-based usability inspection, which focuses on the app's technological mechanisms and features, into a new area in which researchers would have a better way of understanding an app as a sociocultural/technical artifact. In this research, it allowed researchers to understand the app's sociocultural affordances and the possibility of amplifying users' agency and enacting social justice that the app facilitates. Digital rhetoric most simply refers to "the application of rhetorical theories to digital texts and performances" (Eyman, 2015, p. 44). Eyman emphasized that "digital rhetoric must be concerned with understanding all the available elements of document design, including color, font choice, and layout, as well as multimedia design possibilities such as motion, interactivity, and appropriate use of media" (p. 70). Zappen (2005) suggested that research on digital rhetoric focus on a much broader scope, like "identifying characteristics, affordances, and constraints of new media; formation of digital identities; and potential for building social communities" (p. 319). Similarly, Eyman (2015) encouraged scholars to examine the rhetorical function of networks, rhetorical methods used "for uncovering and interrogating ideologies and cultural formation in digital work" and agency that digital work facilitates (p. 44). Other scholars have followed this advocacy and produced scholarship with a focus on empowerment of digital spaces, challenging social and cultural oppression and asserting rhetorical agency (Owens, 2015; Seigel, 2014; Vinson, 2018; Wang, H., 2021, 2022). Their insightful work powerfully demonstrates digital rhetoric, especially digital space, is a potential tool for enacting social justice.

Data Collection and Ethics

This researcher obtained the approval of the Institutional Review Board of Michigan Technological University and then registered and created an account on the app as a user for research purposes. Through the cognitive walkthrough and the perspective-based usability inspection with a focus on the visual appearance and the organization of the app's interfaces, the researcher collected data when completing the representative tasks. The data used in this research cover the app's various functions, interface arrangement, textual content, icons, colors, and users' interactions on the app BabyTree in 2019. The researcher had no interactions with the app users. The app is for the public, and any registered user can read the information on the app. Because all the users use their random-generated usernames or pseudonyms, it is not possible to identify users' private information such as their real names, personal contact information, and their occupations. Because their identities are unknown, it is not feasible to contact users to obtain informed consent.

The posts used in the analysis are translated from Chinese into English, which makes the original Chinese posts difficult to identify. For the direct quotations in the analysis, the original posts are not included, nor are the usernames or pseudonyms of posting individuals.

INSPECTION PROCEDURES

For the cognitive walkthrough, the persona in the inspection was an ideal new user who is an expectant and/or new mother. The inspection focused on whether a new user could choose a step correctly, execute it successfully, and make progress to achieve the goal, and if an error occurred, whether the user could recover on her own; the perspective-based usability inspection performed an expert point of view to check the app's interfaces for efficiency, consistency, and flexibility in facilitating the task completion. Specifically, the inspection focused on possible short-cuts, more selections on a large area associated with the item selected, readability of the text, appropriate organization with the most important information or function listed first and understanding the nature of the error and how to handle the error.

For research purposes, I conducted the usability testing by the two simplified inspection methods independently. In the cognitive walkthrough inspection, I performed an expert use perspective to examine the information design and the visual appearance in supporting task completion. The specific representative tasks, which were developed based on the app's vision and the research goal, are listed in Table 10.1. I used my smartphone to accomplish all the tasks by adopting a cognitive walkthrough method and completed Tasks 2–5 through an expert use point of view.

Table 10.1. The Evaluation Tasks

Number	Task
1	Create and register an account on the app BabyTree
2	Obtain information and knowledge about pregnancy
3	Find a physician for consultation about the expectant mother's high blood pressure
4	Record the weight of a newborn
5	Order one bottle of baby shampoo
6	Log out and cancel the account

Source: Created by the author.

Results

Altogether 10 sessions were completed, and all above tasks were completed successfully without discontinuation. The task completion and the inspector's observation presented a range of uses of the app.

The app has four main interfaces with numerous sub-interfaces. The main interfaces are the Home Page interface (Figure 10.1), the Chat interface (Figure 10.2), the Shopping interface (Figure 10.3), and the "I" interface

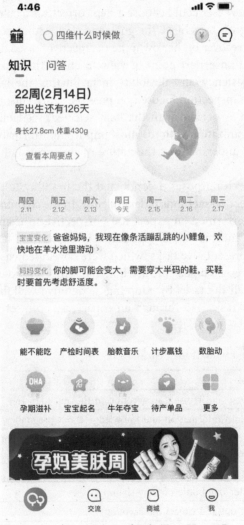

Figure 10.1. Screenshot of BabyTree Home Page.

(the user's Profile page; Figure 10.4). The Home Page interface (Figure 10.1) provides information and knowledge about pregnancy and parenthood. Each icon on the interface connects a sub-interface. The "Chat" interface (Figure 10.2) contains more than 10 group themes such as pregnancy preparation, pregnancy, postpartum, same age, emotion, life, identity, illness, fashion, and same city. Clicking a theme icon, users can find sub-group topics around the theme. The Shopping interface (Figure 10.3) provides various kinds of

Figure 10.2. Screenshot of BabyTree Chat Page.

Figure 10.3. Screenshot of BabyTree Shopping Page.

products related to pregnancy and parenting such as maternity supplies, toiletries during pregnancy, newborn supplies, pregnant women's clothing, nutritional products during pregnancy, baby toiletries, baby products, and home daily necessities. Users can shop online in this area easily. The "I" interface is the user's profile page (Figure 10.4) where users can edit their personal information, join the app's membership, and register for running a business store to earn income.

Figure 10.4. Screenshot of BabyTree User's Profile Page.

The evaluation results are summarized as follows:

- The language used on the interfaces is familiar to the inspector as a new user and is easy to understand.

- The design of all the interfaces keeps consistent in layout, formatting, etc.

- For a new user, the icons with captions on each interface provide users with clear progressive steps to achieve the goal.

- The searching bar on each interface makes the inspector locate information quickly.

- The four buttons on the bottom of the screen make the inspector easily "jump" among the four main interfaces.

- The information on each interface is structured visually and rhetorically for easy navigation.

- When errors occur, such as selecting a wrong item, executing a wrong step, or navigating to a wrong place, users can reverse the side effect by returning the previous interface or clicking one of the four buttons on the screen to make a new start.

The above usability evaluation demonstrates that the technological features of the app facilitate the expected use. Tasks 2, 4, and 5 fulfill users' needs of healthcare, shopping, and child-raising assistance, indicating the app's instrumental and material usages. Task 3 gets users to access high-level medical assistance, affording the app's medical use context. The remaining part of the chapter investigates how the app's interfaces function rhetorically and socioculturally to empower users by increasing users' interactivity and social connections.

Using "Q&A" to Create Multidirectionality

The interfaces on the app BabyTree use the mode of multidirectionality to increase interactivity through the "chat" communication model. On the "Q&A" sub-interface of the Homepage interface, users can ask peers and physicians questions related to pregnancy and parenthood. Peers' answers are quick and free of charge. Usually, a question can be answered by many users within a short time. For example, one user asked the following question: "I gave a c-section birth one and half a year ago, can I have a second child now?" This question was answered by 19 users within 18 seconds. Through the many-to-one mode of communication, the user may obtain a reasonable answer based on these peers' embodied experiences. Importantly, all the users can read the question and all the answers, which shows a many-to-many mode of communication that facilitates the social support use context.

In addition, a user can ask a physician for medical suggestions and assistance. She can ask a question in words with visuals or by phone with payment. Payment for an online consultation depends on a healthcare provider's own pricing. The higher the professional level of a physician is, the more expensive the charge. However, the charge is inexpensive compared with the payment for visiting a physician in a hospital. For example, the average outpatient cost per time in public hospitals in 2019 is approximately RMB 290.8 [$45.4] ("Statistics on the development of health services in 2019," 2020). The range of online charges on the app BabyTree varies from RMB 39 [$6] to RMB 99 [$15]. More importantly, a user can consult with higher-level healthcare providers whom she cannot access in her local hospitals. Physicians can answer the question in a minute or within one hour. The asker is anonymous. All the physicians provide credible information such as their real names, professional level, and hospital where they work. The question is displayed on the interface. If other users want to see the physician's answer, they must pay for access. For example, an anonymous user asked the following question:

> I had a natural delivery in hospital and the baby was born on March 21. I and the baby were discharged on the 23rd. Before the baby was discharged from the hospital, the doctor measured the jaundice index, which was 189 μmol/L [the normal index for newborns is less than 204 μmol/L]. The doctor asked me to come on March 25 to check the newborn. After the examination, the doctor said that the jaundice index was 370 μmol/L, and my baby needed to be hospitalized with blue light treatment. Yesterday, I went to the hospital and the doctor said that the jaundice index had dropped. It was 245 μmol/L. Today the doctor said it dropped to 165 μmol/L, but the doctor refused to let my baby leave the hospital and said that the baby needed to be observed for a while. I'd like to ask if there are any side effects of blue light treatment. Can my baby be discharged from the hospital today? If not, how long will it be hospitalized?

The asker paid RMB 59 (about $9) for the physician's medical assistance, and 40 users paid RMB 0.99 ($0.15) individually to access the answer. Through the "audit" function, this one-to-one mode of communication expands into a one-to-many mode of communication, which not only helps build users' interactivity but also brings benefits to many users in medical care. The interface offers special services, such as mental health counseling through which users can contact psychiatrists by phone or messages. This service can be beneficial to users who are in rural areas or have difficulty accessing mental health counseling. The one-to-one communication mode, which allows other users to access the question and answer through payment, not only provides the user with high-level medical consultation but also invites more users whose agency is constrained by geography, uneven medical resources distribution, or shortage of healthcare providers to access the medical assistance.

Mapping Various Schemas of Social Connections to Engage Users

The app offers the "Q&A" communication mode and social networking functions to invoke social schemas of efficient healthcare support and assistance, increasing interpersonal communication. On the Homepage interface (Figure 10.1), users can seek healthcare support related to pregnancy and parenthood. The peers' "Q&A" interface is like an online chat room: a user can ask a question and others can read the questions and choose those they want to answer. When a user asks a question, the interface reminds her how soon (usually within a few minutes) she will obtain answers; after the question is posted, the interface will show how many people are answering this question within a certain time (usually a few seconds). Importantly, users can have access to all the questions and the answers and gain some useful information no matter where they are in China. The participants interact with one another and build social connections, which increases a high-level interactivity through involvement with the social and spatial schemas facilitated by the interface. To some extent, this virtual communication is better than face-to-face interaction because it allows more users to answer questions and share their answers, which benefits more users.

On the physicians' "Q&A" interface, users can contact physicians from high-level hospitals by messages or phone calls. This communication model erases the spatial distance and establishes a momentary digital space where users can efficiently seek high-quality medical care from healthcare professionals without geographical limits. Through this mode of interaction, users increase their agency to access healthcare services and gain extra help from healthcare professionals before making their medical decisions. The distribution of healthcare enacted by professionals in the digital space can empower users to reclaim their agency and confidence in their healthcare decision making.

Apart from the medical use context, the app supports a social networking expansion context. On the Chat interface (Figure 10.2), a user can join various groups to build connections with other users across the country. In the groups, a user can exchange information with others. The interface enables users to "like," "comment," and "forward" others' posts. A user is both a sender and a receiver. She can also follow others' accounts to build personal relationships. Through this responsive dialogue model of communication, "messages take on a more referential quality" (Carnegie, 2009, p. 168). In addition, the interface provides a "private message" function to users. By means of this communication mode, users can communicate with other users privately after they follow each other, helping users expand social networking.

The Chat interface (see Figure 10.2) also facilitates building various social groups that users can join. More than 70% of users on the app BabyTree are 25–35 years old, and many are new (expectant) mothers who have anxieties about and during pregnancy, confusion about parenting, etc. (Wen, 2012). The Chat interface offers various groups such as "the same age group," "the same city group," "the same hospital group," and other groups with the themes like fashion, traveling, and photography. Users can join those groups to share embodied experiences of pregnancy, childbirth, parenting, or talk about handicrafts, food, travel, beauty care, and other aspects of life. Users may feel connected to one another in the digital place when they share their knowledge and embodied experience. However, a user cannot manipulate new schemas in the digital facilitated communication by the app. Despite the technological constraints, the "Q&A" interface engages with thousands of users every day.

ENCOURAGING ENTREPRENEURSHIP

Unlike most pregnancy and mothering apps that merely focus on maternal health and the fetus/baby development, BabyTree also encourages consumption and entrepreneurship, which can be seen on the two icons (functions) in bright colors under the user's name on the profile page. When the user checks her profile page, she can immediately identify these two colorful icons (functions): a yellow one and a pink one (see Figure 10.4). The yellow one is with the caption "Joining the membership." When the user clicks it, she will be directed into another interface where she can pay to take online classes on pregnancy and parenting from medical professionals. If the user joins BabyTree's membership, she will win perks such as free and discounted online classes. Simply put, the app encourages users as consumers to develop/buy specific scientific and medical knowledge on pregnancy and parenting.

The pink icon (function) pushes users, especially stay-at-home new mothers, to run an online store by the text "being a mom store owner." After clicking the pink icon, the user enters an interface where there are several kinds of promotions with eye-catching headers to encourage users to run an online store. For example, "Purchase any spree, you will become a mom store owner"; "Becoming a mom store owner, you will enjoy super privileges" such as "Save money when making purchases here"; and "Make money when sharing your purchases." At the bottom of this page, there is a section with a header that reads "real store owner's feedback," in which three stay-at-home mothers share their successful business experience. For example, "After getting married, I looked after my kid at home. Now I am taking care of my kid and at the same time I'm an owner of a retail store on BabyTree. I can make money without being restrained, and I can also get bonuses and gifts from the BabyTree company every now and then. Satisfaction!" As discussed above, pregnant Chinese women and new mothers are routinely discriminated against in employment, and many of them lose their jobs due to childbirth. However, BabyTree gives them an opportunity to achieve success and amplify their economic agency.

Discussion

By adopting a cognitive walkthrough and a perspective of expert-use point of view, the usability inspection displays that the app facilitates technological affordance and accommodates a range of user sophistication from novice to expert use. Both new and proficient users can enjoy the functional, material, medical and sociocultural use that the app facilitates, obtaining knowledge of pregnancy and motherhood, social support, and authoritative and credible medical assistance. The app's interfaces with the multidirectional communication mode and the responsive dialogue model enable users to interact with peers and medical professionals through various schemas of social connections. On the functional level, this usability design of the app provides users with a means to "improve and monitor their pregnancies, health, and their children's development and health" (Johnson, 2014, p. 330). On the sociocultural level, engaging with the app helps users build connections with peers and high-level healthcare providers, expanding social networking and obtaining a wider range of social, material, and medical assistance.

To some extent, the digital community that the app facilitates redistributes medical and material resources on pregnancy and motherhood and provides users access to various specific use contexts. In other words, the usability design of the app meets local users' needs and increases users' interactivity. Miller (2007) stated that interactivity is essential for agency. With the increased medical literacy on pregnancy and motherhood, reliable medical assistance, and sufficient social support, it is likely that users are empowered and can reclaim agency in enacting pregnancy and motherhood. Meanwhile, users can take advantage of the technical affordances of BabyTree to reenter job markets by running an online business. Social justice in this context is closely tied to job opportunities and increased income.

The success of the app BabyTree is a good example of the significance of localization usability design fitting into the local context socioculturally and materially, especially for mobile health technologies that aim to improve health interventions, as healthcare involves various sociocultural and material factors through medical resource consumption and interaction among multiple stakeholders like patients, medical workers, patient family members, etc. If mobile health technologies fail to address sociocultural use in health interventions, stakeholders will give them up, especially patients who seek medical interventions. Research shows that many mobile health apps cannot satisfy patients' diverse needs (Sarkar et al., 2016); according to a global study, 83% of mobile health app publishers had less than 10,000 users (mHealth economics, 2017/2018). With smartphones and other mobile digital technologies becoming part of people's everyday lives, users are interested not only in the usefulness of technologies but also in methods to integrate those technologies into their daily routines.

In the context of shifting the functional use to consumption and entertainment, usability design—with emphasis on usable and useful aspects—is outmoded, and it must be localized to meet local users' emotional, aesthetic, pleasant, entertaining, and other sociocultural needs. Understanding such contextual factors may be complex, but it directly determines the usability of a product. Usable technologies are designed and experienced within a sociocultural context that shapes and is reshaped by them. Recognizing the mutual shaping process is significant for localization usability design as it relates to "how a technology is represented, what social identities are associated with, how it is produced and consumed, and what mechanisms regulate its distribution and us" (Du Gay et al., 2013, p. xxxi).

References

Agboka, G. (2012). Liberating intercultural technical communication from "large culture" ideologies: Constructing culture discursively. *Journal of Technical Writing and Communication, 42*(2), 159–181.

Agboka, G. Y. (2013a). Thinking about social justice. *Connexions International Professional Communication Journal, 1*(1), 29–38.

Agboka, G. Y. (2013b). Participatory localization: A social justice approach to navigating unenfranchised/disenfranchised cultural sites. *Technical Communication Quarterly, 22*(1), 28–49.

Agboka, G. Y. (2014). Decolonial methodologies: Social justice perspectives in intercultural technical communication research. *Journal of Technical Writing and Communication, 44*(3), 297–327.

Agbozo, G. E. (2022). Localization at users' sites is not enough: GhanaPostGPS and power reticulations in the postcolony. *Technical Communication, 69*(2), 7–17.

Acharya, K. R. (2019). Usability for social justice: Exploring the implementation of localization usability in Global North technology in the context of a Global South's country. *Journal of Technical Writing and Communication, 49*(1), 6–32.

Bjork, C. (2018). Integrating usability testing with digital rhetoric in OWI. *Computers and Composition, 49*, 4–13.

Carnegie, T. A. (2009). Interface as exordium: The rhetoric of interactivity. *Computers and Composition, 26*(3), 164–173.

Dorpenyo, I. K. (2020). Decolonial methodology as a framework for localization and social justice study in resource-mismanaged context. In *User localization strategies in the face of technological breakdown* (pp. 53–78). Palgrave Macmillan.

Du Gay, P., Hall, S., Janes, L., Madsen, A. K., Mackay, H., & Negus, K. (2013). *Doing cultural studies: The story of the Sony Walkman*. Sage.

Eyman, D. (2015). *Digital rhetoric: Theory, method, practice*. University of Michigan Press.

Gao, L., Larsson, M., & Luo, S. (2013). Internet use by Chinese women seeking pregnancy-related information. *Midwifery, 29*(7), 730–735. https://doi.org/10.1016/j.midw.2012.07.003

Ge, L. (2018, November 15). BabyTree: Maternal and infant community e-commerce platform giant was officially launched globally today. (H. Wang, Trans).

http://finance.sina.com.cn/stock/hkstock/ggscyd/2018-11-15/doc-ihnvukff 1420437.shtml

Haas, A. M., & Eble, M. F. (Eds.). (2018). *Key theoretical frameworks: Teaching technical communication in the twenty-first century*. University Press of Colorado.

Hellerstein, S., Feldman, S., & Duan, T. (2015). China's 50% Cesarean delivery rate: Is it too high? *BJOG, 122*(2), 160–164.

Huang, P., Li, G., Chen, W., Wang, G., & Zhang, Y. (2004). The changing rate and indications of Cesarean section over 12 years (H. Wang, Trans.). *Academic Journal Guangzhou Medicine College, 32*. 57–99.

International Organization for Standardization (ISO) 9241-11:2018(en). (2018). Ergonomics of human-system interaction. https://www.iso.org/obp/ui/#iso:std:iso:9241:-11:ed-2:v1:en

Johnson, S. A. (2014). "Maternal devices," social media and the self-management of pregnancy, mothering and child health. *Societies, 4*(2), 330–350.

Johnson, R. R., Salvo, M. J., & Zoetewey, M. W. (2007). User-centered technology in participatory culture: Two decades "beyond a narrow conception of usability testing." *IEEE Transactions on Professional Communication, 50*(4), 320–332.

Jones, N. N. (2016). Narrative inquiry in human-centered design: Examining silence and voice to promote social justice in design scenarios. *Journal of Technical Writing and Communication, 46*(4), 471–492.

Joyce, A. (2019, October 20). *How to measure learnability of a user interface*. Nielson Norman Group. https://www.nngroup.com/articles/measure-learnability

Ladner, R. E. (2015). Design for user empowerment. *Interactions, 22*(2), 24–29.

Ledford, C. J., Canzona, M. R., Cafferty, L. A., & Hodge, J. A. (2016). Mobile application as a prenatal education and engagement tool: A randomized controlled pilot. *Patient Education and Counseling, 99*(4), 578–582.

Light, B., Burgess, J., & Duguay, S. (2018). The walkthrough method: An approach to the study of apps. *New Media & Society, 20*(3), 881–900.

Li, J. (2018, October 10). BabyTree plans to go public in Hong Kong: Its revenue depends on advertising and e-commerce, saying that monthly active users are 139 million. https://m.yicai.com/news/100037347.html

Liu, X., Rohrer, W., Luo, A., Fang, Z., He, T., & Xie, W. (2015). Doctor–patient communication skills training in mainland China: A systematic review of the literature. *Patient Education and Counseling, 98*(1), 3–14.

Lupton, D., & Pedersen, S. (2016). An Australian survey of women's use of pregnancy and parenting apps. *Women and Birth, 29*(4), 368–375.

mHealth economics. (2017/2018). Connectivity in digital health. *Research 2 Guidance*. https://research2guidance.com/product/connectivity-in-digital-health

Miller, C. R. (2007). What can automation tell us about agency? *Rhetoric Society Quarterly 37*(2), 137–157.

Nielsen, J. (1993). *Usability engineering*. Academic Press Inc.

Owens, K. H. (2015). *Writing childbirth: Women's rhetorical agency in labor and online.* Southern Illinois University Press.

Roy, D. (2013). Toward experience design: The changing face of technical communication. *Connexions: International Professional Communication Journal, 1*(1), 111–118.

Salazar, K. (2022, February 13). *Evaluate interface learnability with cognitive walkthroughs.* Nielson Norman Group. https://www.nngroup.com/articles/cognitive-walkthroughs

Sarkar, U., Gourley, G. I., Lyles, C. R., Tieu, L., Clarity, C., Newmark, L., Singh, K., & Bates, D. W. (2016). Usability of commercially available mobile applications for diverse patients. *Journal of General Internal Medicine, 31*(12), 1417–1426.

Schneider, S. (2005). Usable pedagogies: Usability, rhetoric, and sociocultural pedagogy in the technical writing classroom. *Technical Communication Quarterly, 14*(4), 447–467.

Seigel, M. (2014). *The rhetoric of pregnancy.* University of Chicago Press.

Scott, J. B. (2008). The practice of usability: Teaching user engagement through service-learning. *Technical Communication Quarterly, 17*(4), 381–412.

St.Amant, K. (2017). The cultural context of care in international communication design: A heuristic for addressing usability in international health and medical communication. *Communication Design Quarterly Review, 5*(2), 62–70.

Statistics on the development of health services in 2019. (2020). (Wang, H., Trans). http://www.nhc.gov.cn/guihuaxxs/s10748/202006/ebfe31f24cc145b198dd730603ec4442.shtml

Sun, H. (2006). The triumph of users: Achieving cultural usability goals with user localization. *Technical Communication Quarterly, 15*(4), 457–481.

Sun, H. (2012). *Cross-cultural technology design: Creating culture-sensitive technology for local users.* Oxford University Press.

Vinson, J. (2018). *Embodying the problem: The persuasive power of the teenage mother.* Rutgers University Press.

Walton, R. (2016). Supporting human dignity and human rights: A call to adopt the first principle of human-centered design. *Journal of Technical Writing and Communication, 46*(4), 402–426.

Wang, H. (2021). Chinese women's reproductive justice and social media. *Technical Communication Quarterly, 30*(3), 285–297.

Wang, H. (2022). Rhetorical crocheting: New Chinese moms fighting postpartum depression on social media. In L. Melonçon & C. Molloy (Eds.), *Strategic interventions in mental health rhetoric* (pp. 105–116). Routledge.

Wang, Y. (2021, June 1). *"Take maternity leave and you'll be replaced": China's two-child policy and workplace gender discrimination.* Human Rights Watch. https://www.hrw.org/report/2021/06/01/take-maternity-leave-and-youll-be-replaced/chinas-two-child-policy-and-workplace

Wen, S. (2012, February 2). *For all your parenting questions, BabyTree.com got the answers.* http://finance.sina.com.cn/leadership/mroll/20120202/164311 300571.shtml

Yao, X. Y., Yang, T. T., & Deng, W. J. (2020). Job satisfaction and its impact factors among pediatricians in China: A cross-sectional survey. *Chinese Journal of Public Health, 36*(4), 560–565.

Zhang, Z., Basili, V., & Shneiderman, B. (1999). Perspective-based usability inspection: An empirical validation of efficacy. *Empirical Software Engineering, 4*(1), 43–69.

Zappen, J. P. (2005). Digital rhetoric: Toward an integrated theory. *Technical Communication Quarterly, 14*(3), 319–325.

Zhou, M., Zhao, L., Campy, K. S., & Wang, S. (2017). Changing of China's health policy and doctor-patient relationship: 1949–2016. *Health Policy and Technology, 6*(3), 358–367.

Part Three

Equality, Access, and Social Justice Topics

In Part Three (Chapters 11–14), we include chapters that focus on technology and UX in labor, social, and political movements in TPC. The chapters consider issues of equity, safety vulnerabilities, and civic engagement. Again, the authors are considering original environments, stakeholders, and collaborations but addressing the pursuit of balance in including all users' voices. Chapter 11 addresses design justice, labor, and workload aspects in considering the value of teaching faculty in TC programs. Also considering design justice concerns, Chapter 12 presents a community-centered focus on voter rights per political context and provides an interesting take on the process of integrating a campaign's purpose into its messages and designs. Chapter 13 moves us to consider beekeepers in Lebanon, where gender concerns and climate-smart information technology are helping beekeepers manage their hives and improve honey production and quality. Chapter 14 promotes feminist UX by examining how China's DiDi Hitch app symbolically constructed women users and how the design provided information that led to the rape and murder of two female passengers; the study advocates for balance in design that also prioritizes safety. These four chapters enlarge our view of balance and multiple considerations to bring about *equilibriUX*.

Chapter 11

Design Justice in Technical and Professional Communication

Equity for Teaching Faculty and Graduate Student Instructors

Amy Hodges, Timothy M. Ponce, Johansen Quijano,
Bethany Shaffer, and Vince Sosko[1]

As Technical and Professional Communication (TPC) core concepts, usability and user experience (UX) are taught in many introductory technical writing service courses. It is ironic that the people expected to be facilitators of this important knowledge are an often underrepresented and exploited population: teaching faculty (nontenure track and adjunct instructors) and graduate student instructors (GSIs). Designers of TPC courses rightly position students as users or co-creators (see Crane & Cargile Cook, 2022), but faculty in contingent positions are not often featured as users whose experiences matter.

   As a complex structural issue in all systems of higher education, the contingency of teaching faculty and GSIs is a crucial factor for the design of introductory technical writing service courses and for the efficacy of usability and UX curricula. We agree with the statement that "contingency is a structural issue beyond the control of most departments, and it is a material reality for *all* faculty in composition and TPC" (Melonçon et al.,

2020b, p. 17), and it is a material reality for all designers of introduction to technical writing (ITW) courses. As both local and national coalitions continue to advocate for better educator working conditions, scholar-teachers in TPC can examine the opportunities that exist to develop courses that are usable, equitable, and just.

Using design justice, a "method that centers structural and institutional analysis of power inequality and is interested in root causes" (Costanza-Chock, 2020, p. 190) as a framework, this chapter documents and reflects on our community's initial stages of enacting design justice in course design for introductory technical writing courses, focusing on users who are underrepresented in TPC scholarship: teaching faculty and GSIs.

Design Justice, TPC Pedagogies, and Course Design

Design justice "rethinks design processes, centers people who are normally marginalized by design, and uses collaborative, creative practices to address the deepest challenges our communities face" (Design Justice Network, 2018). Design justice, an aspect of social justice, is critical at every stage of design to consider the relationship between power and design, who is involved in the design practices, and who is impacted by them. Design justice "focuses explicitly on the ways that design reproduces and/or challenges the matrix of domination (white supremacy, heteropatriarchy, capitalism, ableism, settler colonialism, and other forms of structural inequity)" (Costanza-Chock, 2020, p. 23). Table 11.1 lists the principles from the Design Justice Network (2018) and Costanza-Chock that were used to establish our justice design framework for addressing design of TPC introductory courses.

Design justice shares similar orientations and practices to the social justice turn in TPC, as scholars and practitioners have advocated for increased attention to the structural injustices that impact both users and designers of technologies (Jones, 2016b). Design justice and social justice also share the integration of positionality and reflexive dialogue into every design stage and the consideration of multiple, evolving layers of context for designing (Small, 2021; Thinyane et al., 2020; Walton et al., 2019). By repositioning design as emerging from and accountable to communities, the design justice framework enables practitioners to rethink who has expertise, how knowledge is shared, and how knowledge and knowing inform design (Rose, 2016). Like Getto's (2014) model of engaged design,

Table 11.1. Design Justice Network Principles

Principle 1	We use design to sustain, heal, and empower our communities, as well as to seek liberation from exploitative and oppressive systems.
Principle 2	We center the voices of those who are directly impacted by the outcomes of the design process.
Principle 3	We prioritize design's impact on the community over the intentions of the designer.
Principle 4	We view change as emergent from an accountable, accessible, and collaborative process, rather than as a point at the end of a process.
Principle 5	We see the role of the designer as a facilitator rather than an expert.
Principle 6	We believe that everyone is an expert based on their own lived experience, and that we all have unique and brilliant contributions to bring to a design process.
Principle 7	We share design knowledge and tools with our communities.
Principle 8	We work towards sustainable, community-led and -controlled outcomes.
Principle 9	We work towards non-exploitative solutions that reconnect us to the earth and to each other.
Principle 10	Before seeking new design solutions, we look for what is already working at the community level. We honor and uplift traditional, indigenous, and local knowledge and practices.

Source: Created by the author. Available in English, Catalan, Czech, German, French, Hindustani, Greek, Italian, Korean, Mandarin Chinese (traditional and simplified), Polish, Portuguese, Romanian, Spanish, and Turkish at https://designjustice.org/principles-overview.

design justice "treats the cultural lifeways of stakeholders as invaluable assets in the design of communication infrastructure" (p. 45).

We chose design justice as a framework for designing introductory TPC courses at our institution because it emphasizes community sustainability, agency, and health—key factors threatened by the structural injustices of contingency. A design justice framework, Costanza-Chock (2020) reminds us, critically analyzes the narrative of neoliberalism toward continuous innovation in technology overgrowth and reflection of designers and communities. Similarly, many educational institutions prioritize continuous teaching innovations, often asking faculty to document new technologies and techniques used in courses. Few ask for evidence of continuous reflection or instructor contributions to the health and sustainability of their educational community.

We also value design justice's attention to the narratives that users and stakeholders create around technology and design, aligning with Jones (2016a), who advocates for design processes that allow for critically interrogating both voices and silences that are present in users' narratives. Just as popular narratives elide community-led technological developments in favor of the myth of the singular, often White cis male genius who serves as the creator of a design (Costanza-Chock, 2020, p. 109), so too are course designs often attributed to individuals, usually the singular faculty member or select committee members, passing over the contributions of others. This was especially relevant in our case because Amy was asked by administration to design an online course shell and resources for introductory TPC course instructors, for which she would receive financial compensation, credit toward her teaching and service requirements, and attention from other members of the department. As Costanza-Chock (2020) reminds us, "Design generates attention, and attention is an increasingly scarce resource that is not equitably allocated" (p. 133). This structure was set up to favor Amy, the tenure-stream faculty member, over the teaching faculty and GSIs who were positioned as users of her design. Too few teaching faculty and GSIs receive attribution for their contributions to course design, and it is much less often that other users of courses are attributed as course designers, such as students, staff members, or community partners.

When designing courses, TPC instructors often construct a design narrative that focuses on the local, designing what works for our particular students at our particular institution—those who will go on to work in TPC areas. Although considering local contexts is important to course design, "if we never zoom out to the big picture, then we never take on the larger structures that constantly militate toward the reproduction of design inequality" (Costanza-Chock, 2020, p. 124). Thus, TPC introductory course designers should strive to integrate a balance of narratives into our design processes, with a focus both locally on all course users and globally on larger structures of inequity in higher education.

In this way, administrators and instructors should consider balance in course design, recognizing the users' (students') needs, current and future, but also acknowledging that the instructors who design and/or use courses are stakeholders and thus should also have influence in the construct of the courses. This balance, or equilibriUX, provides multiple perspectives and emphasizes all stakeholders' priorities and thus integrates the social justice framework into the curriculum development and delivery.

Authors' Positionality

As part of the design justice framework, the research team first independently and collaboratively considered each member's positionality, a lens "for seeing how people (including one's self) are positioned relative to one another in the social fabric and, relatedly, for identifying our margins of maneuverability for action in pursuit of justice" (Walton et al., 2019, p. 63). Because this research examined contingency and we previously worked or currently work as teaching faculty and GSIs, this ongoing reflection uncovered our experiences and biases that influenced the structure of the project goals, methods, and framework.

In their analysis of positionality in engineering education, Secules et al. (2020) stated that "an examination of positionality can and should disrupt the notion that any of us can remove ourselves entirely from the personal and interpersonal nature of research, education, or equity" (p. 38). Thus, we share these statements to describe our relevant affiliations and purposes for participating in this project and for advocating with teaching faculty and GSIs. These statements also reflect our commitment to Design Justice Principle 4 and the "accountable, accessible, and collaborative process" that facilitates potential changes to the TPC program.

Amy

After receiving my PhD and working as a GSI, I spent 1 year adjunct teaching, 2 years in postdocs, and 6 years as a faculty member in a teaching-track position. In 9 years of contingent labor, I happily took on extra service and teaching roles in my institutions. I rarely said "no" to any opportunity, both because I found those roles enjoyable and impactful but also because I felt pressure to do more. Though I believed that my position was relatively stable, I sensed that I had to constantly prove my worth and advocate for others in a similar position.

Moving to a new institution and to a tenure-track role during the COVID-19 pandemic, I questioned myself as a professional. Because I redistributed my time to focus on research, I saw clearly how I did not meet my own expectations for teaching. I walked away from more than one lesson feeling sorry for my students that they had the tenure-track me instead of the teaching-track me. Yet I could also see how I was given so many privileges: people in the department were clear about my (very few) service commitments, they reached out to mentor me, they invited my input despite my scarce knowledge of the university's culture—all actions I had not experienced in my previous work. I saw more clearly than ever that institutional structures were set up to move people in my position forward, which made it more important that I find existing coalitions and sustain a community around TPC so that we could work toward more equitable conditions.

Tim

As someone who has spent his entire academic career as contingent faculty, I felt energized by this research project—and more broadly, questions about academic freedom and equity for contingent faculty. I have experienced the ways in which "usability" within curriculum development and administration have been dictated rather than co-created. At this point in my career, I inhabit a land of liminality: the head of an academic program but still constrained by the nature of my contingent rank. I'm constantly referred to and pointed out as the "White male" in Diversity, Equity, and Inclusion meetings despite being of Mexican heritage. I'm placed into a system that defines merit based on research metrics despite the fact that research does not appear in my job description. I have found that this straddling of two worlds goes unnoticed and unaddressed because of a system that does not enact or value co-creation. From both this intellectual and material perspective, I approach the current research problem with the hope that many of the invisible inequities that hamper curricular and programmatic development can be brought into the light so that we might address and dispel them.

Quijano

As someone who has worked as contingent faculty in institutions with little academic freedom and even less institutional support, I am passionate about this topic. The expectations of an adjunct—to teach several classes on various campuses while still having to publish and perform unpaid committee work to prove that we belong—can be grueling and take a toll on one's well-being, especially given the lack of security that comes with the position. I have always advocated for contingent faculty, having been an adjunct for more than 10 years, and now that I am tenured I continue to do so. This project is part of that advocacy.

If contingent faculty are to meet department and institutional expectations, they need to be equipped with the resources to succeed. Creating user-centered documents, including syllabi and assignments, and having a robust mentorship network—both of which are within the purview of this project—are key tools they need to succeed. Although this project does not intend to be the endpoint of these discussions, I think that it is an important step toward a cohesive universal course design, increased contingent faculty retention, more robust mentorship networks, decreased adjunct workloads, and an overall quality-of-life improvement in academia.

Bethany

As a lecturer with 10 years in the English Department, I have very much learned how to make myself "retainable." I have worked as an academic advisor, a recruiting specialist, a fraternity advisor, and a teacher throughout this journey. Though my position is pitched as "permanent," I am on a yearly contract with no promise of renewal unless funds are available. As a lecturer, I have learned to teach fully online courses, to develop both hybrid courses and completely new course subjects, as well as to dive into teaching technical writing courses. This project is of special interest to me because I know how expendable contingent faculty are to a university. To prepare for teaching technical writing, I was lucky enough to have mentors assist me throughout my initial course section. However, I know this is not the case for many of my peers when they teach the course for the first time. Finding and creating teacher agency grants me some power where I often feel there is none. This project is important because it requires thoughtful reflection on current issues and attempts to combat them.

Vince

Spending several years as a PhD student while simultaneously upholding my responsibilities as an Enhanced GSI and fulfilling the duties of various service roles across our university, I was immediately drawn to the objective and necessity of this research. Like my fellow researchers and so many working in academia, I felt pressured and obliged to always respond with "yes" to requests for teaching and service positions. I tutored for the writing center and the athletic department, worked as both a teaching assistant and research associate for women's and gender studies, took on numerous editing roles for different university-led projects and proposals, served on committees within my department, and served as assistant director for our department's first-year writing (FYW) program. The many hats I wore while pursuing my PhD and teaching a variety of English courses is an experience so many contingent faculty have in academia, and it is one that comes with both a disorienting sense of purpose and an honor to be trusted by so many in such an array of skill sets. When I moved into a full-time adjunct teaching role that required more of me in the classroom, I began to feel the increased weight of curriculum design and development. Throughout my different positions as a graduate student, I continually negotiated my identity as I faced the dichotomy between internal and external senses of agency. Subsequently, when I was offered to teach a technical writing course, I found myself caught in another negotiation between my focus as a compositionist, my inexperience with technical writing pedagogy, and a standardized curric-. ulum with departmental expectations; and, thus, the process of agency formation began again.

Coauthoring in Community

Reflecting on our experiences with contingency, the coauthors discussed our shared values of equity among instructors and honoring the labor of teaching. As positionality involves "an ongoing re-positioning as a project proceeds, as relationships develop, and as power dynamics emerge" (Small, 2021, p. 287), we returned to these statements throughout the design process. This accountability to each other and to our research participants facilitated a design process that foregrounds the design's "impact on the community over the intentions of the designer" (Design Justice Principle 3).

WORKING CONDITIONS OF TEACHING FACULTY IN TPC

"Faculty working conditions are student learning conditions" is a well-known saying in education. For students in U.S. higher education, this means that their learning conditions are largely dependent on the working lives of a faculty population whose jobs can be described using a variety of terms: "contingent," "non-tenure track," "lecturer," "graduate student," "visiting," "postdoc," "instructor," or "adjunct" (American Association of University Professors, 2018).[2] In 2016, 73% of faculty held positions primarily focused on teaching and were not protected by tenure or tenure-track status (American Association of University Professors, 2018). Although some teaching faculty may have institutional support, some degree of autonomy, and the institution's assurance of ongoing employment, the working conditions of most teaching faculty do not reflect the realities of this small number of individuals who enjoy these privileged positions.

A U.S. national survey conducted from 2016 to 2017 by Melonçon et al. (2020a) contains further information about teaching faculty working conditions. Of 313 contingent faculty (not including GSIs) teaching FYW and introductory TPC, 46% reported earning less than $40,000 per year, often doing so while teaching 4/4 (or greater) loads of writing-intensive courses (Mechenbier et al., 2020). Additionally, the majority (64%) were serving on 1-year contracts, despite 67% reporting being employed at their current institution for more than 4 years (Mechenbier et al., 2020). Additionally, 49% of the survey respondents shared or did not have access to a work computer in 2016–2017, likely requiring such faculty to buy and support their own technology for workplace tasks (Mechenbier et al., 2020).

GSIs in writing programs are in similar precarious positions, with the additional challenge of their often much-lower salary, with a median pay of $11,000–$16,800 (Writing Program Administrators Graduate Organization, 2019). Such unlivable wages are perceived to be justified through covering graduate students' tuition costs and keeping their course loads low, one to two writing courses per semester, often introductory courses in composition or TPC. However, most graduate students in writing programs work over 25 hours per week on instructional duties, greater than the contracted hours (Writing Program Administrators Graduate Organization, 2019). Like teaching faculty, graduate students may face repercussions from negative comments on student evaluations, which can contribute to the marginalization of international graduate students and graduate students of color (Banville et al., 2021). In English departments like ours, graduate students may be pursuing research and teaching specializations in literature or rhetoric and composition, but they choose to or perceive market demands to teach introductory TPC classes, putting them at further risk for negative comments on student evaluations and increasing their workload by encouraging them to "come up to speed" in the field of technical writing.

Because teaching faculty enjoy working with students, they are vulnerable to being exploited in their service requirements and opportunities; few departments "protect the time" of teaching faculty in the same way they do tenure stream. Institutions and departments benefit from the growing "creep" of teaching faculty members' service obligations, such as serving as advisors to student groups; mentoring undergraduate research projects; and writing endless letters of recommendation for jobs, graduate schools, and internships. Even if service is not required of teaching faculty, their frequent contact with students puts them in a position to sign on to service projects and other forms of invisible labor.

For all their teaching, research, and service tasks, teaching faculty and GSIs navigate intersecting oppressive systems, and their experiences often complicate neoliberal concepts of *agency*. What choices do teaching faculty really have? We echo Lynch-Biniek and Hassel's (2018) statement that "unquestionably, contingency complicates agency as it does every element of teaching" (p. 335). Contingent positions limit the choices of teaching faculty to design and implement innovative pedagogies, and they limit the choices of others in the field, who could benefit from those instructors' expertise. Teaching faculty face limited choices when conflicts arise with students, often for fear of repercussions to their reappointment. As Stenberg (2016) observed, GSIs in particular feel pressure to "perform the self as a 'good investment,' [which] inevitably narrows choices about self-representation in (and beyond) the classroom" (p. 191).

Despite these complications, Melonçon et al.'s (2020a) survey showed that most contingent faculty have some satisfaction with their job, mostly related to their passion for teaching and the value that their teaching adds to the lives and career trajectories of their students. If that teaching expertise is valued by departments, teaching faculty can feel more invested in their work, a finding supported by the data from Wilson et al. (2020): "When instructors had control over their syllabi, textbook adoption, and assignments, there was an increase in job satisfaction" (p. 98). Similarly, Cox's (2018) study showed that even when writing programs shifted from primarily part-time to full-time instructors, these positions "were simply not enough to neutralize the frustrations of living daily the dissonance between how they identified as scholar-teachers and how the institution seemed to perceive them" (p. A12), in part due to the constraints of the prepackaged curriculum and the devaluing of their teaching experience. That is, improved working conditions in one area do not necessarily make teaching faculty more agentive in their work or more satisfied with their positions. A design justice approach considers both factors as relevant to the problem scoping process for designing introductory technical writing courses.

Who Teaches Usability in Introductory Technical Writing Courses?

Of the 224 faculty (not including GSIs) who answered Melonçon et al.'s (2020a) survey question about highest degree obtained, most had obtained an MA in English (49%) or a PhD in English (15%), indicating that most teaching faculty did not have significant graduate coursework in TPC (Mechenbier et al., 2020). Although 76% had taken at least one course on how to teach, few (12%) had taken a course on teaching TPC. As a field, TPC has not adequately addressed the question of what qualifications are needed to teach the introductory service course in technical writing, likely because there would be so few instructors left to meet those minimum qualifications. In particular, faculty trained in literature or writing studies are likely to be less familiar with user experience design, as these concepts are further afield from their disciplinary knowledge than other concepts often covered in the service course. As Chong (2016) noted, new instructors of usability and UX may not easily find many resources on effective methods, approaches, and tools in the TPC classroom.

Teaching usability and UX well requires significant time and thoughtful processes that challenge both experienced instructors during semesters or terms, even in cases where predesigned curricula provide guidance for overworked and marginalized faculty. For example, Cleary and Flammia (2012) invited users with disabilities to visit the class, to test out website designs, and to familiarize students with adaptive technologies, teaching their students to be advocates for users. These kinds of service-learning and community-based projects can be sites of agency for teaching faculty, as they can build connections between their lived experiences in the classroom and outside of it. Yet such projects also require labor that is not always compensated for or welcomed by faculty with too many students and not enough flexible time, such as building relationships and coordinating information with community partners, arranging for site visits, scoping the design problem, and managing students' stress (and joys) from working with others on complex societal problems.

Viewing TPC course design through the lens of design justice prompts complex questions about user access, support, and opportunity within these programs. Advocating for a user-centered design approach for writing program administration, Pinkert and Moore (2021) argued that programs "ought to consider how and if users (in all their diversity) can easily access, understand, and use the program (and its documents)" (p.

63). The results of an educational usability inspection (Rodrigo & Cahill, 2009) of a course design could reveal how not only students as users might face challenges but also where teaching faculty and GSI users face barriers of knowledge, support, and labor.

We face hard conversations about teaching qualifications with people who are structurally disadvantaged by higher education, with people who have invested in educational training that has not resulted in the career that their graduate school implicitly or explicitly promised them, with people who may or may not be looking to move up or out into less contingent positions, with people who are not interested in retooling their already robust skillset to teach TPC. A design justice framework would consider the impacts of course design on these users, engage and involve them in the community of designers, and consider how we can move forward and disrupt structural injustices—and, along the way, empower them to teach usability, UX, and other concepts in introductory technical writing courses.

Contingency at UTA's Department of English

Because "the quantity and role of contingency looks sharply different at different kinds of institutions" (Childress, 2019, p. 46), we describe below what contingency looks like in the English department at our institution, The University of Texas at Arlington (UTA), an R1 doctoral university. At UTA, the student users of the ITW course tend to reflect a diverse student population; UTA is designated as a Hispanic-Serving Institution and an Asian American Native American Pacific Islander-Serving Institution. It is highly ranked for serving military students and the top school in Texas for awarding the most degrees to African American students. Most students take ITW as part of the prenursing degree and are highly motivated to do well to enter the competitive Bachelor of Science in Nursing program; other students take ITW for degrees in computer sciences, engineering, history, and psychology.

At UTA, GSIs are required to teach a sequence of FYW courses over their first 2 years in the graduate program, and their workload can be seen in Table 11.2. They take required pedagogy courses in FYW that correspond with their teaching duties, and this education, preparation, and support system of the FYW program provides GSIs with a strong network of information and resources. In their third year, GSIs may begin to take on new positions within the department and begin teaching courses outside the FYW program. Though a literary pedagogy graduate course is offered (infrequently) to help GSIs prepare for teaching literature, the elective technical communication pedagogy graduate course has not been offered in several years due to staffing constraints. If and when a GSI is offered a teaching assignment for ITW, we currently offer no required courses, seminars, orientations, etc. as teaching prerequisites. In essence, they are left to their own devices to consult with experienced ITW instructors and/or syllabi published from previous semesters, in addition to studying the required assignments and committee-chosen textbooks available for teaching the course.

The Department of English also relies on teaching faculty to staff ITW, and their working conditions are described in Table 11.2. Teaching faculty can choose from several introductory courses in FYW and literature in addition to technical writing, although most teaching faculty prefer to teach in their specializations instead of ITW. Depending on enrollment, the department also employs adjunct assistant professors, who can teach between one and five courses per semester. Full-time teaching faculty have their own offices and computers, while part-time teaching faculty share offices with one other part-time faculty member and have access to a shared computer. Because of these different working conditions, experiences, and positionalities, inequity in our institution, and in higher education more broadly, is an inescapable aspect of course design.

Contingency in higher education cannot be solved in this chapter or through this project, but the design justice framework enables coalitions to make collaborative decisions that best serve the diversity of users and stakeholders and to create balance in the design process. Mindful that "user involvement alone is not enough to ensure that design is more equitable" (Jones, 2016a, p. 476), ITW course design processes can and should involve faculty in contingent positions and consider how those processes enable or hinder local and global coalitions toward labor justice in higher education.

Table 11.2. Working Conditions of Teaching Faculty and Graduate Student Instructors at UTA

Position	Teaching Load	Service Expectations
Graduate Student Instructor	2/2	No
Lecturer	5/5	No
Senior Lecturer	4/3	Yes
Assistant Professor of Instruction	4/4	Yes
Assistant Professor (tenure-track)	2/2	Yes

Source: Created by the author.

Ultimately, what all this means for student learning conditions is that faculty who are most often teaching the introductory service course in TPC do not always have extensive experience with usability, UX, or even technical writing, and, as such, they are often excluded from TPC scholarship as users whose voices and perspectives matter. At the same time, few know more about what is working in introductory TPC courses better than the faculty who teach it semester after semester. If the field of TPC is to significantly advance usability and UX curricula, it must also involve teaching faculty and other instructors (graduate students, GSIs, peer tutors/undergraduate TAs, others) as users of the course and include their voices in course design, with the goal of creating more equitable conditions for teaching faculty users.

Methods

Design justice employs an iterative methodology, and one of the principles is "Before seeking new design solutions, we look for what is already working at the community level. We honor and uplift traditional, indigenous, and local knowledge and practices" (Design Justice Network, 2018). In addition to acknowledging teaching faculty and GSIs' successes in the course, our methods were designed to honor their informed decisions about curricula and pedagogy. Asset-based inquiry shifts the power dynamic between designers and community members by acknowledging the agency that community members owned over their decisions and the agency they acquire over current and new decisions being made (Durá, 2018). Costanza-Chock (2020) has also reminded us that "wherever there are problems, those most affected have nearly always already developed solutions" (p. 129). Thus, our first stages of design focused on identifying how teaching faculty and GSIs have succeeded at teaching ITW and how they acted upon their previous professional and personal experiences and the key concepts in the course. Although participants' levels of experience with TPC varied, we began with the orientation that all had something to share about teaching TPC and how students came to learn about TPC through the course.

Participants and Data Collection

Criterion-based sampling was employed to recruit research participants. We invited all teaching faculty and GSIs (n=8) of the in-person sections of ITW in the Spring 2022 semester to participate. To ensure that they had agency over their ideas and were involved in the process, participants were invited to participate at any level and stage of the research process, including data collection and data analysis. To ensure that the membership of the community remained fluid, the process was designed to allow for the inclusion of future faculty members and GSIs teaching TPC.

The interview questions in Table 11.3 were informed by asset-based inquiry (Durá, 2018) and narrative inquiry (Jones, 2016a), where we sought to gather narratives about successes and challenges teaching the TPC course (which would, we believed, be often accompanied by solutions developed by the faculty member). Approval from our institution's Institutional Review Board was sought, and this project was deemed to not meet the definition of "research," as the intent is related to outcomes of a programmatic process rather than contributing to generalizable knowledge.

Table 11.3. Interview Questions

When you began teaching at UTA, did you ask to teach technical writing? Was it assigned to you? Did you teach it as a graduate student? How did you feel about this course at that time?
Have you ever been trained in how to teach technical writing? Did your graduate school offer courses or workshops? Are there any other resources you have used to learn more about technical writing?
Do you have professional experience as a technical writer or a writer outside of academia? How do your experiences as a writer impact how you teach technical writing?
Tell me about how technical writing has been going for you this semester. How is it similar to and different from previous semesters?
What do you think students appreciate about the course? Have you heard from students after the course is over? What do they say?
How have you made technical writing "your own"? Is there a unique strategy, approach, or lesson that you think you'd like to share with others who teach this course?
Of the three signature assignments in technical writing, which do students struggle with the most? Why do you think it's challenging to them?
Which textbook do you use for the course? Why did you choose that one? How do students respond to the readings? Do you think the textbook is a helpful resource for the signature assignments?
What would you like to know about the teaching of technical writing that would help you with your work?
What changes would you like to see in the course design of technical writing?

Source: Created by the author.

Amy conducted semi-structured interviews with five volunteer participants. The five participants included GSIs, full-time teaching faculty (lecturers and senior lecturers), and adjunct instructors. All had taught ITW at UTA for at least one semester. The interview data were transcribed, anonymized, and member-checked before being collaboratively analyzed.

Quantitative Interview Analysis

The interview transcripts were uploaded to Voyant-Tools (https://voyant-tools.org) to analyze the corpus of participants' responses for linguistic patterns. Corpus linguistics tools enable researchers to discover "underlying patterns of doing and meaning" (Geisler & Swarts, 2019, p. 7) that may not be apparent in other forms of analysis. Voyant-Tools generated a list of the most frequently used single words (eliminating filler words, such as "like" and "um"). This software also provided the percentage of relative frequency for each term as compared to the entire corpus. See Table 11.4 in the results section.

N-grams are combinations of words that occur frequently in a corpus, and "n" refers to the number of words in the combination, so 2 words are bi-grams, 3 words are tri-grams, etc. (Collins, 2019). Analyzing both single words and n-grams provided a method for examining linguistic patterns in context, such as when the single word "think" frequently appeared in context with "I." Thus, the transcripts were also studied using the Text Analyzer tool from Online-Utility (https://www.online-utility.org/text/analyzer.jsp). This tool created lists of 2-, 3-, 4-, and 5-word sequences (n-grams) that occurred frequently in the corpus. (See Table 11.5 in the results section.)

The combination of these quantitative methods enabled us to consider the meaning behind the repeated terms and compare trends. Given our close relationship to the participants in this study, we believed that it was useful to have both qualitative and quantitative lenses to interpret meaning from the interview transcripts.

Qualitative Interview Analysis

For the first coding cycle, the research team individually developed inductive, descriptive codes during repeated readings of the transcripts (Miles et al., 2014). These codes were 1- or 2-word phrases that described the interviewee's experience with the ITW course or described factors that impacted their course design. These initial codes were discussed, revised, and collaboratively assembled into a draft codebook (Geisler & Swarts, 2019).

Then, the transcript data were segmented into t-units, or "a principle clause[s] and any subordinate clauses or non-clausal structures attached to or embedded in it" (Geisler & Swarts, 2019, p. 73). A convenience sample of 10% of each interviewee's transcript was recoded according to the draft codebook. This sample coding had an interrater reliability of 75% among five researchers, and we discussed any differences until we reached consensus. The remaining data were coded by individual researchers and pattern codes (Miles et al., 2014) were gathered into two main themes, described below in the results.

Results

As shown in Table 11.4, concepts about critical thinking, knowledge, and education occurred most frequently in the corpus. "Think" appeared 123 times and "know" appeared 120 times. "Writing," "students," and "class" appeared 102, 83, and 58 times, respectively. Other top terms addressed the importance of the technical component of technical writing, with "technical" and "tech" appearing 43 and 42 times respectively, and the importance of including well-crafted "assignments" (38), giving students the "time" (36) they "need" (36) to complete said assignments, and concerns about the course "textbook" (35).

As shown in Table 11.5, both 6- and 4-word frequently used phrases highlighted the resume and cover letter as assignments of utmost importance, with the importance of teaching technical writing also placing unusually high. Frequently repeated 5-word key phrases showed more diverse comments than 6- and -word phrases. In addition to concerns about the resume and cover letter and the need to teach technical writing, repeated 5-letter strings included concerns about workplace conditions, including instructors feeling the need to have "a quiz over every [textbook] chapter" and the importance of economic discourse "like raising the minimum wage." We also noted that instructors frequently voiced concerns about their agency, or lack thereof, in choosing what courses to teach when they stated that "they gave me tech writing." Ironically, they also repeatedly brought up that they have agency in curricular design, as "we have the freedom to" and "made it my own and" were also 5-word

Table 11.4. Most Frequent Words Spoken in User Narratives about Introductory Technical Writing Courses

Word	Number of Occurrences in Interview Corpus	Percentage of Frequency Relative to Entire Corpus	Word	Number of Occurrences in Interview Corpus	Percentage of Frequency Relative to Entire Corpus
Think	123	0.7109	Learn	15	0.0867
Know	120	0.6935	Job	15	0.0867
Writing	102	0.5895	Instructor	12	0.0694
Students	83	0.4797	Example	11	0.0636
Class	58	0.3352	Documents	10	0.0578
Technical	42	0.2427	Conversation	10	0.0578
Tech	41	0.2370	Proposals	9	0.0520
Assignments	38	0.2196	Letter	8	0.0462
Time	36	0.2081	Google	8	0.0462
Need	36	0.2081	Program	8	0.0462
Textbook	35	0.2023	Process	8	0.0462
Helpful	29	0.1676	Literature	8	0.0462
Teach	27	0.1560	UTA	8	0.0462
Assignment	25	0.1445	Framework	8	0.0462
Proposal	24	0.1387	Teacher	7	0.0405
Audience	23	0.1329	Games	7	0.0405
Resume	22	0.1271	Fun	7	0.0405
Teaching	22	0.1271	Textbooks	7	0.0405
Book	21	0.1214	Video	6	0.0347
Classes	20	0.1156	Training	6	0.0347
Project	20	0.1156	Professional	6	0.0347
Instructions	18	0.1040	Presentation	6	0.0347
Group	18	0.1040	User	6	0.0347
Design	17	0.0982	Instruction	6	0.0347
Communication	16	0.0925	Pedagogy	6	0.0347
Course	16	0.0925	Freedom	6	0.0347
Online	16	0.0925	Materials	6	0.0347

Source: Created by the author.

Table 11.5. Repeated n-grams in the Interview Corpus

	N-gram, also known as Word Sequence	Occurrences
2-grams	I think	97
	tech writing	31
	technical writing	28
	in class	16
	the textbook	15
	the proposal	13
	the resume	12
	first year	11
3-grams	first-year writing	7
	in technical writing	6
	the assignments and	4
	semester long project	4
	technical writing	4
	of technical writing	4
	I would say	4
4-grams	resume and cover letter	3
	to teach tech writing	3
5-grams	the resume and cover letter	3
	they gave me tech writing	2
	not have an understanding of	2
	a quiz over every chapter	2
	we have the freedom to	2
	need to teach tech writing	2
	online and face to face	2
	like raising the minimum wage	2
	example when it comes to	2
	you need to teach tech	2
	for the resume and cover	2
	made it my own and	2
6-grams	for the resume and cover letter	2
	what we are working with so	2
	you need to teach tech writing	2

Source: Created by the author.

key phrases that had an unusually high frequency. Further, 3-word key phrases highlighted the importance of knowledge transference from FYW to technical writing. The most repeated 2-word phrases highlighted the importance of technical writing as a field, with it being mentioned as "tech writing" 31 times and as "technical writing" 28 times, and with the proposal and resume assignments appearing 13 and 12 times, respectively.

Qualitative results suggested two major themes: *agency* and *support.* When reflecting on the process of teaching ITW the first time, our respondents had their own ways of creating agency within their classrooms. In speaking about how they came to teach ITW, one respondent noted, "I continued refining my curriculum, and . . . someone had created a standard curriculum. I played around with it and made it my own, and eventually ended up with the class that I'm teaching now." This respondent found ways to own the course through gaming examples and creative grading assignments where students "start at level zero and have to work [their] way through the game/class." Theming or framing ITW was also vital to the success in teaching for another respondent, who mentioned designing all assignments aimed at an audience of a "consulting firm looking for student interns to assist them with a variety of medical and professional clients." With the same base for each student to pull from, this interviewee found teaching the course "to run far smoother than any previous sections because the students had a common place to work from and as the instructor, I had the created space from which to teach." When these participants had the agency to create larger frameworks and themes for introductory courses, they found their TPC teaching more meaningful to themselves and students.

For others, teaching ITW was an opportunity to combine previous work to create agency within a course. One respondent was initially both nervous and excited for this new opportunity and found footing after "tak[ing] all these different syllabi from all these other people and put them together [and asked myself]: *so what's going on here?* and *how can I put this together?* and sort of make it my own." Without any sort of official training, many ITW instructors used syllabi as an unofficial course guide. Because syllabi are easily accessible, they are a useful tool to assist any new instructor. Further, speaking with others who had taught the course aided nearly every interview participant. Finally, using search engines like Google and video resources from YouTube, most participants were able to add specific personal touches to their courses. One respondent found helpful information for their students: "LinkedIn has a lot of good resources" and "YouTube helps with resume work." Thus, our coding process for *agency* examined how interviewees perceived their freedom to go outside of the established framework for ITW course design, and to reject aspects of the framework, based on the possibilities they saw in their own abilities, their location within the institutional structure, and their relationship with students.

Throughout the interviews, several important instances of mentorship and/or reliance upon a collaborative network were observed. For at least three instructors, their mentoring was rooted in a past experience as a graduate student and perhaps was not recognized as mentorship until they had to enter the technical writing classroom as an instructor for the first time. It was at this point that their student experience of technical writing became a source of reflection and led to agential formation as classroom facilitator, in some ways facilitating their own experience with technical communication within the curriculum they prepared for their students: "I went back and looked at a lot of the stuff my instructor did that I still have from when I had taken tech writing, stuff I liked about that, what I didn't like about that experience." The perspective and knowledge offered from past experiences with technical communication can become an integral part of how technical writing instructors perceive their available pedagogical choices.

There was a more active form of mentorship present in a few responses, wherein instructors had the opportunity to work one-on-one with a colleague who offered insight and guidance through their own training and experience with technical writing instruction. For instructors with little to no previous experience with this curriculum, such active networking can be noted to alleviate the worries and doubts that ITW teaching faculty and GSIs may have, which helps permit instructors to discover and understand what agency they have in the technical writing classroom. One participant expressed that "it was definitely out of my comfort zone, not something I'd done before, but honestly [they] were so helpful with the materials; [they] were able to show me what [they] used successfully so that was nice. That really helped with the transition." Another instructor's experience within their department reveals a similar outcome: "I would go sit in [their] office and [they] would lend me what I thought I needed to know a lot more about design and all of the advanced computer software that you would use for design." For instructors in contingent positions who may take on heavy teaching loads across multiple institutions while also performing service work for their department(s), the fortune of meeting individually can be difficult to come by, but the benefits to the instructor's agency seem invaluable.

Discussion

Repositioning teaching faculty and GSIs as TPC pedagogy experts (Design Justice Principle 6) uncovered new layers of their agency as teachers. Stenberg (2016) has advocated for the concept of located agency, where GSIs discover and act upon "the possibilities that emerge from the nexus of their subjectivities, locations, and relations" (p. 192). This agency comes with rewards, as GSIs can challenge student perceptions of themselves—as in one example when a GSI in Stenberg (2016) explained the multifaceted identities held as a Korean nonnative speaker of English, a graduate student in literature, and "a visa-holding non-immigrant who can easily be misunderstood" in terms of anti-Asian stereotypes. However, located agency comes with some risk (Stenberg, 2016, p. 199) when GSIs risk negative student evaluations or "[make themselves] vulnerable to retaliation, as well as how to approach the issue without being seen/labelled as complainers" (Banville et al., 2021, p. 8). Teaching faculty and graduate student instructors develop located agency in drawing upon their professional and personal lived experiences to teach TPC, and so changes to ITW curricula directly impact their agency as instructors (Design Justice Principle 2).

For our participants, agency came from creating. It came from learning from those who came before, finding the right set of resources through very little means provided by the department, and drawing on past experience. Each respondent had taught the FYW sequence of courses and spoke to the pedagogical applications in those courses crossing over to ITW well. The heavily stressed ideas about revision and audience also helped new professors in ITW create and teach content successfully. These past experiences, along with minimal informal training, are what currently exists for our respondents. The biggest request they mentioned was speaking or engaging with those who currently teach ITW to ease the strain on both time and training.

These opportunities to do hands-on work with colleagues and engage in the pedagogical questions and curricular development of ITW appear to be limited for most instructors interviewed. The overwhelming trend among the responses was a plea for more avenues to collaboration and mentorship. In some instances, this plea arose from a comparison to the strong foundation and influence the first-year writing program maintains: "There's not much that brings together [ITW] instructors like there is for other programs, so there's not as much opportunity to collaborate." What we found most often in the interviews are ITW instructors being left to mentor themselves, in a sense, as in the participant who was able to create her own agency by individually searching out and synthesizing the materials of instructors with more training and experience: "I just started to take all these different syllabi from all these other people and put them together, asking 'what's going on here' and how can I put this together and sort of make it my own." Although mentorship and networking certainly require time, energy, and intentionality, we find that the collaborative efforts of more established writing programs (e.g., FYW, writing center), whether in their resource sharing or organizational meetings, provide a possible framework to build networks and mentoring relationships across all levels of technical writing instruction. As another participant reflects, "I wish I had that for tech writing sometimes because sometimes I wonder if everything I know works well for first-year writing is actually the best thing I should be doing here, kind of like some basic tenets; I like pedagogy training so some basic pedagogical tenets would be helpful."

FACULTY DEVELOPMENT WITH TEACHING FACULTY AND GRADUATE STUDENT INSTRUCTORS

Participants expressed a need for further training in TPC, as several were simply asked if they wanted to teach ITW and then scheduled into sections with no training. One interview participant noted that "pedagogy training and basic pedagogy tenets would be helpful," while another noted that they needed "more training in design." Several participants noted that they felt lost teaching some TPC concepts and thus tried to apply their training in rhetoric and composition to the teaching of ITW. Although this transfer of skills was viewed by participants as encouraging their agency as teachers, we acknowledge that this lack of training opened them up to the risk of reduced student success rates and faculty performance.

Although some teaching faculty and GSIs may want "more training in design," such training should not be presented as an expert facilitator bringing them up to par. Instead, training of this kind must demonstrate how their lived experience has equipped them with expertise (Design Justice Principle 6) that has contributed to their successes in ITW. Rather than manufacturing an army of duplicate, teaching faculty, and GSIs who act, think, and teach like the program administrators, we must instead live out the design justice principles, equipping through partnership and co-creation to create "sustainable, community-led and -controlled outcomes" (Design Justice Principle 8).

Thus, we discussed the implementation of a robust training program in which faculty can learn or update their teaching skills based on the latest research in education and teaching of technical communication; however, the idea of asking teaching faculty and GSIs to do more labor on their own time also gave us pause. Nagelhout et al. (2015) highlighted the key effects of professional development programs on faculty in contingent positions: both burnout from too much work and possible solutions that would benefit their work. They explained that "the idealism of overwork appears to be the default. And, sadly, we cultivate this image on an almost daily basis through our stories, our practices, and the awards we bestow on the 'best' teachers" (Nagelhout et al., 2015, p. 142). Full-time lecturers teaching five writing-intensive courses per semester are all prone to burnout because we currently lack the structure to work more effectively and efficiently within the conditions that currently exist.

A design justice framework encourages us to consider how this solution challenges or enables structural inequity. Teaching faculty and GSIs "in the trenches" do not have additional time to sacrifice unless it is going to aid them in their time usage, so faculty development programs that emphasize efficient course design have a better chance at faculty retention, student success, and program advancement. Nagelhout et al. (2015) described such a program in which all faculty development meetings focus on three main criteria: "the state of the program (how is everyone doing?), the future of the program (what does everyone need?) and a focused activity (what can we do better?)" (p. 151). Such development offers an opportunity for time management, effective strategy, and learning what still needs improvement. Minimizing labor on grading and preparing future classes by learning through specific, organized meetings solely focused on improving the local working conditions is a promising solution for our program but does not challenge the idea that these working conditions are unjust to begin with.

We also discussed how professional development could acknowledge the inequity in scheduling courses (Wilson et al., 2020, p. 64), where teaching faculty and GSIs often have inhumane and inflexible course schedules. Providing flexible times for adjuncts and others moving between institutions and between work and home, a professional development solution would ideally be an online, asynchronous course or resources hub that faculty could browse through and complete at their own pace. The resource could include lectures, discussions, and opinions from experienced technical communication faculty and researchers, research articles, and tools that instructors can then choose to use in their courses. Resources on student-centered practices like engaged learning and collaborative learning could be a cornerstone of the module, and once faculty members complete the modules, they would be able to go back and revisit it to find resources. Finally, the module could include a resources pool where course faculty could add advice, recommendations, and activities to share with each other. Though this solution would provide further assistance for teaching faculty and GSIs' workload, it too does not change the material conditions of faculty contingency.

The final suggestion is establishing a robust mentor network. Often in conjunction with their comments on the lack of resources, teaching faculty and GSIs commented that only two experienced TPC faculty members were available as resources. If faculty members in contingent positions are to truly succeed, it would behoove the institution to set up support networks among teaching faculty like the ones that are already in place for both tenure-line faculty and graduate students. The creation of the abovementioned online spaces and resource repositories would be a first step in the right direction.

ITW course designers often think primarily about meeting the needs of local users and to design helpful solutions for local teaching faculty and GSIs who experienced both successes and challenges in teaching TPC. However, to create usable course design for introductory TPC courses and the faculty development requested for success teaching that course, these primary solutions all involved further labor from teaching faculty and GSIs. Reflecting upon this crux and design justice, we were reminded of Costanza-Chock's (2020) argument that "there are many cases where a design justice analysis asks us not to make systems more inclusive, but to refuse to design them at all" (p. 19). That is, designing local TPC support systems is often a choice to support not only faculty but also the systems of inequity that keep some faculty in contingent positions.

Thus, though we continue to think through sustainable professional resources for our institution's teaching faculty and GSIs, we recognize that a design justice approach calls on us to disrupt systems that require more and more labor for faculty in contingent positions. Just as many of us utilized our agency in teaching introductory TPC courses, we can utilize our agency as a coalition to advocate for better working conditions. Melonçon et al. (2020c) called for more scholarship that addresses contingency in TPC: "To improve our situation means we have to rely on local actions and share in more specific ways how those local actions can then impact national conversations" (p. 129). To this we add further questions: How does our TPC course design reproduce existing inequities that disproportionately impact teaching faculty and GSIs? How does it challenge existing inequities that disproportionately impact teaching faculty and GSIs?

We agree with the perspective that "TPC has remained silent on issues of contingent faculty is one of the great ironies of the last several years" (Melonçon, 2017, p. 270), and we hope that conversations with TPC teaching faculty, GSIs, and other users of introductory technical writing course design continue to flourish. We are thankful for all of those working in coalitions to solve course problems and ensure better working conditions for faculty in contingent positions. Employing design justice for introductory technical writing course design has helped us take our collaborative initial steps toward local and global solutions to contingency in higher education.

Notes

1. The authors are listed in alphabetical order by last name, reflecting their equitable contributions to the authorship of this chapter.

2. The American Association of University Professors uses the term "contingent faculty," and the survey by Melonçon et al. (2020a) uses it as well, although noting that some of their interviewees dislike the term: "Mainly, this was because there are so many types of contingent faculty [. . .] and identifying each in turn throughout the articles would weigh down the point of this research: that all faculty off the tenure-track have a story about how their material work life is affected by their contingency" (Mechenbier et al., 2020b, p. 4). In this chapter, we primarily use "contingent" to refer to positions instead of people, while acknowledging that the material conditions of contingency shape faculty identities and positionalities. We tend to use teaching faculty as a catchall term that unfortunately elides the important distinctions between institutions' investment in part-time labor, full-time labor without possibilities for extended contracts, merit, and promotion, and full-time labor with possibilities for extended contracts, merit, and promotion. We also would like to highlight the contingent positions that graduate students often occupy, particularly at institutions like ours where graduate students may

work at local community colleges or be hired as adjuncts for the same courses that they also taught as GSIs. Thus, we primarily use the terms teaching faculty and GSIs to describe the introductory technical writing course user population in this chapter.

References

American Association of University Professors. (2018, October 11). *Data snapshot: Contingent faculty in US higher ed.* https://www.aaup.org/news/data-snapshot-contingent-faculty-us-higher-ed

Banville, M., Das, M., Davis, K., Durazzi, A., Dsouza, E., Gresbrink, E., Kalodner-Martin, E., & Stambler, D. M. (2021). Identity, agency, and precarity: Considerations of graduate students in technical communication. *Programmatic Perspectives, 12*(2), 5–15.

Childress, H. (2019). *The adjunct underclass: How America's colleges betrayed their faculty, their students, and their mission.* University of Chicago Press.

Chong, F. (2016). The pedagogy of usability: An analysis of technical communication textbooks, anthologies, and course syllabi and descriptions. *Technical Communication Quarterly, 25*(1), 12–28. https://doi.org/10.1080/10572252.2016.1113073

Cleary, Y., & Flammia, M. (2012). Preparing technical communication students to function as user advocates in a self-service society. *Journal of Technical Writing and Communication, 42*(3), 305–322. https://doi.org/10.2190%2FTW.42.3.g

Collins, L. C. (2019). *Corpus linguistics for online communication: A guide for research.* Routledge.

Costanza-Chock, S. (2020). *Design justice: Community-led practices to build the world we need.* The MIT Press.

Cox, A. (2018). Collaboration and resistance: Academic freedom and non-tenured labor. *FORUM: Issues about Part-time and Contingent Faculty, 22*(1), A4–A13.

Crane, K., & Cargile-Cook, K. (2022). *User experience as innovative academic practice.* WAC Clearinghouse. https://doi.org/10.37514/TPC-B.2022.1367

Design Justice Network. (2018). *Design justice network principles.* https://design-justice.org/read-the-principles

Durá, L. (2018). Expanding interventional and solution spaces: How asset-based inquiry can support advocacy in technical communication. In G. Y. Agboka & N. Matveeva (Eds.), *Citizenship and advocacy in technical communication: Scholarly and pedagogical perspectives* (pp. 23–39). Routledge.

Geisler, C., & Swarts, J. (2019). *Coding streams of language: Techniques for the systemic coding of text, talk, and other verbal data.* WAC Clearinghouse. https://doi.org/10.37514/PRA-B.2019.0230

Getto, G. (2014). Designing for engagement: Intercultural communication and/as participatory design. *Journal of Rhetoric, Professional Communication, and Globalization, 5*(1), 44–66.

Jones, N. N. (2016a). Narrative inquiry in human-centered design: Examining silence and voice to promote social justice in design scenarios. *Journal of Technical Writing and Communication, 46*(4), 471–492. https://doi.org/10.1177%2F0047281616653489

Jones, N. N. (2016b). The technical communicator as advocate: Integrating a social justice approach in technical communication. *Journal of Technical Writing and Communication, 46*(3), 342–361. https://doi.org/10.1177%2F0047281616639472

Lynch-Biniek, A., & Hassel, H. (2018). Col(labor)ation: Academic freedom, working conditions, and the teaching of College English. *Teaching English in the Two-Year College, 45*(4), 333–337.

Melonçon, L. (2017). Contingent faculty, online writing instruction, and professional development in technical and professional communication. *Technical Communication Quarterly, 26*(3), 256–272. https://doi.org/10.1080/10572252.2017.1339489

Melonçon, L., Mechenbier, M. X., & Wilson, L. (Eds.). (2020a). A national snapshot of the material working conditions of contingent faculty in composition and technical and professional communication [Special issue]. *Academic Labor: Research and Artistry, 4*(3).

Melonçon, L., Mechenbier, M. X., & Wilson, L. (2020b). Introduction to "A national snapshot of the material working conditions of contingent faculty in composition and technical and professional communication." *Academic Labor: Research and Artistry, 4*(3), 1–21.

Melonçon, L., Mechenbier, M. X., & Wilson, L. (2020c). Looking forward: Considering the next steps for contingent labor material work conditions. *Academic Labor: Research and Artistry, 4*(3), 127–151.

Mechenbier, M. X., Wilson, L., & Melonçon, L. (2020). Results and findings from the survey. *Academic Labor: Research and Artistry, 4*(3), 22–38.

Miles, M. B., Huberman, A. M., & Saldaña, J. (2014). *Qualitative data analysis: A methods sourcebook* (3rd ed.). Sage.

Nagelhout, E., Tillery, D., & Staggers, J. (2015). Working conditions, austerity, and faculty development in technical writing programs. In D. Tillery & E. Nagelhout (Eds.), *The new normal: Pressures on technical communications programs in the age of austerity* (pp. 139–156). Routledge.

Pinkert, L. A., & Moore, K. R. (2021). Programmatic mapping as a problem-solving tool for WPAs. *Writing Program Administration, 44*(2), 58–79.

Rodrigo, R., & Cahill, L. (2009). Educational usability in online writing courses. In S. Miller-Cochran & R. L. Rodrigo (Eds.), *Rhetorically rethinking usability: Theories, practices, and methodologies* (pp. 105–133). Hampton Press.

Rose, E. J. (2016). Design as advocacy: Using a human-centered approach to investigate the needs of vulnerable populations. *Journal of Technical Writing and Communication, 46*(4), 427–445. https://doi.org/10.1177%2F0047 281616653494

Secules, S., McCall, C., Mejia, J. A., Beebe, C., Masters, A. S., Sánchez-Peña, M. L., & Svyantek, M. (2020). Positionality practices and dimensions of impact on equity research: A collaborative inquiry and call to the community. *Journal of Engineering Education, 110,* 19–43. https://doi.org/10.1002/jee.20377

Small, N. (2021). Localize, adapt, reflect: A review of recent research in transnational and intercultural TPC. In J. Schreiber & L. Melonçon (Eds.), *Assembling critical components: A framework for sustaining technical and professional communication* (pp. 269–295). WAC Clearinghouse. https://doi.org/10.37514/TPC-B.2022.1381.2.10

Stenberg, S. (2016). Beyond marketability: Locating teacher agency in the neoliberal classroom. In N. Welch & T. Scott (Eds.), *Composition in the age of austerity* (pp. 119–204). Utah State University Press.

Thinyane, M., Bhat, K., Goldkind, L., & Cannanure, V. K. (2020). The messy complexities of democratic engagement and empowerment in participatory design—an illustrative case with a community-based organization. *CoDesign, 16*(1), 29–44. https://doi.org/10.1080/15710882.2020.1722175

Walton, R., Moore, K. M., & Jones, N. N. (2019). *Technical communication after the social justice turn.* Routledge. https://doi.org/10.4324/9780429198748

Wilson, L., Mechenbier, M. X., & Melonçon, L. (2020). Affective investment. *Academic Labor: Research and Artistry, 4*(3), 83–107.

Writing Program Administrators Graduate Organization. (2019). Report on graduate student instructor labor conditions in writing programs. https://csal.colostate.edu/docs/cwpa/reports/wpago-gsi-2019.pdf

Chapter 12

Institutional Transformation

A Critical Analysis Pairing
UX Methods and Institutional Critique

Emma J. Harris, Ruby Mendoza, and Emily L. W. Bowers

Because User Experience Design (UXD) methods are rooted in designing around a specific user's or individual's needs, they are especially situated to support transformational design, "a practice in which an object, system, or relationship is made with open, authentic, and accountable dialog that invokes intention, takes stock of emotional engagement, and the residual consequences of its creation" (Maketrybe, 2018). When speaking specifically to circumstances surrounding the academy (or higher educational institutions and related research organizations and practices), Ohito (2021) noted, "Death can be natural or manufactured, the latter insofar as it can be hastened by circumstances." Ohito (2021) recognized the idea that oppression, "death," is something that can be designed, "manufactured." We propose the opposite is also true, or that people can work to design circumstances, experiences, and institutions that strive to avoid oppression for specific user groups or other participating individuals.

As made evident through the examination of literature, technologies have material, social, and cultural impact on marginalized communities. For instance, Haas (2012) considered the interlocking nature of technology and race, Jones and Williams (2018) reported that technologies have worked toward disenfranchising Black voters, and Costanza-Chock (2020) amplified how UXD practitioners need to be mindful about how participants (marginalized communities) are excluded from the design process.

These critical perspectives inform and motivate our examination of Stacey Abrams and her voting rights advocacy teams' work to showcase how rhetorical and strategic design can create institutional change. As the examination will indicate, Abrams and her teams' campaigns provide accessible and tangible support to those who do not have access to resources that people need to participate in institutional change and transformation. We classify this work as user-centered design. Furthermore, we propose that this type of grassroots civic engagement has been done in a way that pairs user experience (UX) methods with institutional critique and that researchers in other disciplines and circumstances may benefit from practicing UX methods in partnership with institutional critique to employ an imperative approach in developing critical and life-affirming resources for marginalized groups. In the case of Stacey Abrams's work, these resources have taken the form of, for example, systemic improvements to voter registration accessibility and other civic engagement centered around Black American communities in Georgia.

As an example of using UXD and research methods to pursue transformative institutional change, one key component of Abrams, her campaign, and her nonprofit teams' success is their critical attention to interpersonal, structural, and institutional dynamics of oppression as well as opportunities for action. Bringing institutional critique together with UX methods for research and design can support these methods in more effectively pursuing social justice and transformative institutional change because of the essential context that historic and embodied knowledge of systemic oppression brings to UX and technical and professional communication (TPC) design and research. By building this argument, we hope to both encourage and empower professional communicators and UX designers and researchers to "consider more opportunities where they can be change agents and co-conspirators (Love et al., 2019) to support people with disempowered, marginalized identities within institutional process design" (Bowers et al., 2022, pp. 456–457).

This chapter showcases how, when paired with institutional critique, UXD can work within institutional spaces for transformative change within both the digital and physical realm. This chapter first presents a methodological framework for combining UX design and research methods with institutional critique informed by the matrix of domination (Collins, 1990;

Mendoza, 2023; Morgensen, 2011) to pursue institutional change. Next, the chapter analyzes modes of participation and voting rights advocacy throughout Stacey Abrams's political career as an example of how UXD can be used to pursue institutional change through cohesive work at interpersonal, structural, and institutional levels. Finally, the chapter presents implications of enacting UXD toward social justice through institutional and organizational change, including considering this work in connection with curriculum and pedagogy, to demonstrate that UX and TPC work that pursues social justice and institutional change must directly involve "affordances and dysaffordances (Costanza-Chock, 2020) of the most oppressed people within the design process, especially as UX and TPC practitioners are responsible to those who are impacted by their designs" (Bowers et al., 2022, p. 458).

Literature Review

Existing scholarship has addressed the need to embrace social justice and inclusivity (Agboka, 2014; Jones et al., 2016; Walton et al., 2019). UX and TPC have been connected since the 1970s through their focus on user advocacy (Jones et al., 2016; Redish, 2010; Redish & Barnum, 2011; Sullivan, 1989). TPC has taken a significant and important social justice turn (Walton et al., 2019), and we build from this existing knowledge to illustrate how UX and TPC can create institutional change for the betterment of communities. Before expanding on these notions of institutional change, we examine three critical areas that amplify the exigency of the project, including race and technology, UXD and TPC for institutional change, and design justice (Costanza-Chock, 2020). These three areas inform how we consider institutional oppression and transformational design work in three categorical areas: interpersonal, structural, and institutional.

RACE AND TECHNOLOGY

Within the last decade, conversations on race and technology have increased in TPC (Benjamin, 2019; Haas, 2012; Williams, 2017; Williams & Pimentel, 2016). For instance, Haas has discussed that technology and race are not mutually exclusive and that "race affects the ways in which technologies and documents are designed and used, how national and political values can inspire users to transform the work of technologies beyond their designed intent" (p. 281). From this notion, technology is "not neutral or objective—nor are the ways that we use them" (Haas, p. 288). Haas's discussion on the interlocking nature of technology and race exemplifies that the design of technology is intentional, rhetorical, and colonial. From this standpoint, TPC scholars and practitioners are reminded that they have an ethical responsibility to understand race, rhetoric, and technologies and "how these assemblages are constructed and represented in technical communication theories, methodologies, pedagogies, and practices" (Haas, p. 292). This work on addressing technology and race and their impact on institutional, structural, and interpersonal levels of action against marginalized communities has been addressed; for instance, Jones and Williams's (2018) work examined the historical disempowerment on voter literacy technologies, arguing "that texts and technologies are not always designed with goodwill in mind and texts and technologies that are complicit in supporting and promoting oppressive practices have social, cultural, embodied, and material impacts on communities" (p. 371). The material impacts of technology and race indicate a dire need to critically examine voting technologies. From this standpoint, UXD practitioners have the responsibility to interrogate technologies to create institutional change for the betterment of communities.

User Experience Design and Technical and Professional Communication toward Institutional Change

UXD can work within institutional spaces for transformative change in both the digital and physical realm. For example, Skarlatidou et al. (2019) introduced UXD principles and their effects on citizen science through a discussion of design standards, methods, and participant experiences. Also, Hillmann (2021) used a case study of extended reality to detail a UXD process and its implications for "balancing powerful opportunities with new responsibilities" for digital interaction such as a virtual-reality experience of digital art collections and nonfungible token information (p. 201). Lewthwaite et al. (2018) shared a manifesto for accessible digital UX and argued for foundational principles of accessibility that center disabled people at the core of designing accessible experiences. They emphasized education, the complexity of changing organizational processes, and commitments to action based on principles of social justice. Although this manifesto focused on digital accessibility, Lewthwaite et al. (2018) offered valuable insight on designing for institutional change by suggesting that

UX designers critically consider their position as designers through the lens of the community.

DESIGN JUSTICE

Design processes can have extensive impacts on participants' lives, especially individuals who cannot access or participate in the design process. Excluding the people most impacted by decisions in the design process is a problem that motivated the creation of the Design Justice Network principles, which are an ever-adapting set of principles that ground transformational design by "rethink[ing] design processes, center[ing] people who are normally marginalized by design, and us[ing] collaborative, creative practices to address the deepest challenges our communities face" (Costanza-Chock, 2020, p. 6). We use the Design Justice Network principles as the heuristic for determining the success of UXD because the principles are explicitly focused on transformative design (Costanza-Chock, 2020).

Haas's (2012), Jones et al.'s (2016), and Costanza-Chock's (2020) acknowledgments motivate the exigency of this project, especially considering that "the most disempowered populations are often intentionally (or unintentionally but detrimentally) excluded from developing, creating, and implementing design" (Bowers et al., 2022, p. 456). UXD approaches paired with critical consideration of the institutional context in which UX and TPC scholars and practitioners act are instrumental in developing effective products and services for participants. Using institutional critique as our methodological framework for analyzing Abrams and her teams' work for voting rights reform, UX principles can address inequities and pain points (Gibbons, 2021) in all forms of systematic design, including interpersonal, structural, and institutional (Bowers et al., 2022).

Methodological Framework: Institutional Critique Toward Design Justice in UX and TPC

In rhetoric and writing, institutional critique is an essential methodology. Institutional critique is a means of making visible the practices, structures, and context that are often hidden but nevertheless predicate and motivate particular entities.

In this chapter, we use institutional critique to make visible interpersonal, structural, and institutional contexts for inequity and action toward transformative design through the example of Stacey Abrams's campaign and her teams' work for voting rights. We draw connections between the research, engagement, and design methods they use and that the UXD and TPC fields use. We apply the Design Justice Network principles (Costanza-Chock, 2020, p. 6), which apply to UXD and TPC through attention to user advocacy (Jones et al., 2016).

User advocacy is a central component of TPC and UX (Friess, 2010; Jones et al., 2016). However, deciding which users' perspectives will influence design, research, and communication is challenging. Jones et al. (2016) wrote that "many inclusive scholars intentionally shift power toward users, particularly those who are marginalized, creating space for marginalized users' expertise to be recognized as legitimate (e.g., Agboka, 2014; Durá et al., 2013; Johnson, 1998; Jones, 2016; Mukavetz, 2014; Price, Walton, & Petersen, 2014; Walton, 2016)" (p. 218). Shifting power toward users is a key component of design justice (Costanza-Chock, 2020) and often involves connecting with and uplifting users' knowledge of practices, structures, and context that come from their lived experience. Institutional critique is one way to make these practices, structures, and context visible in a way that UX and TPC practitioners can apply to their work.

Combining institutional critique, UXD, and TPC can be key to advocating for users inclusively through design. An important part of the work that Abrams and her advocacy teams do is making visible and then acting from knowledge of these historic and current structures of oppression and opportunities for transformative design. Working at these multiple levels of oppression and action, and being aware of the present and historic context of the communities that UX and TPC practitioners are working within are both key to transformative justice work in the field.

When considering how institutional critique works to support design justice and transformative UXD, we reference Mendoza's (2023) argument: Considering the matrix of domination (Collins, 1990) alongside institutional critique is critical to enacting that critique with intersectional complexity. Institutional structures are inherently tied to colonizing and Eurocentric ways (Morgensen, 2011), which must be considered to use the methodology of institutional critique in a way that pursues transformational change: "Institutional critique is an inherently decolonizing approach that embodies intersectional feminism (Collins, 1990), queer approaches (Muñoz, 1999), disability justice (Berne et al., 2018; Mingus, 2017; Schalk, 2017), and trans*/formative approaches (Patterson & Spencer, 2020) that can provide rhetorical agency to those who are continuously and negatively impacted by violent Eurocentric tendencies" (Bowers et al., 2022, p. 457). We practice this intersectional approach to institutional critique (Mendoza, 2023) because of the importance of identity and positionality in TPC and UX (Jones et al., 2016). Because identity markers are "co-constructed and shaped by . . . positionality, privilege, and power" (the 3Ps), we amplify Jones et al. (2016) by acknowledging that "an awareness of how the 3Ps are articulated and inscribed in our work is a necessary step toward increasing the inclusivity of our research and practice" (p. 212) in TPC. Therefore, to frame an analysis of Abrams's Georgia gubernatorial campaign in 2016 as a formative example of how UXD methods can be used toward institutional change, we forward an approach to institutional critique that recognizes its implications in colonizing ways and builds its existing potential to further amplify the need and opportunity to look to current and past social movements as examples for enacting UXD toward social and institutional change.

As stated previously, we do not hold that UXD is either inherently socially just or unbiased. Processes like persona development, user interviewing, surveying, focus groups, and usability, all of which focus on human-centric design, do not remove bias. For UXD to pursue social and institutional change, XD methods can be paired with methodologies that practice transparency and accountability for design decisions. By pairing UXD with institutional critique, practitioners are not designing the universal fit for all user groups or all scenarios but are basing design around specific, institutional shortcomings to improve design for a specific group. In the case of Abrams's campaign and nonprofit teams, pairing institutional critique of the United States' democratic government, specifically the election system within Georgia, with UXD-adjacent methods for institutional transformation demonstrates this focused design.

Scholars have critiqued institutions for their oppressive effects on TC and how institutions are intentional and rhetorical (Moeggenberg et al., 2022; Skinnell, 2019). Moeggenberg et al. (2022) discussed how "power and knowledge are deeply situated in the relationship between technical documents and bodies," particularly for trans people (p. 418). They share a theoretical framework for identifying oppression toward trans people in technical documentation by evaluating those documents' rhetoric through an intersectional lens. Skinnell (2019) argued that rhetoricians need to build more awareness of the ways institutions affect their scholarship, and

Mendoza (2023) and Moeggenberg et al. (2022) have built on this work, advocating to generate institutional critiques and rhetorics through an intersectional framework through personal and sexualized embodiments.

An example of institutional critique in TPC to further institutional transformation, Nash (2019) wrote that the term "intersectionality" has been co-opted by many groups throughout its history. Nash's explanation of how universities conflated the term with "diversity" is one example of institutional critique (pp. 11–12). Her critical attention to ways that institutions are influenced by human design is imperative: "It highlights how particular bodies—White cis-gender heteronormative individuals—hold positions of power which often develop and implement policies that impact intersectional multi-marginalized communities" (Bowers et al., 2022, p. 457).

As scholars such as la paperson (2017) have argued, universities (institutions) represent colonizing places that operate like machines, driven toward colonial desires for the future (Bowers et al., 2022). When practitioners better understand the ecologies or layers of interlocking influence of these machines, they can more effectively work toward UXD that pursues social and institutional change. For example, in *Design Justice*, Costanza-Chock (2020) argued for the connection of designers to principles of social justice through considering the overlapping nature of structures and systems being designed (p. 68).

The data for this study come from media coverage of Abrams's campaign and nonprofit teams (Green, 2019; Klein, 2020; Krieg & McKend, 2022; Traister, 2020) and Abrams's documentary (Garbus & Cortes, 2020), memoir (Abrams, 2021), and campaign and nonprofit websites (Abrams, n.d.; New Georgia Project, 2022a, 2022b). Narrative is a valid source of information for research that has value in TPC and UX (Jones, 2016; Jones & Williams, 2018): "Technical communicators can look for possibilities and opportunities for resisting discrimination through texts" (Jones & Williams, p. 371).

Guiding Example Analysis: Stacey Abrams's Gubernatorial Campaign and Voting Rights Reform Work

Abrams's 2016 gubernatorial campaign and related community work to reform voting rights in Georgia effectively implements UXD principles with institutional critique working toward social justice and institutional change. Though Abrams and her teams do not explicitly define their work as UXD, their work and methods overlap with those of UX and TPC.

In her memoir, Abrams (2021) has detailed her family history with voting rights in the broader context of the U.S. Constitution to describe her personal motivation for pursuing voting rights in her work. About the foundation of U.S. democracy, she said, "Let me be clear here: the codification of racism and disenfranchisement is a feature of our lawmaking—not an oversight. And the original sin of the U.S. Constitution began by identifying Blacks in America as three-fifths human: counting Black bodies as property and their souls as non-existent" (p. 28). This codification began with the Constitution but has continued in many iterations throughout America's existence. One key mechanism of power within a democracy is the right to vote. This gives people power—the "right to be seen, the right to be heard, the right to direct the course of history and benefit from the future" (pp. 28–29), and voter discrimination has historically and presently targeted marginalized and disempowered people (Jones & Williams, 2018). Abrams brings this personal and historical knowledge into the work she does with her campaigns and reform work (Abrams, n.d., 2021; Klein, 2020).

Abrams and her teams envision a democratic United States in which people can advocate for their goals and needs, and everyone has equitable access. As it stands, the United States has many barriers to this participation, even in basic elections; Abrams noted, "When entire communities become convinced that the process [of civic participation] is not for them, we lose their participation in our nation's future, and that's dangerous to everyone" (Garbus & Cortes, 2020, 0:06:28).

Abrams's and her teams' reform work serve as an example of using UX and rhetorical strategies toward social justice. In particular, their work aligns with Jones and Walton's (2018) specific attention to disadvantaged communities: "Social justice research in technical communication investigates how communication broadly defined can amplify the agency of oppressed people—those who are materially, socially, politically, and/or economically under-resourced. Key to this definition is a collaborative, respectful approach that moves past description and exploration of social justice issues to taking action to redress inequities" (p. 242). As Bowers and colleagues (2022) mentioned in their work, "Abrams and her team worked to increase accessibility to information and participation, solicit participation and support of the most marginalized and under-resourced communities in Georgia, and build the capacity of her constituents through education and voter registration" (p. 458). Abrams and her advocacy teams have sustained these actions by forming a coalition (Prasad, 2021). Abrams's and her teams' work includes action at multiple levels, specifically interpersonal, structural, and institutional categories of action.

Interpersonal

We begin with interpersonal because interpersonal dynamics influence every other dimension of systematic oppression and ground institutional change. More specifically, Abrams enacts her knowledge of the importance of interpersonal connection and communication through direct engagement with community members in her campaign and nonprofit work. For Abrams and her teams to create institutional change, they know that they must engage on a personal level with the communities they hope to transform. As Sasha Costanza-Chock (2020) shared in a discussion about how the matrix of domination (Collins, 1990) applies to design justice: "Black feminist thought emphasizes the value of situated knowledge over universalist knowledge. In other words, particular insights about the nature of power, oppression, and resistance come from those who occupy subjugated standpoints" (p. 22). Abrams herself has been an actively engaged Georgia resident both personally and professionally for most of her life. For instance, she publicly discusses her childhood experience moving to Georgia; there, she received most of her education and had a memorable experience of the effects of systemic racism when she was denied entrance to the banquet for valedictorians at the governor's mansion (Garbus & Cortes, 2020). Professionally, she served as a minority leader in the Georgia state legislature, seven years before running for governor in 2016. From her experience as an active member of the Georgia community, Abrams believes that being situated in the community she is working with professionally is vital (Bowers et al., 2022) "because we have to care about people, and people's lives aren't just politics" (Traister, 2020).

Abrams extends this valuing of situated knowledge (Costanza-Chock, 2020) to the hiring of her campaign and nonprofit teams. Abrams's campaign and nonprofit teams specifically call for leaders and volunteers in their organizations to be members of the communities they are working with. The New Georgia Project, for example, has "previous experience in organizing volunteers towards a common goal, especially in the state of Georgia" as a preferred qualification for a volunteer organizer position (New Georgia Project, 2022b).

Localized knowledge is irreplaceable (Costanza-Chock, 2020; Jones et al., 2016), and Abrams and her advocacy teams recognize this in the approach to participant engagement that they enact: "We talked to people. We met them at their doors. We met them at their churches, their mosques, their temples, their shrines, their synagogues. I went to DragonCon and one music fest. I also talked about real issues. I talked about my brother and his fight with drug addiction and incarceration. But I also had solutions, and I talked about the things that people care about every single day" (Garbus & Cortes, 2020, 1:22:03).

Involving people who directly experience life in the communities we are working with as UXD and TPC scholars and practitioners—both the effects of injustices and the assets of and relationships within communities—is both an ethical approach to doing work that involves particular communities and a means of recognizing the value of lived experience as expertise. People know themselves and their communities best. This is why Abrams, herself a longtime Georgia resident, feels motivated to work toward voting rights advocacy in Georgia (Traister, 2020) and why the nonprofits she has founded seek to hire people or find volunteers who are local to the areas where they will be working (New Georgia Project, 2022b). As Abrams clearly stated in her interview with Traister (2020), "I understand Georgia, but I only understand Georgia because I worked with Georgians who were here before I got here and who will be here after I'm dead. There is absolutely a necessity to build in place."

We can recognize the value of localized knowledge in UXD by using our resources to meet participants where they are, give them multiple avenues to participate in different spaces and contexts, and create transparency and access to the people who can make changes. One way that transparency and access can be built is from equitable education.

The New Georgia Project, a nonpartisan organization founded by Abrams that is focused on registering and engaging voters across Georgia, was created with the goal of empowering the most marginalized lives through interpersonal action. When building the New Georgia Project, Abrams and her advocacy teams centered their efforts on institutional transformation by wielding their resources—the material, social, economic, and emotional support—to create accessible and engaging voter registration and education opportunities (Bowers et al., 2022). Abrams framed her motivation for this approach by saying, "My belief is [voter registration] has to be registration plus education. [With just registration,] it's the equivalent of giving someone keys to a car but never teaching them how to drive" (Traister, 2020). Abrams and her team worked in coalition to create education and voter registration workshops that were informed by their historical, contemporary, and embodied knowledge of ways people are informed about and are barred from voting (Traister, 2020). The goal of these workshops is to build community capacity toward more sustainable engagement with these issues over time, particularly for individuals and communities Abrams and her advocacy teams are working to support (Bowers et al., 2022). This approach directly connects with Design Justice Principle 7: "We share design knowledge and tools with our communities" (Costanza-Chock, 2020, p. 7). Although not explicitly *design* knowledge and tools in all instances, the education and registration workshops provide insights that community members and Abrams's teams can use to better inform the design of their campaigns and advocacy work and can lead to a more interconnected, coalitional approach. Coalition, as Prasad (2021) described in connection to Chávez's (2013) work, "is something more than merely strategic alliances towards a shared political goal; it also offers "differently positioned groups" to "imagine the conditions of their politics" differently (p. 81). Chávez, in this way, does not think of coalitions as fixed entities—she prefers instead the concepts of coalitional possibilities or moments, which capture the idea of coalition as an enduring practice with material implications rather than a series of neat intersections and initiatives." An example of Abrams and her teams' coalitional work exists in their continued advocacy beyond the context of Abrams's 2016 gubernatorial campaign. Two specific examples of bringing "differently positioned groups" together to create an "enduring practice with material implications" that Abrams and her campaign and nonprofit teams participate in are hiring to eliminate potential barriers to participation (e.g., having American Sign Language [ASL] interpretation at every event; Krieg & McKend, 2022) and committing to education as a part of their campaign and nonprofit work (Traister, 2020).

Coalition is relational; it comes not only from responding to events after they have happened but also from being involved in a collective support system that is committed to acting consistently to pursue more "livable li[ves]" (Chávez, 2013, p. 147) for those involved and even beyond these groups (Prasad, 2021). This is why practices like hiring and building community capacity (Simpson et al., 2003) through education, connections that can endure past moments or mere initiatives, showcase Abrams's and her voting rights advocacy teams' interpersonal and relational work in connection to their UXD and TPC work.

Their cause—voting rights—also speaks to a need for a coalitional approach. Voting in and of itself can be a life-affirming practice because of its connection to elements of Americans' daily lives. As Abrams told Klein (2020), "When you have a robust democracy that is fully engaged and that is fully accessible to those who are eligible, what you then see are actual changes in the outcomes of lives." This type of support builds the capacity (Simpson et al., 2003) of the communities involved with the campaign and nonprofit organizations.

Even in their public communication about their work (e.g., Abrams's documentary *All In: The Fight for Democracy* [Garbus & Cortes, 2020]), Abrams and her advocacy teams share practical steps that people can take in the process of voting—creating a sort of journey map for the possible advantages and pitfalls to spread awareness of our rights (1:34:16). This is one example that overlaps with methods we find in UXD (i.e., journey mapping), and this short, powerful list of steps can make a big difference in what people understand their voting rights and process to be.

Whether through being an involved member of the community, consciously engaging in public communication and education, practicing more inclusive hiring and communication practices, or applying other means of interpersonal action, Abrams's example highlights how strengthening interpersonal relationships with community members is a vital component to expanding knowledge about the historical systematic oppression multimarginalized folks implicitly face. By developing interpersonal connections, Abrams and her nonprofit and campaign teams also develop a better understanding of their constituents' goals and needs to create their opportunity to design more responsive products and services for the people they are working to support, which directly aligns with the work of UX and TPC practitioners and scholars (Bowers et al., 2022).

For example, an amendment was passed to expand access to voting across the state of Florida, specifically for people who were formerly incarcerated, through participants' direct action with a ballot initiative (Garbus & Cortes, 2020). Desmond Meade in Abrams's documentary frames this work by asserting that when "politicians did not have the political strength or courage to actually change the policies, then the next best thing was to let the citizens take it into their own hands" (0:28:30). Through the nonprofit organization, the Florida Rights Restoration Coalition, volunteers and organizational leaders enacted change by knocking on doors, making phone calls, and sharing a simple but consistent message that voting is a fundamental right. When the amendment passed with a wide margin at 64%, the coalitional action enacted institutional change through policy (Garbus & Cortes, 2020). Importantly, this action centered people who were impacted by disenfranchisement and who have been disenfranchised as a result of historical discrimination through the criminal justice system. From these examples of personal community support work, Abrams and her advocacy teams illustrate that institutional change begins with interpersonal relationships leading the action.

Structural

Abrams's and her team's work engages action at structural levels of institutional change by educating the public on the ways historical, structural racism was written into and continues in U.S. policies and legislation. Structural oppression has existed and prevailed through particular bodies, identities, and perspectives (predominantly White men) holding the majority of positions of power (Garbus & Cortes, 2020).

As cited in Bowers et al. (2022), Abrams and her team work from the understanding that, "in George Washington's election, only six percent of the people living in the U.S. could vote—white, property-owning men, which indicates that the U.S. historically neglected to include multi-marginalized bodies, identities, and perspectives (Abrams, 2021; Jones & Williams, 2018)" (p. 459). To bring more people to understand the context that Abrams and her campaign and nonprofit teams knew they were working in, they take the opportunity to share the history and context that informs their work in much of their public messaging (Abrams, n.d., 2021; Garbus & Cortes, 2020; New Georgia Project, 2022a; Traister, 2020).

As Abrams shared in her 2021 memoir, voting rights have "had to be purchased by blood and protest in each generation" (p. 28). Abrams continued to describe several iterations of policy and legislation that establish the long history of discrimination within U.S. democracy and human rights. This history motivates her voting rights advocacy work. She established how voting rights in the United States, specifically Georgia, are affected in particular by the U.S. Supreme Court decision in *Shelby County v. Holder* (Abrams, 2021).

The Voting Rights Act's protections were extended and expanded until 2013, when the U.S. Supreme Court decided in *Shelby County v. Holder* that, essentially, "racism had ended in election practices and so too must federal oversight" (Abrams, 2021, p. 39). The *Shelby County v. Holder* decision opened the floodgates for voter suppression to expand, which informed and motivated Abrams and her teams in continuing work. Narrative and qualitative data cited in Abrams's memoir (2021) also illustrated how current government processes are still predominantly developed, created, and implemented by White men.

Because these structural inequities are still present in U.S. democracy, Abrams has spent much of her career working to reduce voter disenfranchisement and uplift the perspectives and experiences of multimarginalized people who are devalued, ignored, and disempowered (Abrams, n.d.;

Garbus & Cortes, 2020; Walton et al., 2019). When people feel that their voices do not matter structurally, trust-building needs to happen before people may feel comfortable enough to participate, even if the resources to do so are widely available: "There's always a debate on, 'why aren't our communities participating?' The most important part of voter registration is, that human connection, and being able to understand why that person does not trust" (Alejandra Gomez in Garbus & Cortes, 2020, 0:55:01).

Specifically, Abrams is creating positions within her own campaign team to mitigate these historic and present exclusionary practices and work to build trust for participants in the system they're participating in through voting. In an interview with CNN in 2018, Abrams's spokeswoman highlights the implications of Abrams hiring a full-time ASL interpreter to her campaign team, a job traditionally only filled after officials become elected, "as a choice example of [the campaign's] intent to immediately engage more deeply in underserved communities around the state" (Krieg & McKend, 2022). This is an example of ways that structural inequities like ableism extend to the interpersonal.

The structural and interpersonal overlap when, as Abrams (2022) wrote, actions furthering structural inequities like voter suppression are framed as user error. Not having an ASL interpreter at all events makes participation in any political campaign inaccessible for people who are deaf or hard of hearing. This is not the fault of deaf or hard-of-hearing people but of the structure that assumes that people participating are nondisabled, therefore, not designing campaigns from the start to include disabled people. Hiring an ASL interpreter for all campaign events is a first step to remedying this inequity. Using knowledge of structural inequities (like institutional critique can help develop) in design decisions helps Abrams's campaign better pursue inclusion and transformative design.

For UX designers and researchers, paying attention to ways structural inequities are enacted through products and services is imperative for improving user experiences and pursuing transformative institutional change. In Abrams and her nonprofit teams' work, that service happens to be voting. Voting is suppressed even today by a monetary poll tax for formerly incarcerated people (in the form of fines and fees) and by an indirect poll tax for people to take unpaid time off from work and to wait in line for hours to vote, on top of potentially putting their employment at risk for taking time off. Voting is also inhibited when polling places close, no public transportation is available to the other locations, and forms of ID required to vote are made specific and limiting (Klein, 2020).

In a discussion about these examples with Klein (2020), Abrams said, "Those are examples of how bureaucratic rules take on the veneer of logic but have the most heartless effect, because they distract from the responsibility of the state to engage in providing the right to vote. They also convince citizens that it's either too hard, or that they were not worthy enough, and that they didn't work hard enough." In response to this knowledge, Abrams and her campaign team are advocating to eliminate bureaucratic barriers like specific and limiting ID requirements and decreased access to absentee ballots (Abrams, 2021).

We can build the understanding of structural inequities that are connected to the products and services we're researching and designing by learning more about their historical context, thinking about the structures they are connected to that may be hidden, and pursuing change in our approaches to design and research that are more directly engaged with the specific communities we aim to support.

Engagement with the specific communities we are working to support often allows for opportunities to specialize platforms that we are creating to better suit user needs and goals (Skarlatidou et al., 2019). Abrams founded several nonprofits in support of community voting rights based on participants' varied needs and her teams' understanding of the structural context of voter advocacy. After losing the 2018 gubernatorial election, Abrams continued to support promises that she made during her campaign and responded to specific requests from community members (Klein, 2022). Through her response, she and her advocacy teams align with Design Justice Principle 8: "We work towards sustainable, community-led and controlled outcomes" (Costanza-Chock, 2020, p. 7).

This multidirectional approach shows Abrams's and her team's understanding that these structural inequities in voting are interlocking parts and that working toward institutional change means that important work must be done that encompasses all of those parts at once.

Institutional

The interpersonal, structural, and institutional categories of action converge in many ways (Bowers et al., 2022). Institutional action focuses on addressing specific, discriminatory systems that have been normalized as part of public institutions, like U.S. democracy, in this case. "Abrams and her team use participatory design and research informed by historical and embodied knowledge of systemic oppression in our governing institutions to create more accessible and life-affirming opportunities for civic engagement" (Bowers et al., 2022, p. 459). Similarly, UXD and TPC practitioners need to create more accessible engagement practices that uplift and empower particular and often-ignored bodies. People who are marginalized and disempowered will continue to be excluded from civic engagement opportunities if UX and TPC practitioners, researchers, and educators do not acknowledge institutions as a major contributor to systemic oppression.

Working to stop voter suppression is one of the ways that Abrams and her advocacy teams work to create civic engagement opportunities that are more accessible and life-affirming (Bowers et al., 2022). Abrams clearly articulated that voter suppression goes beyond Georgia—it is an institutional issue: "Voter suppression is often targeted at communities that are seen as non-normative, and we saw this come into sharp relief in North Dakota, where a law that required a residential address on drivers' licenses seemed benign. But the reality was that for those communities, getting a residential address on a reservation required that the state or local government grant that to you, and it didn't happen" (Garbus & Cortes, 2020, 1:08:04). By engaging directly with marginalized communities who were experiencing voter suppression through means associated with normalized bureaucratic procedures (i.e., requiring an address on a driver's license), Abrams and her team were able to determine specific institutional barriers to civic participation that could affect people all over the United States. This is just one example of using a method commonly practiced in UX and TPC research—interviewing—to gauge institutional barriers more directly.

Abrams and her voting rights advocacy teams also practiced participatory action research throughout the creation and engagement of these organizations. In 2013, Abrams established the New Georgia Project, an organization that "trained poor folks in South Georgia to be quasi-navigators, deploying them across 39 counties to help folks sign up for the Affordable Care Act" (Traister, 2020). Understanding the institution of voting to have many contributing elements of participation—a complicated or inaccessible registration process being one of them—Abrams also founded Fair Fight to eliminate barriers to voter registration, focusing on the 800,000 Black and Brown Georgians who were not yet registered (Traister, 2020). These programs were community-led and focused on empowerment through education and resources, as well as building from the connections and knowledge participants already had to effectively spread access to healthcare and voting rights to communities across Georgia. By collecting information about pain points (Gibbons, 2021) within the process of voting and civic participation from people across Georgia as part of this programming, as well, Abrams and her team were able to "map" the process and contributing factors to people's disempowerment and use their resources to find ways to increase participants' agency in the voting process. Aligning with many of the design justice principles in their approach to encouraging civic participation, Abrams and her advocacy teams' work was successful in pursuing a social justice agenda for UX and TPC.

The New Georgia Project also established an internship program to train young people to better understand and work with politics and policy—building their capacity to create civic change through voting and voting rights advocacy. Within TPC scholarship, Jones and colleagues (2016) shared that "user advocacy is not fully enacted by merely making objects easy to use but also includes respecting users enough to convey effects of use so they can make informed decisions (Johnson, 1998)" (p. 218). This internship program helps Abrams and her New Georgia Project team facilitate the education of community participants to "sustain, heal, and empower" (Costanza-Chock, 2020) people within Georgia communities to engage in political decisions that impact them. Through coalition and collaboration, Abrams and her team worked to empower people who have been disempowered, particularly by limits to education and resources.

Abrams and her advocacy teams' work can be used within TPC and UX as an example of using UXD principles and practices in conjunction with institutional knowledge and critique to pursue transformative change. Participatory design and research are some of the most valuable means of understanding community members' existing knowledge (Moore & Elliott, 2016; Spinuzzi, 2005) that we have as UX and TPC scholars and practitioners. To pursue more accessible and equitable civic engagement opportunities through our work, UX and TPC practitioners can learn from Abrams's example and engage in participatory research and design that is guided by the historical and embodied knowledge of community members (Bowers et al., 2022).

Abrams and her team center their work in historical, embodied knowledge by meeting people where they are in the communities they hope to serve and giving everyone multiple options for engagement, including varying the contexts and people they interact with so that they can choose what's most comfortable for them. They also practice inclusive UXD by "being transparent about their intentions and actions so that they can be accountable to those promises, and constituents can feel more empowered to engage with the people and policies making changes that matter to them (Abrams, n.d.; Garbus & Cortes, 2020; Walton et al., 2019)" (Bowers et al., 2022, p. 459).

UX and TPC practitioners and scholars have tools, such as journey mapping, interviews, and participatory design that we can use to pursue institutions that more effectively meet the needs and goals of participants. Using institutional critique in conjunction with UXD and research methods, we can learn from systems and how they interact with interpersonal, structural, and institutional elements to inform our design and research.

Implications: User Experience Design toward Institutional and Organizational Change

Abrams's and her advocacy team's work shows us that today's government institutions are not monolithic (Bowers et al., 2022). Governmental institutions can be designed in a way that prioritizes certain groups and creates manufactured death (silence) of others (Ohito, 2021). But this also means that the institutions can change. By critically considering the work and positions of UX designers and researchers in connection to the institutions and actions that contribute to inequities in our society, practitioners can better hold themselves accountable for design choices by creating transparency and context around which users who are centralized in design decisions and why. A universally just design does not exist for every context—for instance, it could be futile to strive to create a government system that centers all users in the same way. However, it is possible to clearly articulate the focus on designing experiences in a particular way for a particular group to pursue institutional change toward social justice.

Abrams and her team's social justice–oriented coalitional approach exemplify critical awareness of opportunities and inequities in interpersonal, structural, and institutional design to enact institutional change that empowers multimarginalized populations. We use Abrams and her team's efforts in this chapter as *a* formative example, not *the* formative example of how UX and TPC informed by institutional critique can be used to effect institutional change. However, as UXD continues toward institutional change, Abrams's and her campaign and nonprofit teams' work acts as a significant example of the potential that enacting UXD methods and methodologies in specific ways has to create civic change that empowers and supports multimarginalized individuals impacted by systematic oppression (Bowers et al., 2022).

When Abrams and her nonprofit and campaign teams do design and research work for voting rights with a critical awareness of how inequities and opportunities for social justice exist at multiple levels, they exemplify how these designs and methods can further transformative institutional change. Although the situation and context in which UX and TPC scholars and practitioners might use these methods vary, TPC practitioners can still learn from and apply this critical awareness (of the various levels of inequity and action) to the work they do in pursuit of social justice.

There are many ways in which Abrams's example can support UX scholarship and pedagogy. One is that, as UX scholars and practitioners, "we can look to Abrams's work as an example of how to be accomplices in supporting marginalized, especially multi-marginalized, identities. Abrams uses her position as an integral part of the community to work WITH the community she aims to serve" (Bowers et al., 2022, p. 459). In academic and industry contexts, we must work to better understand our own positions within existing power structures, communities that we are a part of, and structures of power that we are involved in through the institutions in which we are participating. Similarly, if we aim to be UX researchers and community practitioners pursuing institutional transformation, we must value communities as critical participants in this work. We must immerse ourselves in the communities we want our designs to support and change our mindset from designing *for* to designing *with* community partners. By engaging UX methods and methodologies in ways that Abrams and her advocacy teams exemplify as using UX toward social justice, "we can work towards being co-conspirators (Love et al., 2019) with underserved and marginalized communities in pursuing institutional critique, trans-formation, and change" (Bowers et al., 2022, p. 460).

References

Abrams, S. (n.d.) *Meet Stacey*. Stacey Abrams Governor. https://staceyabrams.com/about

Abrams, S. (2021). *Our time is now: Power, purpose, and the fight for a fair America*. Henry Holt and Company.

Agboka, G. Y. (2014). Decolonial methodologies: Social justice perspectives in intercultural technical communication research. *Journal of Technical Writing and Communication, 44*(3), 297–327. https://doi.org/10.2190/TW.44.3.e

Benjamin, R. (2019). *Race after technology: Abolitionist tools for the new Jim code.* Polity Press.

Berne, P., Morales, A. L., Langstaff, D., & Invalid, S. (2018). Ten principles of disability justice. *Women's Studies Quarterly, 46*(1), 227–230. https://doi.org/10.1353/wsq.2018.0003

Bowers, E. L., Harris, E. J., & Mendoza, R. (2022). UX methods as transformative institutional change: Stacey Abrams' Georgia campaign as a formative example. In *Proceedings of 2022 IEEE International Professional Communication Conference (ProComm), 456–461.* https://doi.org/10.1109/ProComm53155.2022.00090

Chávez, K. R. (2013). *Queer migration politics: Activist rhetoric and coalitional possibilities.* University of Illinois Press.

Collins, P. H. (1990). *Black feminist thought: Knowledge, consciousness, and the politics of empowerment.* Routledge.

Costanza-Chock, S. (2020). *Design justice: Community-led practices to build the worlds we need.* The MIT Press. https://doi.org/10.7551/mitpress/12255.001.0001

Durá, L., Singhal, A., & Elias, E. (2013). Minga Perú's strategy for social change in the Peruvian Amazon: A rhetorical model for participatory, intercultural practice to advance human rights. *Rhetoric, Professional Communication, and Globalization, 4*(1), 33–54.

Friess, E. (2010). The sword of data: Does human-centered design fulfill its rhetorical responsibility? *Design Issues, 26*(3), 40–50. https://doi:10.1162/DESI_a_00028

Garbus, L., & Cortes, L. (Directors). (2020). *All in: The fight for democracy* [Film]. Amazon Studios. https://www.amazon.com/All-Fight-Democracy-Stacey-Abrams/dp/B08FRQQKD5

Gibbons, S. (2021, May 16). Three levels of pain points in customer experience. *Nielsen Norman Group.* https://www.nngroup.com/articles/pain-points

Green, A. (2019, August 16). Stacey Abrams is playing the long game. *The Atlantic.* https://www.theatlantic.com/politics/archive/2019/08/stacey-abrams-voting-rights/596206

Haas, A. M. (2012). Race, rhetoric, and technology: A case study of decolonial technical communication theory, methodology, and pedagogy. *Journal of Business and Technical Communication, 26*(3), 277–310. https://doi.org/10.1177/1050651912439539

Hillmann, C. (2021). *UX for XR: User experience design and strategies for immersive technologies.* Apress. https://doi.org/10.1007/978-1-4842-7020-2_6

Johnson, R. R. (1998). *User-centered technology: A rhetorical theory for computers and other mundane artifacts.* State University of New York Press.

Jones, N., Moore, K., & Walton, R. (2016). Disrupt the past to disrupt the future: An antenarrative of technical communication. *Technical Communication Quarterly, 25*(4), 211–229. https://doi.org/10.1080/10572252.2016.1224655

Jones, N., & Walton, R. (2018). Using narratives to foster critical thinking about diversity and social justice. In A. M. Haas & M. F. Eble (Eds.), *Key theoretical frameworks: Teaching technical communication in the twenty-first century* (pp. 241–267). University Press of Colorado. https://doi.org/10.7330/9781607327585

Jones, N. N., & Williams, M. F. (2017). The social justice impact of plain language: A critical approach to plain-language analysis. *IEEE Transactions on Professional Communication, 60*(4), 412–429. https://doi.org/10.1109/TPC.2017.2762964

Jones, N. N., & Williams, M. F. (2018). Technologies of disenfranchisement: Literacy tests and black voters in the US from 1890 to 1965. *Technical Communication, 65*(4), 371–386.

Klein, E. (2020, November 6). Stacey Abrams on minority rule, voting rights, and the future of democracy: In 2020, democracy is on the ballot. *Vox.* https://www.vox.com/ezra-klein-show-podcast/21540804/stacey-abrams-2020-biden-trump-election-voter-suppression-laws-republicans

Krieg, G., & McKend, E. (2022, May 23). Stacey Abrams' playbook faces a new test in second run for Georgia governor. *CNN.* https://www.cnn.com/2022/05/23/politics/stacey-abrams-georgia-governor-democratic-primary/index.html

la paperson. (2017). *A third university is possible.* University of Minnesota Press.

Lewthwaite, S., Sloan, D., & Horton, S. (2018). A web for all. In K. Ellis, R. Garland-Thomson, M. Kent, & R. Robertson (Eds.), *A manifesto for critical disability studies in accessibility and user experience design* (pp. 130–141). Routledge. https://doi.org/10.4324/9781351053341

Love, B., Sealey-Ruiz, Y., & Akinnagbe, D. B. (2019, March 19). *We want to do more than survive: Abolitionist teaching and the pursuit of educational freedom* [Book talk], Schomburg Center for Research in Black Culture, New York, NY.

Maketrybe. (2018). What is transformative design? http://www.maketrybe.org/transformativedesign

Mendoza, R. (2023). *A rhetorical, coalitional, and decolonial critique on cistematic academic scholarly practices: Mobilizing Queer and Trans*/formative BIPOC resistance.* [Unpublished doctoral dissertation]. Michigan State University.

Moeggenberg, Z. C., Edenfield, A. C., & Holmes, S. (2022). Trans oppression through technical rhetorics: A queer phenomenological analysis of institutional documents. *Journal of Business and Technical Communication, 36*(4), 403–439.

Moore, K. R., & Elliott, T. J. (2016). From participatory design to a listening infrastructure: A case of urban planning and participation. *Journal of Business and Technical Communication, 30*(1), 59–84. https://doi:10.1177/1050651915602294

Morgensen, S. L. (2011). The biopolitics of settler colonialism: Right here, right now, settler colonial studies. *Settler Colonial Studies, 1*(1), 52–76. https://doi.org/10.1080/2201473X.2011.10648801

Mukavetz, A. M. R. (2014). Towards a cultural rhetorics methodology: Making research matter with multi-generational women from the Little Traverse Bay Band. *Rhetoric, Professional Communication, and Globalization, 5*(1), 108–125.

Nash, J. C. (2019). *Black feminism reimagined: After intersectionality.* Duke University Press.

New Georgia Project (2022a). *About New Georgia Project.* https://newgeorgiaproject.org/about

New Georgia Project (2022b, September 12). *Peanut gallery lead organizer.* Indeed. https://www.indeed.com/viewjob?t=peanut+gallery+lead+organizer&jk=2a9be155d11d507d&_ga=2.235632193.1195639042.1663014037-336646079.1658772545

Ohito, E. O. (2021). Some of us die: A Black feminist researcher's survival method for creatively refusing death and decay in the neoliberal academy. *International Journal of Qualitative Studies in Education, 34*(6). https://doi.org/10.1080/09518398.2020.1771463

Prasad, P. (2021). "Coalition is not a home": From idealized coalitions to livable lives. *Spark: A 4C4 Equity Journal.* https://sparkactivism.com/volume-3-introduction

Redish, J. (2010). Technical communication and usability: Intertwined strands and mutual influences. *IEEE Transactions on Professional Communication, 53*(3), 191–201. https://doi:10.1109/TPC.2010.2052861

Redish, J. G., & Barnum, C. (2011). Overlap, influence, intertwining: The interplay of UX and technical communication. *Journal of Usability Studies, 6*(3), 90–101.

Simpson, L., Wood, L., & Daws, L. (2003). Community capacity building: Starting with people not projects. *Community Development Journal, 38*(4), 277–286. http://www.jstor.org/stable/44258892

Skarlatidou, A., Ponti, M., Sprinks, J., Nold, C., Haklay, M., & Kanjo, E. (2019). User experience of digital technologies in citizen science. *Journal of Science Communication, 18*(1). http://doi.org/10.22323/2.18010501

Skinnell, R. (2019). Toward a working theory of institutional rhetorics. In R. Skinnel, J. Holiday, K. Gerdes, & A. Alden (Eds.), *Reinventing (with) theory in rhetoric and writing studies* (pp. 69–82). Utah State University Press. https://doi.10.7330/9781607328933.c005

Spinuzzi, C. (2005). The methodology of participatory design. *Technical Communication, 52*(2), 163–174.

Sullivan, P. (1989). Beyond a narrow conception of usability testing: A rationale for broadening usability. *IEEE Transactions on Professional Communication, 32*(4), 256–264. https://doi.org/10.1109/47.44537

Traister, R. (2020, November 19). Stacey Abrams on finishing the job in Georgia: "It can be undone just as quickly and as effectively as we did it." *The Cut.*

https://www.thecut.com/2020/11/stacey-abrams-on-flipping-georgia-blue.html

Walton, R. (2016). Supporting human dignity and human rights: A call to adopt the first principle of human-centered design. *Journal of Technical Writing and Communication, 46*(4), 402–426. https://doi.org/10.1177/0047281616653496

Walton, R., Moore, K. R., & Jones, N. N. (2019) *Technical communication after the social justice turn: Building coalitions for action.* Routledge. https://doi.org/10.4324/9780429198748

Williams, M. F. (2017). *From black codes to recodification: Removing the veil from regulatory writing.* Routledge. https://doi.org/10.4324/9781315224541

Williams, M., & Pimentel, O. (2016). *Communicating race, ethnicity, and identity in technical communication.* Routledge. https://doi.org/10.4324/9781315232584

Chapter 13

The Beekeeper's Companion

Enabling Ag-Extension through Localizing and Customization of a Climate-Smart Technology for Women in Lebanon

SARAH BETH HOPTON, LAURA BECKER, MAX RÜNZEL,
AND JAMES T. WILKES

As global warming and anthropogenic climate change threatens, disrupts, or devastates natural ecosystems, human rights to clean food and water, economic stability, and personal health and safety also become jeopardized. Although many climate-mitigation policies and climate-smart technologies treat biodiversity loss, global warming, and human rights as independent of each other, they are not; disruption in one system disrupts the other systems to which they are connected. Thus, climate-smart technical solutions should be designed holistically, with consideration given to global warming, biodiversity loss, and human rights impacts collectively (Pörtner et al., 2021).

Recognizing that climate change disproportionately affects women; children; and Black, Indigenous, and people of color (BIPOC; Boyd, 2021), climate-smart technologies—especially those designed in Majority World contexts for Majority World users—should be localized, customized, and designed using participatory design and design thinking methods

(Pope-Ruark et al., 2019; Rose et al., 2018; Sun, 2006; Sun & Getto, 2017; Tham, 2021) that share power with users and incorporate indigenous knowledge without appropriating or exploiting it (Bogale, 2009; Haas, 2012; Koerber, 2000; Salvo, 2002), thus creating balance and equilibrium or, as is the theme of this book, what scholars Lancaster and King (2023) have termed "equilibriUX."

This chapter presents a case study of the *AI-Driven Climate-Smart Beekeeping (AID-CSB) for Women Advisory & Extension (AID-CSB A&E)* project, which worked with beekeepers in Lebanon to co-design and localize the "Beekeeper's Companion," a climate-smart information communication technology for development (ICT4D) app designed to support beekeepers' hive management practices and improve honey production. The project also co-designed a web portal for extension workers that pulls data from the app for better monitoring and engagement of beekeepers and bee health. Findings from *AID-CSB A&E* will be compared and contrasted with a similar project started a year earlier in Uzbekistan and Ethiopia and will highlight the ways that developers and project leaders attempted to balance the goals of designers with the needs of users.

Commonalities between projects include the goals of developing and customizing *Beekeeper's Companion* (detailed below) application for women beekeepers in all three countries, with the goal of improving hive management, boosting productivity, and improving resilience in the face of climate change and its inequitable distribution of burden on women by increasing women's economic opportunity, autonomy, and digital literacies. Similarities in beekeeping challenges across both studies were also identified. A core difference with the *AID-CSB A&E* project in Lebanon was its new focus on aiding extension workers in monitoring beekeeping practices and bee health remotely through a customized web portal. This proved to be of particular importance, as Lebanon faces a major economic crisis exacerbated by COVID and resulting in fuel shortages and power outages. These on-the-ground conditions highlighted the importance of supporting women in rural areas and enabling agricultural extension services to manage access to knowledge remotely and efficiently.

By using participatory design methodologies that included customization and localization of the app, we hoped to achieve what scholars Lancaster and King (2024) have called *equilibriUX*, which is the central focus of this collection. Linguistically, *equilibriUX* is a portmanteau, a combination of the words "equilibrium" and "user experience." In practice, it is both a goal and a methodology that integrates "reactors of character and influence to establish balance and respect diversity in design" (Lancaster & King, 2024). Lancaster and King explained that achieving equilibriUX requires us to design and develop technologies in ways that account for our own expertise and cultural competence *and* that of our user. This case study is an example of equilibriUX in practice and our attempt to answer Lancaster and King's (2024) call to "extend UX design practices beyond translating and tailoring for local users to broader global users, while considering the diversity of user uniqueness, customization desires, all stakeholders, and social needs" (p. xiii).

Launched in 2010, the first HiveTracks app was designed to help beekeepers manage hive data that had traditionally been recorded by hand and often difficult to assess longitudinally and, therefore, of limited value both to the beekeeper and the scientific community. Originally designed as a web app for North American beekeepers, the words, symbols, hive types, queen types, pest species, and logics of use were design-centric to the United States, despite this version of the app being used in 150 countries. Through the *AID-CSB* and the *AID-CSB-AE* projects, the new version of the app—the Beekeeper's Companion—underwent iteration to localize the app for target users in the three countries studied: Uzbekistan, Ethiopia, and Lebanon. This process required extensive secondary research and user research to understand and later customize the user interface and application functions across four different languages (Uzbek, Russian, Amharic, and Arabic); four different climates; and two different user profiles (beekeepers themselves and extension workers working with beekeepers) and to accommodate different forms of Indigenous knowledge regarding beekeeping practices.

The Beekeeper's Companion is mobile only and offline capable, and it creates a true digital companion for beekeepers, aimed at generating season and location-based information, reminders, and decision-making aids that beekeepers need to protect hives from weather and climate risks, pests, disease, and nutritional deficiencies and malpractices. Though the beekeepers own their data and those data are protected—thus protecting privacy of all users—in the future, scientists and researchers will be able to access nonidentifiable aggregated data to understand the variables associated with honeybee health and the local conditions exacerbating pollinator declines, while giving voice to women whose voices in agricultural policy and technological production have been historically silenced. The app is a step toward the project's overarching goal: to build data-driven, user-centric, and discrimination-aware solutions and to monitor, evaluate, and manage biodiversity impacts on and of the honey supply chain while creating value for the individual beekeeper and extension worker.

Many global digital projects fail because development firms too frequently design *at* users not *for* them (Hopton et al., 2022). This is perhaps especially true of projects in Majority World contexts because many technologies are designed with Western users in mind, such as the direction of navigation for left-to-right languages (Acharya, 2019). Majority World users are often left out of design decisions, and usability research is difficult to conduct in times of political turmoil and global health crises, which interrupt access to communications needed to conduct tests, to say nothing of the disruption to participants' lives (Currie et al., 2022; Gonzales et al., 2022). Despite these difficulties, which we detail later in this chapter, the *AID-CSB-AE* project team employed participatory design methodologies that involved stakeholders, designers, researchers, and end users in the customization and localization process, ensuring the final product met the needs of its audience—not the will of the developer—and was ethically designed (as supported by technical/professional communication [TPC] and design researchers; see Acharya, 2018; Bannon & Ehn, 2013; Spinuzzi, 2005; Steinke et al., 2022; Zachry & Spyridakis, 2016).

Though the use of the Beekeeper's Companion app at all three sites is ongoing, preliminary findings suggest that use of the app among *AID-CSB A&E* participants will help beekeepers keep more complete records and access information that will help them better manage their hives. Furthermore, smartphone trainings and usage of the app narrow the gender digital divide by increasing women's digital literacy and raising awareness around women's technology use. In the long term, we expect the localized app to support low-barrier economic activities for women that also improve local biodiversity, crop yields, and the health of bee populations and thus support the human rights to safe work, shelter, food, and environmental health. Specifically in Lebanon, the cocreation of a web portal for extension workers will enable the monitoring of honey bee health and beekeeper activities remotely at a frequency and scale previously not possible.

This research project addresses issues such as how to extend localization practices and principles to increase user agency and support user advocacy in expanded global contexts, as well as how designers engage and empower user groups to reflect and respect diverse voices, thus building unifying experiences for diverse user groups (Lancaster & King, 2024) and creating *equilibriUX*. This case study also provides valuable contributions to the TPC literature. Existing literature on designing for underrepresented voices has included social justice and UX research in areas of localization, voter registration, accessibility and universal design, environmental issues, and risk and safety, to name a few. More recently, TPC's social justice turn has called upon those in design fields to advocate for users—to ensure their voices are heard in designing technologies that cater to their needs (see Lancaster & King, 2022, 2024; also Hitt, 2018; Hopton, 2021; Sackey, 2020; Shirley, 2019; Shivers-McNair, 2017). This study offers a new perspective on the challenges of designing for global users in specific and localized contexts and addresses underrepresented users (women and people of color) in the unique context of the beekeeping profession (an agricultural lens).

As a final note of clarification, because the methodologies used across all three sites are similar, small segments of this work (particularly in the findings and methods section) resemble content in a previous publication entitled, "Localizing Climate-Smart Applications Through Participatory Design: A Case Study of the Beekeeper's Companion" (Hopton et al., 2022). Additionally, some information has been adapted from *AI-Driven Climate-Smart Beekeeping for Women 2021 Report* (Becker et al., 2022), which was collectively written by the authors and delivered to our granting institutions. However, the work highlighted in this case study is original material from a new site of study (Lebanon). As the use of the application is ongoing, some of our findings and implications in the "Discussion" sections are preliminary and/or speculative.

Cultural and Contextual Background: Agricultural Labor, Human Rights, and Inequities for Women and People of Color

Nearly half of the world's food and agricultural resources depend on women's labor (World Bank Group, 2017). Despite this, they do not enjoy the various rights that should accompany their agriculture work, including financial autonomy, equal pay to men, land ownership, and environmental health. Compounding these disadvantages, cultural norms around women's use of technology often make it difficult for women to access and use technologies that could improve their livelihoods or make labor conditions, which require a delicate balance between work in and outside the home, more equitable.

Beekeeping is unique among agricultural labor options for women in that it does not require land ownership but can still provide women with a source of income and a kind of labor opportunity that can be performed at home while managing other domestic responsibilities. Home-based or small-scale beekeeping can also give women voice in local or regional climate-mitigation policy (United Nations, n.d.) and when facilitated by new technologies, can normalize women's use of information-communication tools and services that, in many cases, can also significantly improve poverty outcomes (Kakar et al., 2012) and even reduce mortality (Murphy et al., 2009).

CLIMATE CHANGE AND HUMAN RIGHTS

Though basic human rights include those to safe work, shelter, food, and environmental health (United Nations General Assembly, 2019b), these rights are not enjoyed universally. Vulnerable populations, children, BIPOC, and women are disproportionately affected by climate change, which exacerbates existing inequalities (Boyd, 2021). During extreme weather events, for example, women are more likely to die than men, owing to differences in socioeconomic status and in financial and decision-making power to access resources and life-saving information (United Nations General Assembly, 2019a). The economic stress induced by extreme weather events can result in spikes of domestic violence against women or trigger famine and forced migration (Baada & Najjar, 2020). On a day-to-day basis in many parts of the Majority World, women are responsible for labor that depends on the climate, such as harvesting coffee in Ethiopia. This becomes more difficult in the face of rising temperatures that enable the spread of destructive pests and decrease the coffee quality (Asegid, 2020), which drives down the value of such products and thus diminishes livelihoods.

The relationship between biodiversity and human rights is not always obvious; consider the right to a clean, healthy, and sustainable environment. Clean water is supported by diverse plants and animals that help remove pollutants and improve water flow by reducing runoff; the reduction of infectious zoonotic diseases is associated with higher biodiversity, and the maintenance of mental health is supported through exposure to nature, which requires biodiversity. Loss of biodiversity, therefore, exacerbates existing vulnerability and marginalization of Indigenous populations dependent on natural resources, and they often lack the capital or political power necessary to replace lost resources or compete with wealthy global corporations intent on exploiting native lands (United Nations General Assembly, 2019b). But perhaps the most obvious consequences of biodiversity loss and human rights are visible in our food systems. Because diverse plants and crops enrich the nutrients available in soil and mitigate farmer risk in case one crop fails, biodiversity and healthy agricultural practices are a critical link in protecting and strengthening food systems as well as people's livelihoods.

Supporting all of these human rights through biodiversity are pollinators. Pollinators help plants reproduce and flourish—in this case, plants that become medicine used in both Indigenous and Western cultures. Healthier plants root deeper and live longer, which, depending on whether those plants are in soil or water, affects oxygen levels and food supplies for microorganisms essential to biotic life. Strong pollination services can shorten the time between flowering and fruiting, thus reducing risks associated with pests, disease, and weather while concurrently decreasing the need for application of agro-chemicals, which disrupt and damage ecological systems and improve yields, thus ensuring livelihoods dependent on farming (Aktar et al., 2009). Ultimately, pollinator services can improve a beekeeper's productivity, preserve and restore the biodiversity of the habitats they rely on to provide these services, and in turn protect the Indigenous knowledge required to maintain healthy hives.

THE POTENTIAL FOR WOMEN'S ECONOMIC EMANCIPATION AND AUTONOMY THROUGH APICULTURE

Women make up almost half of the agricultural workforce in the world today; and in some parts of the world their role in agriculture is growing, as they take on additional responsibility when men migrate for work or when they become heads of households (Baada & Najjar, 2020). Nonetheless, women are often unpaid and are frequently categorized as "home" producers or assistants and not accorded the dignity of being considered farmers, apiarists, or income-producers in their own right. As with agriculture generally, beekeeping has historically been viewed as a men's activity, but women have always played a significant role in the daily labor required to maintain hive health or to trade and sell honey and hive products (Olana & Demrew, 2018). Due to women's growing role in apiculture, designing and working with women on innovative technology-driven agriculture projects can have significant positive effects on communities and agri-food systems and, thus, they should not be overlooked as a key user group for technology development (United Nations General Assembly, 2019a).

CLOSING THE DIGITAL DIVIDE

The most recent conclusions from the Commission on Status of Women (CSW) emphasize that women and girls should be enabled to participate in climate change solutions; in part by strengthening women's access to education and technology. By some estimates, doubling the number of women online in developing countries has the potential to increase global gross domestic product of 9% (Antonio & Tuffley, 2014). Yet, close to half the world remains offline, and the majority of those without access to Internet connections in developing countries are women (Iglesias, 2020).

For the past 30 years, experts working at the intersections of gender and ICT4D theorized that digital literacy and skills, cost, and availability of delivery mechanisms (like cell phones) were the principal reasons for the gender digital divide (Steele, 2019). Certainly, these factors contribute to the divide, but new research suggests that norms have an equal and sometimes greater impact on women's access to and use of technology (Sterling et al., 2020; Sterling & Koutsky, 2018). Many women across the Majority World have cell phones or access to communication technology because they are *allowed* to, and millions of women do not have access to tech because of family, community, legal (Chandran, 2016), or religious regulations (Grant, 2014) that order women to protect themselves or preserve their dignity by abstaining from use (Primo, 2003).

Across all three study sites, women beekeepers expressed the need for access to more beekeeping knowledge and resources. To localize a tool for knowledge transfer, *AID-CSB* and *AID-CSB A&E* used a co-design approach with women beekeepers to gather their input on pilot climate-mitigating features within the Beekeeper's Companion app, such as notifications on blooming plants and weather, and integrated their feedback into newer iterations of the app: this led to improved digital literacies, supporting women's economic autonomy, and demonstrating the powerful and positive effect women equipped with technology can have on their families and local communities. In Lebanon, women beekeepers will further benefit by connecting their apps to extension worker advice via the web portal, and in the future, this additional feature will also be implemented in Uzbekistan and Ethiopia.

Degendering the digital divide and technology solutions requires practitioners and scholars to improve digital literacy, technology skills, decrease cost of access, customize delivery mechanisms, consider and—where possible—incorporate or disrupt the normative values and attitudes associated with gender and technology (Feeney et al., 2021). Though technology is not a panacea, ICT4D solutions can catalyze and empower women and BIPOC communities, especially in agriculture. Coupling women's existing knowledge about agri- and apiculture with responsively and appropriately designed technology could result in a force-multiplier effect in mitigating climate-related problems like pollinator decline. But, for this multiplier effect to emerge, designers and developers must prioritize participatory design, localization, and the degendering of technology in their designs. Balance and equilibrium of design must become standard operating procedure. As the world becomes increasingly technology dependent, the consequences grow for those who remain without access or understanding of how to use new technologies. Responsively designed technology and the standardization of participatory design processes can ensure that fewer and fewer women are left behind.

Designing the Beekeeper's Companion App: Considerations and Predesign Goals

AVOIDING BIAS IN DESIGN

Despite its potential, bias concerns exist when designing technology, even technology that is codesigned using participatory design methods. "Smart" apps, including the Beekeeper's Companion app, revolve around data-driven rules and algorithms that pave the way for more advanced statistical methods, such as machine learning. Research has shown that rules and algorithms contained within smart apps tend to inherit the bias of developers (Suresh & Guttag, 2019). This is particularly the case when inputs, operations, and training mechanisms are neither visible nor auditable by third parties and are considered a "black box." Biased outcomes implicit in the decision-making process may not be the intent but can certainly be the by-product of development (Borgesius, 2018). For example, when operating in an openly or subconsciously patriarchal society, machine-learning algorithms could also learn and proliferate this bias.

To develop fair artificial intelligence (AI) and discrimination-aware, data-driven apps, we employed several steps. In line with the six phases for nondiscrimination by design, and by developing and localizing the Beekeeping Companion, our app mitigates the unintended consequences of its data-driven and automated functionalities as much as possible. Starting at the problem and use case definition of the model, the phases of our app development included (1) data collection, (2) data preparation, (3) model selection, (4) implementation, and (5) evaluation.

Following the best practices of design thinking and usability, and with the goal toward achieving "equilibriUX" (i.e., balance in design; Lancaster & King, 2024), we employed steps that localized, customized, or incorporated user feedback; from there, prototypes were rapidly produced and users again tested the design in an iterative cycle. For example, after defining the purpose and necessity of potential statistical models alongside their impact on all stakeholders and determining required success criteria, data were selected based on quality, size (which affects storage), and origin (to minimize biased inputs). Similarly, data preparation followed a transparent and thoughtful process that included defining inclusion and exclusion criteria and determining specific integration and aggregation criteria. Premodeling, model selection, and technical and user tests were the focus thereafter. Ultimately, our user testing approach was only applied to implementation and evaluation phases, but the participatory model of application development required ethical consideration of users at every step of the process, and model alterations relied on continuous, iterative user evaluation and feedback as we strove for balance between design and usability.

To ultimately arrive at a sound and responsible implementation of AI, the first two steps were the focal points of this stage of the project, with a particular focus on data collection, which proved to be a key enabler for future implementation of statistical models. In line with this process, the app development followed the principles that aim to limit the bias inherent in the design of technology associated with data collection by using a gender-responsive codesign process detailed in the following sections. Starting in Lebanon, the Admin Portal (discussed and illustrated in the Case Study Methods) plays a significant role. Notably, agricultural extension experts will be able to send recommendations and messages to beekeepers based on collected data points on bee behaviors, pests, and diseases. Logging these recommendations and messages and overlaying weather and blooming data allows us to discover patterns that can then be codified and turned into algorithms. Having these data-driven rules and algorithms audited and studied will be key to avoiding design bias in future app iterations and will be a central part of future development efforts.

Protecting User Privacy

The right to privacy is enshrined in Article 12 of the Universal Declaration of Human Rights (United Nations General Assembly, 1948); Article 17 in the legally binding International Covenant on Civil and Political Rights (United Nations General Assembly, 1966); and Article 16 of the Convention of the Rights of the Child (United Nations General Assembly, 1989). Also, many privacy laws are designed to protect personal data from state actors and corporations and the right to privacy intersects with many other human rights, such as freedom of expression; the right to seek, receive, and impart information; and freedom of association and assembly. Privacy also helps ensure accountability should data be stolen; and incorporating strict data privacy protections helps build trust between user and developer. Thus, our app protects user data and ensures privacy through compliance with best practices as outlined by the Bee XML Data Standardization group (BeeXML Apimondia Working Group, 2021), including rounding location data to the nearest 3 km (about 1.8 miles) and aggregating beekeepers' data at regional levels, which further anonymizes data, thus protecting identity and privacy and improving safety for women.

Case Study Methodology

To bring innovative global solutions to a local context, the project partnership included ICARDA's MEL team (with regional research expertise), HiveTracks (with an international technology solution), and LARI (with relationships with beekeepers for local implementation). In addition to the strengths of bridging local, regional, and global partners, *AID-CSB A&E* leveraged the public-private-partnership (PPP) model because PPPs transfer certain risks to the private sector and provide incentives for assets to be maintained while lowering infrastructure costs to public organizations by reducing development and project life-cycle expenditures. Additionally, PPPs can introduce private sector technology and innovation with the speed and scale that most public services cannot and are often more flexible in their methodological decisions, which means they have the time, resources, and—importantly—the political *will* to enact equilibriUX design principles and methods like multiple user tests, translations, and focus groups that encourage a multiplicity of user voices during more phases of design and development.

ICARDA (https://www.icarda.org) is an international nonprofit organization that undertakes applied research-for-development aiming at reducing poverty and enhancing food, water, and nutritional security and environmental health in the face of global challenges. HiveTracks (https://www.hivetracks.com) is a small US-based private company that has developed software for beekeepers since 2010 and has supported 38,000 beekeepers across 150 countries since then. HiveTracks works to create a positive impact on bee health, beekeeper livelihoods, and biodiversity by connecting people, nature, and data. LARI (https://www.lari.gov.lb) is an independent governmental agricultural research institute under the supervision of the Minister of Agriculture in Lebanon. The institute conducts applied and basic scientific research for the development and advancement of the agricultural sector in Lebanon. LARI started its work with beekeeping in 2018, it conducts activities related to laboratory tests for major bee diseases and pests as well as tests for honey quality. LARI also provides extension services to beekeepers across the country.

The 6-month *AID-CSB A&E* project in Lebanon was funded by the GIZ Innovation Challenge.

SELECTION OF PARTICIPATING BEEKEEPERS AND IN-COUNTRY PARTNER: LEBANON BEEKEEPER PROFILE

Given the short (6-month) timespan of the *AID-CSB A&E* project and ideal functionality of the Beekeeper's Companion app to work with beekeepers with some experience and a small number of hives, the project selected 21 beekeepers that already had smartphones and typically had less than 15 hives. Given the project's focus on women's economic empowerment and decision-making for hive management, the project worked with a majority of women (12 beekeepers). We also worked with men (9 beekeepers) to increase the geographical diversity of participants, understand any gender-differences in beekeeping challenges and app feedback, and consider their feedback, as the localized app will be freely available to all beekeepers following registration. Beekeepers' level of experience ranged from less than a year to 20+ years, with the majority of beekeepers (43%) holding 2–5 years of beekeeping experience.

Research Design

Both the *AID-CSB* and the *AID-CSB-AE* projects leveraged a 3-step process to localize and customize the Beekeeper's Companion app that independently supported the mitigation of bias and protection of user privacy:

1. Secondary research was conducted to gather both a general understanding of the honey sector and beekeeping practices, as well as collect specific data points that would be used later on in the app development, such as flowering calendars.

2. Preliminary user interviews were conducted to understand the day-to-day beekeeping experience, practices, and challenges of project beekeepers to design the app prototype.

3. Prototype testing enabled women beekeepers to interact directly with the app prototype to identify problems and areas of improvement early, so designers could make the necessary changes prior to development and build products that meet user needs and expectations. During user testing, which was conducted via Zoom, the user experience/user interface (UX/UI) designer requested the beekeeper to complete certain tasks or test certain functions of the app, and we observed both their real-time navigation choices, while collecting verbal feedback. The participants were not exposed to any version of the app or to the wireframe prototypes prior to testing to minimize bias. For simplicity, to avoid creating stress for the participants, and to include those users who did not yet have access to a smartphone, the prototype(s) were displayed on a laptop and screen-shared during testing. This process promoted feedback on the user interface and sharing of technical beekeeping information in users' native language(s), which is one way to be gender-inclusive during technology design processes.

Though Uzbekistan and Ethiopia were two other sites we included in our grant-funded project, they are not the primary focus of this chapter's case study. However, we offer comparisons and contrast to establish design practices and decisions for the app going forward, though, for the sake of brevity, we do not describe the methods used at those sites. They can be found, however, in the full report (Hopton et al., 2022). All interviews and training sessions were conducted remotely by the HiveTracks UX/UI designer via Zoom to establish a personable connection between the international staff and local beekeepers. Though the same remote methodology was originally planned for Lebanon, given the high COVID-19 case rate and fuel crisis in Lebanon in December 2021 and early 2022, LARI field assistants were trained by the UX/UI designer instead and provided with questionnaires to conduct the preliminary interviews and prototype tests themselves. Though this setup did give the UX/UI designer and international staff less insight than the interviews and less connection with participating beekeepers, this methodology built the capacity of local staff to conduct UX/UI interviews, and most importantly, was the safest option for project participants.

The Admin Portal consisted of six main features: a dashboard, an apiary map, communications, reports, accounts, and settings (see Figures 13.1, 13.2, 13.3, 13.4, and 13.5).

Figure 13.1. The Admin Portal Dashboard.

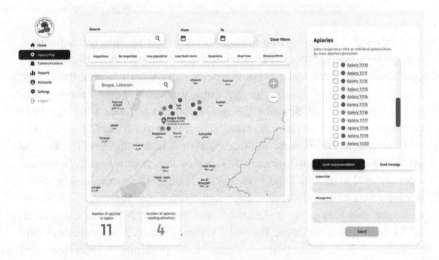

Figure 13.2. The Admin Portal Map View.

Figure 13.3. The Admin Portal Map View, Recommendation.

Figure 13.4. The Admin Portal Map View, Varroa Mite Infestation.

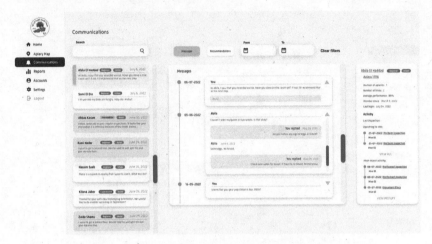

Figure 13.5. The Admin Portal Communication Interface.

Comparative Results

Key findings from the preliminary interviews with beekeepers in Lebanon revealed top challenges of beekeepers as Varroa, pests (particularly wasps and other insects), diseases, agricultural pesticides, cost/expense of beekeeping equipment and treatment, lack of knowledge on beekeeping, and weather/humidity. One-third of beekeepers already kept records on their hives, a practice that is digitized by the Beekeeper's Companion app. The majority of beekeepers do not have access to an advisory body (i.e., mentor, beekeeping association, or other beekeeping organization), emphasizing the need for an equitably accessible solution that connects extension workers to beekeepers remotely. In terms of challenges specific to women beekeepers, only 3 of the 11 women beekeepers reported the same gender-specific challenge: that beekeeping can be physically demanding, which was also reported by women beekeepers in Uzbekistan and Ethiopia. But other than this, there seemed to be few barriers specific to women beekeepers in Lebanon.

Once the Arabic prototype was ready, beekeepers participated in prototype testing, overall revealing that the app was intuitive and would be a useful decision-support tool. With the exception of a few too literal or academic translations, the overall translation, including the localization (i.e., making the app available in Arabic and mirroring the entire app experience from right to left instead of left to right), was successful.

The prototype enabled beekeepers to better understand the project and usefulness of beekeeping-related record-keeping, as well as to learn better beekeeping practices.

Several challenges were also uncovered by the prototype testing; despite the majority of beekeepers finding navigation easy, several felt less comfortable with using mobile applications. Beekeepers also expressed the need for adding the ability to create large numbers of hives, to-dos, and records in bulk, as well as the ability to perform apiary-level inspections for beekeepers with more hives than the pilot group participants (currently the app is best suited for beekeepers with a few hives).

This challenge was also reported by beekeepers in Uzbekistan and Ethiopia; hence, the app is being changed to ease the processes for beekeepers with 20+ hives. Finally, the bee species in Lebanon is known to be aggressive and gloves are a requirement when visiting hives, which can affect the usability of a mobile app for a means of keeping records while at the apiary. To address this last challenge, the project is testing future use of smart-gloves and stylos.

Key findings from the Admin Portal design consultation emphasized the value extension workers place on the ability to monitor the health of hives remotely, and the perception that the Admin Portal will be a useful tool to do so. Extension workers believe they will benefit from the ability to easily communicate information with beekeepers (individually and as a group) about weather, blooming information, healthy beekeeping practices, etc. using the portal's messaging and recommendation feature. Interviews and feedback sessions with extension workers (two women, two men) on the Admin Portal revealed that the key data points to track are

- weather;

- location of apiaries (especially proximity to agricultural regions/farms using pesticides);

- number and frequency of inspections;

- number of hive deaths (and the relationship with proximity to agricultural areas);

- diseases and pests recorded;

- status of the queen and brood (including number of capped and uncapped frames); and

- application of mite treatments.

In terms of communication needs, extension workers reported the importance of communicating the following information to beekeepers in the form of alerts/notifications and recommendations:

- what is the weather (alerts of storms, cold weather, humidity levels, etc.);

- what is blooming and when (the beginning and end of honey flows);

- when are pesticides being applied (so that beekeepers in close proximity to farms know in advance to move their hives);

- when to harvest honey (to prevent harming the wellbeing of the colonies);

- when to winterize;

- when to requeen;

- when to perform seasonal beekeeping tasks (e.g., feeding bees, first inspections, last inspections); and

- what are shared modern and vetted beekeeping advice and practices.

The *AID-CSB A&E* project introduced an innovative advisory and extension approach for beekeepers: for the first time, a hive management app and extension support system were customized for Lebanon. Together, the smartphone app and extension Admin Portal will enable constant access to resources and information and remote knowledge transfer.

Comparing and Contrasting Results from Lebanon, Uzbekistan, and Ethiopia: Designing for Glocal Users

Glocalization design is the concept of designing for groups of users who share some local concerns and needs for design, but in a slightly larger (i.e., more global) context. Scholars in TPC have defined glocalization as "an approach that considers audience expectations related to both local and global concerns" (Breuch, 2015, p. 113) and "the process of localizing documents or materials to local contexts of use" (Saru & Wojahn, 2020, p. 187). Considering design decisions, we combined our larger project datasets and findings from design testing to develop next iterations of the Beekeeper's Companion app. The next section summarizes key findings that apply to the glocal user population for our app. We offer comparisons and contrasts of our three sites' user populations to situate future design considerations (i.e., how we might redesign for enhanced equilibriUX).

UZBEKISTAN AND ETHIOPIA: IMPORTANT OUTCOMES FOR FUTURE DESIGN

As cited in Hopton et al. (2022), of all three countries, beekeepers in Uzbekistan keep the most hive records; all beekeepers interviewed—except one—recorded their observations in paper journals and kept notes on each hive. Some recorded additional data during inspection, and some recorded other types of information in their heads (especially schedules of what to do when). The most pressing challenges for beekeepers in Uzbekistan generally were disease, pesticides or chemicals from nearby manufacturing plants or farms, and pests. Combatting the varroa mite, Foulbrood, and Nosemosis (nosema) were the top disease concerns. Pesticides were deeply concerning to all interviewees and composed a significant part of every interview because they risk the hives and contaminate honey, which means that Uzbek beekeepers cannot meet the import standards of other countries, limiting sales channels and economic potential. Several beekeepers expressed the need for more and better-quality laboratories for honey composition testing and help communicating the risks of pesticide use to farmers. When asked about challenges specific to women beekeepers, like Lebanon, interviewees remarked that beekeeping is physically demanding.

Additionally, the importance of the one-on-one user interview approach was validated during mixed-gender beekeeper discussion groups at the final workshops in Uzbekistan, where men dominated the conversation and women felt less comfortable speaking up. In one-on-one user interviews, women were more comfortable and vocal to voice their opinions.

The effects of climate change were clearly visible shortly before the final project workshops, as Tashkent experienced its worst dust storm on record due to drought (NASA, 2021). When project staff visited a beekeeper's apiary a few days later, he expressed his concern about the potential effects of the dust storm on his honey bees' health and the quality of honey but said would not know how harsh the consequences would be until he opened his hives in the spring.

In Uzbekistan, the most common communication between beekeepers occurs in a large Telegram group to exchange information on markets, share beekeeping photos, ask questions, and sell honey and hive products. Beekeepers also communicate and access information through membership in beekeeping organizations and knowledgeable family members and mentors. Despite the grassroots-level organization of the Telegram group, several types of information (such as pesticide use) and key decisions (such as honey pricing) flow top-down. Details like this are important to maximize networks of people effectively, and they could also have significant impacts on adoption and use of the application.

Unlike Uzbekistan and Lebanon, Ethiopian beekeepers do not keep written records of hive inspections or plan beekeeping tasks. Building these habits will be critical to the successful adoption of the Beekeeper's Companion app and are a key advantage to its introduction. Despite differing from their current practices, beekeepers expressed the helpfulness of the app in recommending what to do next depending on selections made during the inspection process. A greater factor in Ethiopia than in Lebanon and Uzbekistan is the gender digital divide, in which rural, low-income women are less likely to be able to access and use digital technologies.

In terms of gender-specific challenges, a unique challenge reported by Ethiopian beekeepers is that the practice of catching swarming colonies is difficult for women. Additionally, as the subspecies of *Apis mellifera* bee in Ethiopia is more aggressive than their European or American counterparts, apiary visits and inspections need to be conducted in the evening hours when bees are more relaxed, but doing so increases danger to women because their apiaries are often situated in remote areas. Furthermore, most hives are traditional and thus are located on tall trees, and some do not believe the practice of climbing the tree or accessing and controlling the beehive is an appropriate activity for women.

In Ethiopia, traditional gender roles do not encourage women to speak up. During the smartphone training (mixed-gender small groups of five to nine beekeepers with family members, and two field assistants), some women beekeepers were not comfortable actively participating in group discussion. This challenge was anticipated and mitigated by including many peer-to-peer smartphone exercises in which women could practice directly with one another or family members, thus increasing their agency but also normalizing their use of technology.

Ethiopia has no mainstream peer-to-peer communication channels. Instead, trusted research centers (such as Holeta Bee Research Center) and projects (such as YESH) are the main channels for advisory services but are not adequate for the estimated 1 million beekeepers across Ethiopia, emphasizing the need for alternative channels of communication and support, such as the Beekeeper's Companion app.

Particularly in light of COVID-19 and the conflict in Northern Ethiopia, the app will provide a way to evaluate the challenges and successes of beekeepers when in-person visits are challenging. The *AID-CSB* and *AID-CSB A&E* projects demonstrated that it is not only possible to conduct an inclusive, multicountry, multilanguage agricultural UX/UI development project, it should be the new standard. There are several lessons learned and recommendations for other agriculture development projects and future *AID-CSB* and *AID-CSB A&E* work.

Discussion

In the *AID-CSB A&E* project, we successfully developed an infrastructure that supports beekeepers, extension workers, and honey bee health, consisting of the localized Beekeeper's Companion app for hive management and the Admin Portal for extension workers that aggregates and visualizes data from the app. Through close collaboration with local beekeeping experts and researchers, the Beekeeper's Companion app was successfully translated, localized, and made available to a subset of beekeepers, due to excellent working relationships and improvements to the application as beekeepers provide their feedback in a trusted environment. Notably, best practices as well as regional challenges can now be targeted to improve the organization of documenting beekeeping activities that in turn allow for timely advice on required activities, such as hive inspections or disease treatment guided by pooled knowledge and experiences. Further, first steps toward the assimilation of detailed scientific information such as relevant flowering data have been made to better notify on key beekeeping events in due time.

During the short six-month *AID-CSB A&E* project in Lebanon, we had less time than hoped for beekeepers and extension workers to pilot the technologies, which would have gathered more data, and enabled further iterations to the technology based on beekeeper feedback. This is in part due to late receipt of project funds (early February) which prohibited the hiring and payment of select staff and services, along with technical app delays to ensure beekeepers could access the app locally. We furthermore wanted to ensure a level of quality and stability before sharing the app with beekeepers; recognizing that sharing an underdeveloped product may generate mistrust and low level of confidence in the technology.

Despite this, the beekeepers who are already piloting the app are actively using it and have provided positive feedback about the overall beekeeping support provided by the app, as well as suggestions for enhancements of select features (e.g., ability to record data at apiary-level).

The Way Forward: EquilibriUX

To achieve the overarching design goal of equilibriUX (balance and equilibrium between designer and user; Lancaster & King, 2024), stakeholders have to work closely together to carefully evaluate the tension between localization and customization, on the one hand, and scalability on the other hand; a tension that is exacerbated by project, financial, and technology constraints. But it is essential that users feel heard and understood. For our project, users' engagement over time will continue to validate or invalidate our design and use assumptions. At the same time, to be sustainable in the long run, the final product designed and developed must be maintainable across countries of users. Scale is a perennial challenge for developers, especially those designing for balance and equity, because scale and localization/customization are at odds with one another as design principles. Perhaps a solution is to conceptualize the idea of scale differently, to consider scale not as volume of users, for example, but as function and features, which can be customized and localized with the help of ever-advancing AI and commitments to participatory design. For the *AID-CSB* and *AID-CSB A&E* project, future scaling opportunities include tapping the marketplace for data-authenticated honey and pollination services outlined above, which have the potential to pave the way toward local and global financial sustainability. In Lebanon, app users will continue to have free and unlimited access to the app, which will be maintained by HiveTracks. The project team hopes to secure another grant in the near future to further expand and enhance the digital system to further address user needs. The PPP approach chosen by the project team seems well-suited to effectively solve this issue.

Finally, it will be key to understand the depth, quality, and frequency of data that can be collected while safeguarding the users' privacy. For our project, this has the potential to enable myriad new data-driven policies that protect pollinators and local ecosystems across Uzbekistan, Ethiopia, and Lebanon (and other areas of the globe). In so doing, we hope the Beekeeper's Companion and the Admin Portal developed for LARI will become a beacon of smart technology use that protects vulnerable groups of people, indigenous knowledge systems, and the environment alike.

References

Acharya, K. R. (2019). Usability for social justice: Exploring the implementation of localization usability in Global North technology in the context of a Global South's country. *Journal of Technical Writing and Communication, 49*(1), 6–32. https://doi.org/10.1177/0047281617735842

Acharya, K. R. (2018). Usability for user empowerment: Promoting social justice and human rights through localized UX design. In *Proceedings of the 36th ACM International Conference on the Design of Communication* (pp. 1–7). http://dx.doi.org/10.1145/3233756.3233960

Aktar, M. W., Sengupta, D., & Chowdhury, A. (2009). Impact of pesticides use in agriculture: Their benefits and hazards. *Interdisciplinary Toxicology, 2*(1), 1–12. https://doi.org/10.2478/v10102-009-0001-7

Antonio, A., & Tuffley, D. (2014). The gender digital divide in developing countries. *Future Internet, 6*(4), 673–687.

Asegid, A. (2020). Impact of climate change on production and diversity of coffee (Coffea Arabica L) in Ethiopia. *International Journal of Research Studies in Science, Engineering and Technology, 7*(8), 31–38. http://www.ijrsset.org/pdfs/v7-i8/5.pdf.

Baada, N., & Najjar, J. (2020). A review of the effects of migration on the feminization of agrarian dryland economies. *Journal of Agriculture, Gender, and Food Security, 5*(2), 1–12. https://dx.doi.org/10.19268/JGAFS.522020.1

Bannon, L. J. & Ehn, P. (2013). Design matters in participatory design. In J. Simonsen & T. Robertson (Eds.), *Routledge handbook of participatory design* (pp. 37–63). Routledge.

Becker, L., Rünzel, M., Hopton, S. B. (2022). *AI-Driven Climate Smart Beekeeping for Women 2021 Project Report Brief*. International Center for Agricultural Research in the Dry Areas (ICARDA). https://hdl.handle.net/20.500.11766/67515

BeeXML Apimondia Working Group. (2021). *BeeXML data standardisation*. BeeXML. https://beexml.org

Breuch, L. A. K. (2015). Glocalization in website writing: The case of MNsure and imagined/actual audiences. *Computers and Composition, 38*, 113–125.

Bogale, S. (2009). Indigenous knowledge and its relevance for sustainable bee-keeping development: A case study in the Highlands of Southeast Ethiopia. *Livestock Research for Rural Development, 21*(11), 184. http://www.lrrd.cipav.org.co/lrrd21/11/boga21184.htm

Borgesius, F. Z. (2018). *Discrimination, artificial intelligence, and algorithmic decision making.* Council of Europe. https://rm.coe.int/discrimination-artificial-intelligence-and-algorithmic-decision-making/1680925d73

Boyd, D. R. (2021). *Report of the special rapporteur on the issue of human rights obligations relating to the enjoyment of a safe, clean, healthy and sustainable environment.* Office of the High Commissioner United Nations Human Rights. https://www.ohchr.org/sites/default/files/Documents/Issues/Environment/SREnvironment/Report.pdf

Chandran, R. (2016). Indian villages ban single women from owning 'distracting' mobile phones. *Reuters.* https://www.reuters.com/article/us-india-women-phone/indian-villages-ban-single-women-from-owning-distracting-%20mobile-phones-idUSKCN0VZ1AA

Currie, H., Harvey, A., Bond, R., Magee, J., & Finlay, D. (2022). Remote synchronous usability testing of public access defibrillators during social distancing in a pandemic. *Scientific Reports, 12*(1), 1–10. https://doi.org/10.1038/s41598-022-18873-7

Feeney, M. K., Fusi, F., Gasco-Hernandez, M., Nesti, G., Cucciniello, M., & Gulatee, Y. (2021). A critical analysis of the study of gender and technology in government. *Information Polity, 26*(2), 115–129. https://doi.org/10.3233/IP-200303

Food and Agriculture Organization. (2011). *The state of food and agriculture: Women in agriculture—Closing the gender gap for development.* Food and Agriculture Organization of the United Nations. https://www.fao.org/3/i2050e/i2050e.pdf

Gonzales, L., Lewy, R., Cuevas, E. H., & Ajiataz, V. L. G. (2022). (Re)designing technical documentation about COVID-19 with and for Indigenous communities in Gainesville, Florida, Oaxaca de Juárez, Mexico, and Quetzaltenango, Guatemala. *IEEE Transactions on Professional Communication, 65*(1), 34–49. https://doi.org/10.1109/TPC.2022.3140568

Grant, M. (2014, September 1). Egyptian Islamic authority issues Fatwas against selfies and chatting online. *Newsweek.* https://www.newsweek.com/egyptian-islamic-authority-issues-fatwas-against-selfies-and-chatting-online-%20267832

Haas, A. M. (2012). Race, rhetoric, and technology: A case study of decolonial technical communication theory, methodology, and pedagogy. *Journal of Business and Technical Communication, 26*(3), 277–310.

Hitt, A. (2018). Foregrounding accessibility through (inclusive) universal design in professional communication curricula. *Business and Professional Communication Quarterly, 81*(1), 52–65.

Hopton, S. B. (2021). The tarot cards of tech: Foretelling the social justice impacts of our designs. In G. Agboka & R. Walton (Eds.), *Equipping technical communicators for social justice work: Theories, methodologies, pedagogies* (pp. 158–178). University of Colorado Press.

Hopton, S. B., Rünzel, M., & Becker, L. (2022). Localizing climate-smart applications through participatory design: A case study of the Beekeeper's Companion. In C. Stephanidis, M. Antona, & S. Ntoa (Eds.), HCI International 2022 Posters. *Communications in Computer and Information Science, 1581.* Springer. https://doi.org/10.1007/978-3-031-06388-6_45

Iglesias, C. (2020). *The gender gap in internet access: Using a women-centred method.* World Wide Web Foundation. 10. https://webfoundation.org/2020/03/the-gender-gap-in-internet-access-using-a-women-centred-method

Kakar, Y. W., Hausman, V., Thomas, A., Denny-Brown, C., & Bhatia, P. (2012). Women and the web: Bridging the internet gap and creating new global opportunities in low and middle-income countries. *Intel.* https://www.intel.la/content/dam/www/public/us/en/documents/pdf/women-and-the-web.pdf

Koerber, A. (2000). Toward a feminist rhetoric of technology. *Journal of Business and Technical Communication, 14*(1), 58–73.

Lancaster, A., & King, C. S. T. (Eds.). (2022). Localized usability and agency in design: Whose voice are we advocating? [Special issue]. *Technical Communication, 69*(4), 1–6.

Lancaster, A., & King, C. S. T. (Eds.). (2024). *Amplifying voices in UX: Balancing design and user needs in technical communication.* State University of New York Press.

Murphy, S., Belmonte, W., & Nelson, J. (2009, September). *Investing in girls' education: An opportunity for corporate leadership.* Harvard Kennedy School. https://www.hks.harvard.edu/sites/default/files/centers/mrcbg/programs/cri/files/report_40_investing_in_girls.pdf

NASA. (2021, November 5). *Dust blankets Tashkent.* NASA Earth Observatory. https://earthobservatory.nasa.gov/images/149067/dust-blankets-tashkent

Olana, T., & Demrew, Z. (2018). The role of women in beekeeping activities and the contribution of bee-wax and honey production for livelihood improvement. *Livestock Research for Rural Development, 30*(7), 118.

Pörtner, H. O., Scholes, R. J., Agard, J., Archer, A. E., Arneth, A., Bai, X., Barnes, D., Burrows, M., Chan, L., Cheung, W. L., Diamond, S., Donatti, C., Duarte, C., Eisenhauer, N., Foden, W., Gasalla, M. A., Handa, C., Hickler, T., Hoegh-Guldberg, . . . Ngo, H. T. (2021). *Scientific outcome of the IPBES-IPCC co-sponsored workshop report on biodiversity and climate change.*

Intergovernmental Science-Policy Platform on Biodiversity and Ecosystem Services (IPBES). https://zenodo.org/record/5101125

Pope-Ruark, R., Tham, J., Moses, J., & Conner, T. (2019). Introduction to special issue: Design-thinking approaches in technical and professional communication [Special issue]. *Journal of Business and Technical Communication*, *33*(4), 370–375.

Primo, N. (2003). *Gender issues in the information society*. UNESCO Publication for the World Summit on the Information Society. https://unesdoc.unesco.org/ark:/48223/pf0000132967

Rose, E. J., Edenfield, A., Walton, R., Gonzalez, L., McNair, A. S., Zhvotovska, T., Jones, N., Garcia de Mueller, G. I., & Moore, K. (2018). Social justice in UX: Centering marginalized users. In *SIGDOC '18: Proceedings of the 26th ACM International Conference on the Design of Communication*. *21*, 1–2. https://doi.org/10.1145/3233756.3233931

Sackey, D. (2020). One-size-fits-none: A heuristic for proactive value sensitive environmental design. *Technical Communication Quarterly*, *29*(1), 33–48.

Salvo, M. J. (2002). Critical engagement with technology in the computer classroom. *Technical Communication Quarterly*, *11*(3), 317–337.

Saru, E. H., & Wojahn, P. (2020). "Glocalization" of health information: Considering design factors for mobile technologies in Malaysia. *Journal of Technical Writing and Communication*, *50*(2), 187–206. https://doi.org/10.1177/0047281620906131

Shirley, B. (2019). Working toward social justice by engaging other disciplines in engaging communities: A technical communication scholar's role. In *Proceedings for 2019 CPTSC Annual Conference*.

Shivers-McNair, A. (2017). Localizing communities, goals, communication, and inclusion: A collaborative approach. *Technical Communication*, *64*(2), 97–112.

Spinuzzi, C. (2005). The methodology of participatory design. *Technical Communication*, *52*(2), 163–174.

Steele, C. (2019). *What is the gender divide?* Digital Divide Council.

Steinke, J., Ortiz-Crespo, B., Etten, J., & Müller, A. (2022). Participatory design of digital innovation in agricultural research-for-development: Insights from practice. *Agricultural Systems*, 195, Alliance of Biodiversity International and CIAT, Digital Inclusion, Montpellier, France.

Sterling, R., Grubbs, L., & Koutsky, T. (2020, December 17). Breaking through the gender digital divide: Technology, social norms, and the WomenConnect challenge. TPRC48: The 48th Research Conference on Communication, Information and Internet Policy. http://dx.doi.org/10.2139/ssrn.3749643

Sterling, R., & Koutsky, T. (2018, March 16). Understanding the gender digital divide: Social norms and the USAID WomenConnect Challenge. TPRC46: The 46th Research Conference on Communications, Information, and Internet Policy. https://ssrn.com/abstract=3142049

Sun, H. (2006). The triumph of users: Achieving cultural usability goals with user localization. *Technical Communication Quarterly*. *15*(4), 457–481.

Sun, H., & Getto, G. (2017). Localizing user experience: Strategies, practices, and techniques for culturally sensitive design. *Technical Communication*, *64*(2), 89–94.

Suresh, H., & Guttag, J. V. (2019). A framework for understanding unintended consequences of machine learning. *Equity and Access in Algorithms, Mechanisms, and Optimization*, https://doi.org/10.1145/3465416.3483305

Tham, J. C. K. (2021). Engaging design thinking and making in technical and professional communication pedagogy. *Technical Communication Quarterly*, *30*(4), 392–409.

World Bank Group. (2017). Help women farmers "get to equal." *The World Bank*. https://www.worldbank.org/en/topic/agriculture/brief/women-farmers-getting-to-equal

United Nations. (n.d.). Women as agents of change. *United Nations Climate Action*. https://www.un.org/en/climatechange/climate-solutions/womens-agents-change

United Nations General Assembly. (1948). *Universal Declaration of Human Rights*. Refworld. https://www.refworld.org/docid/3ae6b3712c.html

United Nations General Assembly. (1966). *International Covenant on Civil and Political Rights*. *United Nations, Treaty Series*. Refworld. https://www.refworld.org/docid/3ae6b3aa0.html

United Nations General Assembly. (1989). *Convention on the Rights of the Child, United Nations, Treaty Series*. Refworld. https://www.refworld.org/docid/3ae6b38f0.html

United Nations General Assembly. (2019a). Analytical study on gender-responsive climate action for the full and effective enjoyment of the rights of women. https://undocs.org/A/HRC/41/26

United Nations General Assembly. (2019b). Human rights obligations relating to the enjoyment of a safe, clean, healthy and sustainable environment. https://undocs.org/en/A/76/150

Zachry, M., & Spyridakis, J. H. (2016). Human-centered design and the field of technical communication. *Journal of Technical Writing and Communication*, *46*(4), 392–401. https://doi.org/10.1177/004728161665349

Chapter 14

"It Killed That Girl"

Toward Safer Ridesharing Experiences
through Feminist UX Design

Lin Dong and Elizabeth Topping[1]

Gender bias in digital products can be hard to perceive. A recent series of tragedies involving a Chinese ridesharing application demonstrates that overlooking the role of gender in design can cost lives.

DiDi was China's first licensed online ride-hailing servicer ("DiDi's milestones," 2015). One of its businesses, DiDi Hitch, was a C2C, inter-city platform that matched private car owners and commuters to share transportation resources. Hitchhikers paid a small amount to the driver based on DiDi's algorithm. DiDi started the Hitch service in 2015 as a move into the sharing or gig economy. In the *2014–2020 China's Ride-sharing Industry Development Bluebook* (Dida Travel, 2020), ridesharing is acknowledged as a "green travel means that is supported by the government policy" because it can "promote transport supply, ease traffic congestion, reduce the emissions, and harmonize interpersonal relationship" (p. 1). This bluebook also names the essential traits of ridesharing: nonbusiness, nonprofit, nonservice, but mutual help oriented. By the time of publication, DiDi Hitch had 15 million drivers and 150 million hitchhikers registered—that is, 1 in 10 Chinese use DiDi Hitch. Chinese law affirms that the intermediary service platform DiDi has legal obligations when matching the driver and the hitchhiker.

In 2018, two female passengers were raped and killed by their Hitch drivers. The public outcry centered around the user design, which minimized female rider agency. In the first case, a female flight attendant called the Hitch service at 11:55 p.m. for a ride to the train station but encountered sexual assault when the driver aborted the service, logged out of the app, and murdered her on a waste dumping ground ("Flight attendant," 2018). Three months later, a 20-year-old female hailed a Hitch car for a 1-hour afternoon journey, but the driver deviated from the requested route and drove to a remote mountain road where the driver robbed, raped, killed, and dumped her body ("8/24 Leqing girl," 2018).

People blamed DiDi for its flawed safety design, ineffective emergency responses, and delayed crisis communication. For example, both drivers had a bad record (traffic accident or gambling) but passed the background check by using fake identities; no alarms were sent to the victims when they were taken off the scheduled path; and the user interface did not have an emergency button. Customer service was also criticized because its bureaucratic communication style seriously delayed a potential rescue of the victim in the second case.

These two heinous cases urgently call designers to review safety vulnerabilities in ridesharing applications. In China alone, 4 homicide cases and 13 other sexual assault cases related to Didi have been reported,[2] and frequent similar crimes targeting females have occurred around the world. In this study, we focus on the domain of user experience design (UXD) to see how ridesharing technology design failed and continues to fail female users. Lack of attention to the role of gender in ride-sharing application design creates an unequal and unsafe user experience for women. As female passengers ourselves, we realize our own vulnerability, and therefore as researchers we study UXD from a feminist perspective: that is, we investigate gender issues, analyze their technical embodiment, and suggest design heuristics for improving user advocacy in ridesharing's online and physical spaces. We use the term UXD to refer to the union of technology, design, product, and service that includes these dimensions: information architecture, human-computer interaction (HCI), interface design, visual design, usability, and customer service. These elements comprise the complete system of product design and the entire experience of usage.

A feminist approach to UX design means that we view gender as an element of design (see historical examinations: Bardzell, 2010; Berg & Lie, 1995; Gurak & Bayer, 1994; Rode, 2011; van Oost, 2003). Offline gender situations extend to and are reflected in online space. Designers will create a deficient digital product if they do not consider gender issues in real life (e.g., gender identity, relation, interaction, participation), just as the DiDi cases suggest. Technology itself is not gender neutral; rather, it is inherently gendered. Studies have argued that computer-mediated communication enabled by information technologies does include gender bias in many aspects, including access to technology, dominance of discussions, use of language and rhetoric, information content, and misogynist attitudes (Gurak & Ebeltoft-Kraske, 1999; Herring, 1999). We believe that the idea "technology is gendered" should be understood from two sides: on the one hand, gender does not—and should not—be consequential to the use of technology; on the other hand, gender experiences in real-life settings should be incorporated into technology design. Parameterizing gender differences and biases in technical products and spaces is detrimental. A feminist approach will help technical designers be more aware of gender biases as one of many variables necessary to consider in design to achieve equality for users.

Creating a more friendly, inclusive, and empathic technical culture both online and offline for users requires designers and researchers revisit past calls of integrating feminism into technical communication (TC; e.g., E. A. Flynn, 1997; J. F. Flynn, 1997; Gurak & Bayer, 1994; Lay, 1989; Smith & Thompson, 2002), reexamine the vitality of early feminist theories and models, track the development of new approaches to feminist design, and put new design insights and opportunities into practice. Our study is an attempt to concretize this blueprint.

We first rhetorically analyze the business strategies and UXD of DiDi, primarily through marketing material, to find its gender biases and design epistemology. Then, this study puts forward a heuristic for effective safety design in ridesharing user interfaces. This chapter shows that feminism helps demonstrate objectivity, interpretation, values, and social consequences in UXD. We hope through this study, user experience (UX) designers, product managers, and educators in Technical and Professional Communication (TPC) will consider participatory or user-centered design from a new perspective of feminist UX. We also hope this feminist UX methodology will contribute to constructing more empathic, inclusive, and functional localized products and services worldwide, a potential benefit for TC's rising interest in social change (see Agboka & Dorpenyo, 2022).

Feminism in Technical Communication

APPROACHES TO GENDER AND TECHNOLOGY

TC is not the only field to discuss gender and technology. HCI and information and communications technology (ICT) also engage with gender. Although there are various ways of addressing, or not addressing, gender and technology in communication design, these three fields bear much similarity in theorizing feminism and approaching gender. This similarity is a natural result of three positions on the relevance of gender and technology—disregarding, overemphasizing, and somewhere in between. Below, we describe these three mainstream feminist positions in TC, HCI, and ICT.

LIBERAL FEMINISM—"TECHNOLOGY IS GENDER-FREE"

Liberal feminism claims that women have been long excluded from the history of technology (E. A. Flynn, 1997). To resolve women's underrepresentation in male-dominated technical institutions, liberal feminists argued that women deserve equal opportunities to credentials and technical jobs (Gurak & Bayer, 1994). In this perspective, technology is both a job and knowledge tool to "develop their intellectual capacities" (E. A. Flynn, 1997, p. 315) and leads to women's emancipation. Centering around the appeal of "equality for women" (Gurak & Bayer, 1994, p. 264), liberal feminism minimizes or even denies biological gender differences and accordingly the relevance of gender to technology, assuming that technology itself is gender neutral (Gurak & Bayer, 1994; Rode, 2011).

In TC and HCI, holding this position results in the dismissal of gender as a design element (Rode, 2011). Designers do not view themselves as having the agency to narrow the gender gap since technologies are not the source of gender differences—social problems are. One representative for gender-free design, Cassell (2002) promoted "undetermined design" (p. 12), which means giving users the freedom to construct their own gender identity through their use of gender-blind technologies.

RADICAL FEMINISM—"TECHNOLOGY IS GENDER SPECIFIC"

In challenging the beliefs of neutral technology, radical feminism emphasizes biological and psychological gender differences in task performance and separates the spheres within which gender performances operate (E. A. Flynn, 1997; Gurak & Bayer, 1994). However, this attempt to correct liberal feminism's neutrality easily oscillates to a male/female dichotomy.

As Rode (2011) pointed out, technology designers in HCI treat gender as a variable and parameterize gender differences based on repeated experiments. Feminist technology embraces feminine subjectivity and intuition (Gurak & Bayer, 1994, p. 264) and thus becomes gender-specific when it targets female users. Examples of radical feminism studies include Tan et al. (2003) on gender differences in spatial navigation ability, Kelleher et al.'s (2007) design for teaching girls to program, and Laurel's (2003) game Purple Moon, specifically designed for pre-adolescent girls.

Postmodern Feminism—"Technology is Gender Polynary"

Postmodern feminists repudiate radical feminism's binary male/female thinking and its focus on women's segregation from men and instead recognize gender as socially constructed (E. A. Flynn, 1997; Gurak & Bayer, 1994), aiming to understand nuanced differences within the general category of "female." E. A. Flynn (1997) suggested a fluid consideration of gender to provide men an opportunity to participate in feminism.

Postmodern feminism brings to TC and HCI a new approach to gender and technology—as we describe it, "gender-polynary," equivalent to "pluralism" (Bardzell, 2010) or "a situated activity" (Rode, 2011). This approach means situating gender in historical and material circumstances where gender and technology mutually shape each other in everyday interaction. It means to resist any single, totalizing, or universal point of view on technology design (Bardzell, 2010).

Acknowledging "gender is polynary" in design is a rejection of "gender is universal." Bardzell (2010) warned about the fallacy of "universal usability" built on Western technological norms and practices, saying it is "dangerous" to "demote cultural, social, regional, and national differences in user experiences and outlooks" (p. 1305). Instead, she proposed a human-centered UXD to meet diverse and specific needs of non-Western female users.

Taken together, these three feminist approaches of defining technology vary drastically. Liberal feminism refutes gender inequality inside the milieu of technology design process. Radical feminism parameterizes gender in design. Postmodern feminism situates gender design in social contexts and encourages a constructive engagement of cultural differences.

Which approach holds the most promise? Our study indicates the pitfalls of liberal feminism's view of technology and the failure to account for the lived experience of gender in design. Instead, our design heuristic is rooted in polynary thinking.

GENDER IN TECHNICAL COMMUNICATION: PAST, PRESENT, AND FUTURE

Research on feminist TC began with Mary Lay's 1989 article "Interpersonal Conflict in Collaborative Writing: What We Can Learn from Gender Studies." Drawing perspectives from gender studies, linguistics, philosophy, social psychology, managerial communication, and rhetoric and composition, Lay illustrated that knowing conflict sources enables TC scholars to predict and ease collaborative conflicts. After more than three decades, scholars from TC and the broader field of rhetoric and composition developed the study of feminist TC, but not as substantially or as advanced as the passage of time suggests. Rather, we describe the overall scholarship in this area as homogeneous and stalled. Our claim is based on four comprehensive literature studies on feminism in TC: Smith and Thompson (2002), Thompson (1999), Thompson and Smith (2006), and White et al. (2015). These four reviews examine feminism studies in TC from 1989 to 2012 published as journal articles[3] and textbooks.

First, the number of articles published pertaining to gender and women steadily declined from 1989 to 2012. From 1989 to 1997, 40 articles about gender and feminism were published, averaging four articles per year (Thompson, 1999). From 1998 to 2004, 21 articles in total and 3.5 articles per year were published (Thompson & Smith, 2006). From 2005 to 2012, 22 articles were published, meaning 3.14 articles per year (White et al., 2015). Although these data seem insufficient to generate a precise numerical trend, they alarmingly indicate that feminism has become less popular in TC.

Second, the research content seems homogenous. Thompson (1999) found that feminist articles in 1989–1997 "share[d] a common concern for the inclusion of women" (p. 163). This common theme of *inclusion/ exclusion* calls for valuing women's talents, skills, and historical contributions in homes, workplaces, and classrooms. To realize inclusion, feminist TC scholars suggested (1) eliminating sexist language, (2) providing equal opportunity in the workplace, (3) valuing gender differences, (4) recovering women's historical contributions to TC, and (5) critiquing previously uncontested terms and concepts (p. 164). Smith and Thompson (2002) found two thematic changes from 1998 to 2003: first, "sexist language" was replaced by "the implicit masculine bias in value-neutral technology (such as in Durack, 1997; Ranney, 2000), and second, the theme of equal opportunity was dropped. White et al. (2015) revisited the scholarship, listing four common themes: studies of history and recovery of female rhetorical practices, embodiment and rhetorics of the female body, technology and design considerations for female audiences, and gender problems in the workplace. White et al. (2015) also investigated nine TC textbooks on the treatment of gender and found comparable topics as Thompson's (1999) categories. Overall, most topics about gender and technology in TC have remained the same for decades, which suggests that research on feminist TC has stalled.

Third, the epistemologies behind the publications have not changed much. The most common research topic, inclusion/exclusion of women in history and practice of TC, has prevailed for decades, which reflects an epistemology of liberal feminism that sees technology as "value-neutral" or gender-free (discussed earlier). Although early scholars argued that TC had to be expanded to include women "as technologists" (Gurak & Bayer, 1994) or "as users" (Durack, 1997) of technology, their work carried a strong radical feminist note to underscore women's ways of knowing, speaking, or doing technologies—the feminine qualities traditionally attributed to women (e.g., Durack, 1997; Lay, 1994; Lind, 1999; Lippincott, 1997; Sauer, 1994; Tebeaux, 1990).

Although feminist TC scholarship has not changed much since the 1980s, new developments and turning points are materializing. In the broader field of social science, a trend to emphasize gender and Internet/information technologies is emerging (Royal, 2005). In TC, we see new research topics, such as embodiment (Koerber, 2000; Wang, 2021; Zdenek, 2007); information design and usability centered on female users (Hallenbeck, 2012; Stenstrom et al., 2008; Zdenek, 2007); agency and empowerment (Novotny & Hutchinson, 2019); and fertility and other women healthcare technologies (Frost, 2020). New feminist methodologies are also appearing, such as antenarrative in feminist historiography (J. F. Flynn, 1997; Petersen & Moeller, 2016) and human-research design (Jones, 2016). In terms of epistemological development, studies increasingly use (such as Barker & Zifcak, 1999; Tong & Klecun, 2004) a postmodern feminist lens to view gender in a panoramic context. Doing so responds to E. A. Flynn's (1997) call for understanding gender as socially constructed, Thompson and

Smith's (2006) suggestion on turning research "from inclusion to critique" (p. 196) and Herrick's (1999) proposal of examining local enculturation.

By adopting a postmodern feminist perspective to investigate human body/technology interactions, female embodiment, and user agency in a non-Western country, our study will fill a theoretical, practical, and contextual gap in feminist technical communication. We call this gap *feminist UXD*.

Our Theorization of Females in Traveling Apps

The tragic circumstances in which women were victimized while using ridesharing technology illustrate that technology is rarely, if ever, neutral. Thus, attention to how technology obscures underlying ideologies follows Donna Haraway's call in her 2016 *Manifestly Haraway*: "*pleasure* in the confusion of boundaries and *responsibility* in their construction" (p. 7, emphasis in original). A service nominally intended to bring together individuals to share transportation seems a neutral interaction. However, more exists in this cybernetic space. Using Haraway's exploration of the idea of cyborg—"a hybrid of machine and organism, a creature of social reality as well as a creature of fiction" (p. 5)—this chapter explores the ways in which the technology constructs gender (Herrick, 1999) as a lived body experience (Young, 2005). This section develops the theoretical framing of our analysis of Didi's marketing material.

BOUNDARIES AND INTERSECTIONS

Haraway (2016) has identified the relationship between men and women as one part of the "border war" between organism and machine, in which women's bodies "can be dispersed and interfaced in nearly infinite, polymorphous ways with large consequences for women and others" (p. 31). The tragedies summarized in this chapter's introduction suggest that women's bodies are interpellated as objects in the rideshare technological ecosystem rather than being locally constructed as autonomous individuals. The mediated interaction between driver and user places the identity of the individuals within "a cybernetic information system" (Balsamo, 1992, p. 11). Because our identities interweave with technology, the technology establishes functional connections and boundaries linking individuals through the app. Thus, the application mediates how we read individuals, their bodies, and their gender. These boundaries are partially arbitrary (Balsamo, 1992). However, the responsibility of constructing those boundaries falls to design choices.

We exist in a web of symbolic relationships that construct our reality (Stob, 2008). These symbol systems are influenced by and impact our social relationships, which have personal and political consequences. In the case of app technology, the symbolic system linking riders and drivers uses marketing materials to construct identities long before the ride begins. The gendered systems within us are mediated further by the visual architecture of the app. Opting to believe that technology is gender neutral or gender specific can further constrain design choices in ways that deliberately or implicitly fail to recognize how gender is socially and locally constructed. However, recognizing how gender operates and is interpellated through the technology can allow designers to choose a path that responsibly attempts to avoid the worst outcomes.

SYMBOLIC CONSTRUCTION OF USER-APP INTERACTION

Haraway's (2016) cyborg model also helps to explain how the symbolic construction of ridesharing technology neglects the boundaries the application itself places on the interaction. Often, design and programming choices focus on making the service work. Social aspects are overlaid on the economic purpose. Safety concerns appear in this second layer, or sometimes not at all. On its surface, the rideshare app brings together individuals who would not normally otherwise interact; the driver shares the car with the rider, who needs transportation service. During the journey, the two can establish a brief interpersonal relationship. Hailing a ride is first and foremost an economic transaction. The user inputs a destination and chooses a predetermined option. The request goes to a nearby driver's queue. Driver and user do not know each other's identities until the ride has been confirmed. The interaction and negotiation between rider and driver are indirect, inviting our preexisting internal symbol systems into the interaction. This rhetorical space allows users to locally construct their relationship with each other, their genders, and local power dynamics (Herrick, 1999).

Before the ride is booked, app users are directed to think of the other as a service provider—the one who drives and the one who pays. The individuals become defined by their roles. The app-mediated interaction positions both rider and driver "in systems of evaluation and expectations which often implicate their embodied being" (Young, 2005, p. 17). Each participant depends on social expectations that prioritize the economic transaction and safety over other desires and behaviors. When marketing and other business materials outside the application direct attention toward the female body and app design choices do not center rider agency, social norms serve as the last safety check. For traveling apps, that assumed social contract is too frequently broken.

Public outcry has led to considerable safety improvements in ride-sharing applications. Though the response is a welcome change, the technology must be designed in ways that center rider autonomy and agency. If the application is truly meant to be social, then designing for social relationships that recognize how gender is constructed in the moment should not be an afterthought.

Research Methods

Conceptualizing the DiDi app as a social/rhetorical space that shapes interactions between drivers, riders, and the platform itself, we wanted to examine the gendered features of DiDi and ask what UX design problems contribute to sexual crimes in ridesharing. We posed the following research questions:

1. Is DiDi a gender-free, gender-specific, or gender-polynary technical product?

2. How does DiDi embody gender?

3. What gendered features can or cannot be found in its UX design?

UX studies are usually based on product and/or user analysis. However, this method is not applicable for DiDi because its previous version was removed from the app store after the 2018 crimes. Without access to either the product or the users, we took an indirect approach to UX—we started from examining business materials, drew implications about business and marketing requirements, and deduced user requirements. To be specific, we used rhetorical criticism to study DiDi Hitch's advertisements to analyze its creation and circulation of sexual rhetoric and then evaluated the product design based on those analytical results.

The rationale behind this approach is a common belief and established practice that good UXD aligns both user goals and business goals (Marsh, 2016). Product teams collect requirements from a variety of sources (e.g., sales, marketing, customers, end users) and use the information to determine the functionality of the product. Some business, marketing, and sales requirements do represent a form of end user requirements (Baxter et al., 2015). Product logic (such as user positions, user needs/wants/benefits, context of use) can be found in advertisements. (We found that Hitch's advertisements carry strong eros-appealing and gender-biased messages, so we believe this indicates the same tendency in designers' interest in and perspective of users, as well as designers' values and ideologies behind product making.)

The research data come from a complete collection of 40 Hitch advertisements published in 2015–2018 (after the launch of service and before the homicide cases). Using Saldaña's (2016) coding methods for qualitative research, we followed a research process with first- and second-cycle coding: (1) coding visual and textual data in 40 pictures in words or short phrases, (2) categorizing the codes, and (3) theming the categories to assertions in extended phrases or sentences. For the first-cycle code, we used descriptive (documenting the material products and physical environments in a word or short phrase), action (describing observable activities), in vivo (drawing people's words from their own language), and values (reflecting people's needs and wants) codes. Our second-cycle coding applied a theoretical coding method, looking for central themes that incorporate opposing norms and patterns between genders. To generate richer codes and ensure coding validity, two researchers independently coded all the data. Their intercoder agreement was 89.28%.

Findings: Sexist "Gender-Free" UX Design

After the first-cycle coding, we separated female-driver advertisements from male-driver ads after finding striking differences in gender representation. We then focused on advertisements containing male drivers and female riders. The coding charts in Table 14.1 and 14.2 (see Appendix 14.A) document the codes and categories from both cycles of coding.

In comparing female- and male-driver advertisements, we saw similarities and differences. In number, there were 35 male versus 5 female driver advertisements. Targeting men as drivers and women as passengers/

riders reinforced physical and social discrepancies among genders. In male-driver advertisements, female riders possess feminine attractiveness in appearance (pretty, young, and thin), whereas men were attractive in other aspects (e.g., kind, punctual, reliable, good driving skill). In female-driver advertisements, women were mature, educated, and independent but feeling empty inside, whereas males were nonexistent in the picture but implied as a riding and romantic mate for whom the female driver was looking. Both types of advertisements emphasize the social benefits of ridesharing (e.g., beautiful meetings, exciting journeys). Doing so transfigures mutual-help hitchhiking into romantic traveling opportunities, which invites the dangers of sexual crimes.

Extracting the themes from second-cycle coding, we conclude how DiDi identifies genders. DiDi's theoretical construct of genders—the "cyborg identities" of women and men are as follows:

Core Theme: Women are sexual subjects and men are sexual predators.

Supporting Themes

- Women are attractive and feel empty inside.

- Women are active and want a riding/romantic companion.

- Women are evaluated, chosen, gazed on, served, gifted, and discussed before, during, and after trips.

- Women are usually riders who have no control of the car or the relationship.

- Female drivers are viewed as "different" and hope for a return to traditional feminine roles.

- Males are usually drivers who have power to pick up female riders, choose the route, and reward female riders with economic benefits.

DiDi's sexist perspective is rooted in a patriarchal culture in which women exist only because men made them so, and woman as a self never exists. The DiDi case demonstrates Haraway's (2016) criticism of the cyborg as offspring of patriarchal capitalism that intensifies male domination (p. 9) through mobile media and communication, where this new space of human living does not annihilate "conventional inscription of the gendered, race marked body" (Balsamo, 1992, p. 26). On the contrary, it breeds, nurtures, amplifies, and solidifies sexism.

DiDi's ads delineate target consumers/users of the app and their characteristics. Identifying target users and analyzing their needs/wants is the bedrock principle for product design. Using the results of ad analysis, we retrodicted the user persona, user goals, and context of use, all of which are basic to user design. Doing so helps underline DiDi's embedded sexism in its UXD. Figures 14.1 and 14.2 (see Appendix 14.A) are two user personas we created for DiDi Hitch based on the findings of ad analysis.

The two user personas demonstrate DiDi Hitch's conception of user goals, needs, and expectations—that is, typical users like Lily Wang (Figure 14.1) and Dave Tong (Figure 14.2) want the app to be cost effective and socially supportive. We do not think it problematic to design a ridesharing service to meet economic and social needs. We think the problem is meeting *merely* these needs but neglecting other equally or more critical issues, such as safety design, emergency response, and privacy protection. These are life-matter issues that pose great danger for ridesharing users. Though we did not have access to Hitch's original interface design, user personas illuminate that safety and privacy protection were not primary concerns when determining user needs, which verifies design problems (e.g., no emergency button, no alert sent to the passenger or their contacts) confirmed by regular users. The emphasis on finding meaningful dating relationships required female riders to give up their privacy if their male driver wished to continue contact—an expectation built into the user personas and marketing materials. Such a presumption positions women's security and safety as unimportant in contrast to the goal of growing the business, a strategy that can be fatal for both users and the company. Valuing the brand's business goals over user needs can be a detrimental UX strategy.

Another UX problem shown in the personas is designing the same social expectations for all genders (i.e., an interesting companion in commute, an enjoyable ride together). This is gender-free design. It becomes malevolent when Hitch narrowed "social needs" to sexual ones by pairing up a male driver and a female passenger and portraying these interactions as potential romantic relationships. This app created a sexual scenario for users that began inside the car and expanded beyond the ride to other social situations for sexual purposes. In other words, Hitch was not designed for transportation services but for users seeking lovers.

DiDi violated basic UX principles on user personas and scenarios. DiDi also designed other functions to support the development of romantic relationships beyond the ride, such as making driver and passenger evaluation tags public (especially eros-appealing tags for females) as well as providing visible contact numbers of passengers, free photo-beautifier, and substantial coupons for successful dates. This sexist "gender-free" design idea contributed to the loss of users' lives.

Feminist UX Design Heuristics

The DiDi case propels us to ask how to design safety for ridesharing? From a feminist UX perspective, we propose a set of ruling principles and specific suggestions that runs through the lifecycle of UXD.

BALANCE AND EQUALITY IN UX

Designers must balance diverse design variables to achieve equality for all users (King & Lancaster, 2024). Our study explores this concept in ridesharing applications and explains from a feminist perspective.

UX designers should balance user goals and business goals. A good alignment of the two will help businesses generate benefits when users reach their goals (not vice versa). Such an alignment is built on extensive collection of requirements for product development drawn from not only business and marketing analysts but also user researchers. The latter should include all genders to ensure their needs are represented.

UX designers should also balance the realization of social, economic, and safety needs. Platt (2016) found that, although users value security, they were unwilling to put forth extra effort. This means designers should proactively consider security needs and features for users.

Moreover, UX designers should balance needs of different genders when facing various and even conflicting needs. Instead of prioritizing one gender, we endorse a human-centered design that integrates characteristics and expectations of different gender groups, which helps achieve a dynamic balance of user needs.

Suggestions for Feminist UX Practice

In balancing design and creating equilibriUX, designers should consider the following for best practices.

Designers should engage with gender diversity by taking a gender polynary approach to technology. This includes understanding gender in its historical and cultural circumstances and studies the uniqueness of gender experience at the individual level. It requires designers to focus on a particular user's lived body experience in the user's specific contexts. Such a critical, wide-range, and nuanced techno-cultural examination of gender and its relevant matters is a must for feminist UX design.

Designers should also share design authority with users through participatory approaches. For example, designers should build an empathic relationship with underrepresented genders, encourage users to share stories for use as design variables, or convert to a user role to experience the real conditions and limitations of usage. Designers and users can collaboratively decide design principles, frame questions, analyze data, choose methods, and so on.

Technology companies should offer gender-related designing courses and learning sources for designers to better engage with and empower users. Harper (2020) recommended that companies go beyond solely implicit bias training to address how they are perpetuated in harmful ways. We would like to adapt this suggestion for feminist UXD. Courses that address women's historical subordination and the subtle ways in which that status is reinforced and reproduced would provide tools to make different design decisions, allowing companies to be proactive rather than reactive.

Researchers can promote user empowerment and advocacy using an integrated approach. First, examine how technologies/products construct and perpetuate social construction of gender and gender biases. Second, integrate feminism in a more intellectually rigorous way into technical communication and interaction design studies. Third, develop and legitimize a feminist UXD research agenda.

Users should be more aware of their technical identity. Technologies change humans into users, create our technical identities, regulate our online behaviors, and parameterize genders for business benefits. Our study demonstrates how Didi Hitch contributed to an objectification that promoted the potential for harm. Therefore, we strongly encourage users to examine their digital identities and strengthen safety measures. Regardless of gender, all users should be agents of their own lives.

Conclusion

Our analysis of DiDi's ads and discussion of its safety design conclusively reveal that DiDi leveraged eros appeals. The company symbolically constructed women as objects of desire who otherwise lack agency in their own lives. This design cost lives and threatened users. Because those symbolic choices construct social realities (Stob, 2008), women in general cease to be autonomous individuals worthy of respect and safe rides in this situation. Thus, in the actual ride, the rider is positioned in a system "of evaluation and expectations" (Young, 2005, p. 416) that foregrounds the rider's potential as an object of desire, placing the individual at significant risk. Responsible design choices would acknowledge this Lived Body Experience (Young, 2005) and seek out the subtle ways that disempowering and objectifying narratives continue to influence design choices and women's experiences. We believe gender-polynary thinking will make possible cooperative, compassionate changes in communications technology design.

In an era of "social media feminism" (Megarry, 2020), the link between social media (or tools for social networking) and a resurgence of attacks on women and on women's rights is clearly seen in the DiDi case and other events such as the #MeToo movement. It is unclear if this new social development will lead to more hostility or equality for women. Moreover, deeper questions remain regarding if and how technology will address gendered power dynamics and help prevent sexual crimes. After studying the DiDi case, we want to ask a bigger question: in the social media era, how should we design technology, products, or services to secure our safety in physical and online worlds? This question is not within the inquiry scope of this study, but we hope our readers would consider it.

Without access to DiDi Hitch's previous version or the real users, we were unable to fully analyze its UX features. Further research would expand beyond marketing material to include users. Future projects would also integrate more fully the changing nature of ridesharing technology and its global presence, allowing greater understanding of how these new technologies are constantly reconstructing our identities and ways in which users can take control. We expect such investigation will yield further principles that would allow ridesharing companies to be successful businesses while ensuring their users' safety.

Appendix 14.A

Table 14.1. Coding Chart of Male-Driver Advertisements

First Cycle Coding		
	Female Riders	Male Drivers
Descriptive Codes and Categories	*Characteristics*: young (20s), good-looking, thin, unmarried college girls and female office workers	*Characteristics*: young or middle-aged, average-looking but polite, kind, punctual, reliable men with clean cars, great driving skills, and low-profile personality; unidentified professions
	Setting: inside the car, train station, office building, plaza, amusement park, seashore, college campus, grocery store, city street, home	
	Occasion: daily life, Valentine's Day, World Sleep Day, Spring Festival	
Action Codes and Categories	*Before the trip*: pulling luggage; holding a grocery bag, umbrella, coffee cup, or puppy; placing a ride order; beautifying profile picture; editing profile nickname; pouting; standing in the dark and feeling lonely; jumping for joy	*Before the trip*: holding a rose in the mouth, holding flowers, holding an ad board, reading passenger's profile tags
	During the trip: Sitting in the passenger's seat, giving snacks and talking to the driver, flinging hair, leaning toward the driver, smiling, enjoying the ride	*During the trip*: Sitting in the driver's seat, holding the car door, taking snacks from and talking to the rider, smiling, enjoying the ride
	After the trip: kissing a man, sitting on a sofa, chatting with a driver online, adding evaluation tags for a driver, making another appointment with the driver	*After the trip*: waving and smiling, waiving the fare, checking the appointment schedule, sitting on the sofa at home, chatting with a frequent rider, kissing a woman, chatting with a girl on the app, confessing his love to a female rider

continued on next page

Table 14.1. Continued.

First Cycle Coding		
In-Vivo Codes and Categories	*From Women:* "Sooner or later, I will be yours." "You are the reason why I am not single anymore." "He is my superman." "10 mins waiting is worthwhile for a lifelong journey together." "I will still be your sweet rider beyond the journey." "You look like my future boyfriend." "I want to be drunk, not worry, and leave the city."	*From Men:* "My warm car can drive your chill away." "Happy to waive her fare." "Push one button and I will take you home." "Share a vacant seat on the way home and earn yourself an opportunity to fall in love." "I'm fed up with seeing many beautiful legs here and eager to see different types of girls in other places." "Wet? Tight? Hard?"
	To Both/Unisex: "Let's date!" "I am afraid of nothing but no dates." "Beautifying your profile picture will make the car hailing faster." "DiDi Hitch makes us meet." "We seem destined to meet." "It was a stroke of luck that I found you here, but not for our next meet." "Sharing a ride with you is the best way home." "How lucky we are to have the same route and the same taste." "You and me traveling together is a wonderful thing." "Dating is the right purpose on DiDi." "Having the same way home is the best trick to date." "Not only you are on my way home, but also on my way to find love." "Being sexy is the universal password for city travel." "The night breeze will help me sleep (with you)."	
Values Codes and Categories	*Benefits to social life:* fun, romantic, warmhearted, beautiful meet; being part of the biggest ever "mobile blind date"; wonderful journey; companionship	
	Benefits in other aspects: economical; earning extra money; comfortable, relaxing, eco-friendly, quick and simple, no trouble	

First Cycle Coding	
Second Cycle Coding	
Central Themes	Stakeholders: clearly identified women (young, good-looking girls with limited social experience) riders vs. vaguely identified men (young/middle-aged, average looking, but kind and reliable) drivers
	Actions: women waiting for a man (not only a car), women receiving (attention, car ride, fare waive, romantic relationship) vs. men picking up girls, men giving (ride, fare waive, gifts, invitation to meet)
	Issues: taking a ride vs. looking for a date; rider-driver relationship vs. romantic relationship; women's power/initiative vs. men's power/initiative

Source: Created by the author.

Table 14.2. Coding Chart of Female-Driver Advertisements

First Cycle Coding	Female Drivers	No Passengers
Descriptive Codes and Categories	*Characteristics: mature* (30s), nicely dressed (in business casual), warm, unmarried, educated, independent, feeling void, longing for a potential partner	
	Setting: inside the car/beside the car	
	Occasion: daily life	
Action Codes and Categories	*Before the trip:* standing in front of the car, leaning against the car	
	During the trip: sitting in the driver's seat, steering the car, looking forward, looking at the back seat, smiling, talking to a female passenger	
	From Women: "Not many people could persuade me (sleep with me, homophones in Chinese) in my car." "Happy to let you win me over." "I am Yu Feifei. I like DiDi Hitch. I like doing something different." "I support the project of sharing vacant seats."	
	To Women: "Sharing a vacant seat on the way home will earn you an opportunity to meet your Mr. Right." "Sharing a vacant seat will help an enterprising girl to become soft and home-centered."	
Values Codes and Categories	*Benefits to social life:* romantic, beautiful meet exciting journey, better to travel with a companion; helpful with finding love	
	Benefits in other aspects: economical; earning extra money; good to be unique; becoming less manly, softer, and more sensitive	

First Cycle Coding	
Second Cycle Coding	
Central Themes	*Stakeholders*: clearly identified women (young, nicely dressed, educated, white collar, independent, but longing for a partner) rider vs. nonexistent but implied men (caring, persuasive, stable; an ideal long-term partner) driver
	Actions: women waiting for a man as traveling and life companion
	Issues: taking a ride vs. looking for a date; rider-driver relationship vs. romantic relationship; women's power/initiative vs. men's power/initiative

Source: Created by the author.

LILY WANG	#commuter #budget-conscious #social

Age 23
Location Beijing
Occupation Secretary
Income Low
Status Single

Needs and Wants
- Wants to save money on her monthly budget
- Wants the comfort & convenience of a personal vehicle
- Wants to broaden social circles in a big city

Behavior
- Travels on the Beijing Metro most of the time
- Uses shared cars when tired, in emergency, or in bad weather

Pain points
- Using public transport messes up appearance
- Feeling awkward or bored in conversations with cab drivers

Expectations
- A companion in commute who is reliable, funny, and has common topics
- A low-price ride but enjoyable journey

Background
Likes traveling to new places around the city; love shopping; open to develop a serious relationship; hopes to gain a foothold in work and in the big city.

Figure 14.1. Female Persona. *Source*: Created by the author.

DAVE TONG	#responsible #budget-conscious #social

Age 31
Location Beijing
Occupation Computer Engineer
Income Medium
Status Single

Needs and Wants
- Wants to avoid monotony during his journey
- Wants to save money from his monthly budget
- Wants to broaden social circles beyond a boring workplace

Behavior
- Drives to work and back home, but gets tired
- Worried about the rising gas price

Pain points
- Feeling irritated by the congestion
- Hard to find reliable and regular co-riders

Expectations
- A companion in commute who is reliable, funny, and has common topics
- An economical ride but enjoyable journey

Background
Likes to travel out of city for leisure; talkative when being with friends; has a clean car; eager to develop a serious relationship.

Figure 14.2. Male Persona. *Source*: Created by the author.

Notes

1. This work was supported by the Fundamental Research Funds for the Central Universities in UIBE under Grant No 19QD13.

2. On the government website for public court papers, China Judgments Online, we searched using "DiDi" and "crime" as the key words and found 13 rape cases and 4 homicide cases (including the two cases discussed above) since DiDi started its service in 2015.

3. *IEEE Transactions on Professional Communication, Journal of Business and Technical Communication, Technical Communication, Technical Communication Quarterly.*

References

8/24 Leqing girl Hitch murder case. (2018). *Baidu Baike.* https://baike.baidu.com/item/8·24乐清女孩乘车遇害案/22835678?fr=aladdin

Agboka, G. Y., & Dorpenyo, I. K. (2022). The role of technical communicators in confronting injustice—Everywhere. *IEEE Transactions on Professional Communication, 65*(1), 5–10. https://doi.org/10.1109/TPC.2021.3133151

Balsamo, A. (1992). The virtual body in cyberspace. *Research in Philosophy and Technology, 13*, 119–140.

Bardzell, S. (2010). Feminist HCI: Taking stock and outlining an agenda for design. In *Proceedings of the SIGCHI Conference on Human Factors in Computing Systems, 1301–1310.* https://doi.org/10.1145/1753326.1753521

Barker, R. T., & Zifcak, L. (1999). Communication and gender in workplace 2000: Creating a contextually-based integrated paradigm. *Journal of Technical Writing and Communication, 29*(4), 335–347. https://doi.org/10.2190/648j-e1vu-4je6-jwwc

Baxter, K., Courage, C., & Caine, K. (2015). *Understanding your users: A practical guide to user research methods* (2nd ed). Elsevier.

Berg, A. J., & Lie, M. (1995). Feminism and constructivism: Do artifacts have gender? *Science Technology and Human Values, 20*(3), 332–351. https://doi.org/10.1177/016224399502000304

Cassell, J. (2002). Genderizing HCI. In J. Jacko & A. Sears (Eds.), *The handbook of human-computer interaction: Fundamentals, evolving technologies, and emerging applications* (pp. 402–411). Lawrence Erlbaum Associates.

Dida Travel. (2020). *2014–2020 China's Ridesharing Industry Development Bluebook.*

DiDi's milestones. (2015). *DiDi Official Website.* https://www.didiglobal.com/about-special/milestone

Durack, K. T. (1997). Gender, technology, and the history of technical communication. *Technical Communication Quarterly, 6*(3), 249–260. https://doi.org/10.1207/s15427625tcq0603_2

Flight attendant was raped and killed by the Hitch driver. (2018). *Wangyi News.* https://www.163.com/dy/article/H560E5K40550TA0M.html

Flynn, E. A. (1997). Emergent feminist technical communication. *Technical Communication Quarterly*, *6*(3), 313–320. https://doi.org/10.1207/s15427625tcq0603_6

Flynn, J. F. (1997). Toward a feminist historiography of technical communication. *Technical Communication Quarterly*, *6*(3), 321–329. https://doi.org/10.1207/s15427625tcq0603_7

Frost, E. A. (2020). Ultrasound, gender, and consent: An apparent feminist analysis of medical imaging rhetorics. *Technical Communication Quarterly*, *30*(1), 48–62. https://doi.org/10.1080/10572252.2020.1774658

Gurak, L. J., & Bayer, N. L. (1994). Making gender visible: Extending feminist critiques of technology to technical communication. *Technical Communication Quarterly*, *3*(3), 257–270. https://doi.org/10.1080/10572259409364571

Gurak, L. J., & Ebeltoft-Kraske, L. (1999). Letter from the guest editors: The rhetorics of gender in computer-mediated communication. *The Information Society: An International Journal*, *15*(3), 147–149. http://dx.doi.org/10.1080/019722499128457

Hallenbeck, S. (2012). User agency, technical communication, and the 19th-century woman bicyclist. *Technical Communication Quarterly*, *21*(4), 290–306. https://doi.org/10.1080/10572252.2012.686846

Haraway, D. (2016). *Manifestly Haraway*. University of Minnesota Press. https://doi.org/10.5749/minnesota/9780816650477.001.0001

Harper, K. C. (2020). *The ethos of black motherhood in America: Only white women get pregnant*. Lexington Books.

Herrick, J. W. (1999). "And then she said": Office stories and what they tell us about gender in the workplace. *Journal of Business and Technical Communication*, *13*(3), 274–296. https://doi.org/10.1177/105065199901300303

Herring, S. C. (1999). The rhetorical dynamics of gender harassment on-line. *The Information Society: An International Journal*, *15*(3), 151–167. http://dx.doi.org/10.1080/019722499128466

Jones, N. N. (2016). Narrative inquiry in human-centered design: Examining silence and voice to promote social justice in design scenarios. *Journal of Technical Writing and Communication*, *46*(4), 471–492. https://doi.org/10.1177/0047281616653489

Kelleher, C., Pausch, R., & Kiesler, S. (2007). Storytelling Alice motivates middle school girls to learn computer programming. In *Proceedings of the SIGCHI Conference on Human Factors in Computing Systems* (pp. 1455–1464). https://doi.org/10.1145/1240624.1240844

Koerber, A. (2000). Toward a feminist rhetoric of technology. *Journal of Business and Technical Communication*, *14*(1), 58–73. https://doi.org/10.1177/105065190001400103

Lancaster, A., & King, C. S. T. (Eds.). (2024). *Amplifying voices in UX: Balancing design and user needs in technical communication*. State University of New York Press.

Laurel, B. (2003). *Design research: Methods and perspectives.* The MIT Press.

Lay, M. M. (1989). Interpersonal conflict in collaborative writing: What we can learn from gender studies. *Journal of Business and Technical Communication, 3*(2), 5–28. https://doi.org/10.1177/105065198900300202

Lay, M. M. (1994). The value of gender studies to professional communication research. *Journal of Business and Technical Communication, 8*(1), 58–90. https://doi.org/10.1177/1050651994008001003

Lind, M. (1999). The gender impact of temporary virtual work groups. *IEEE Transactions on Professional Communication, 42*(4), 276–285. https://doi.org/10.1109/47.807966

Lippincott, G. (1997). Experimenting at home: Writing for the nineteenth-century domestic workplace. *Technical Communication Quarterly, 6*(4), 365–380. https://doi.org/10.1207/s15427625tcq0604_1

Marsh, J. (2016). *UX for beginners.* O'Reilly Media, Inc.

Megarry, J. (2020). *The limitations of social media feminism: No space of our own.* Springer Nature. https://doi.org/10.1080/14680777.2022.2048598

Novotny, M., & Hutchinson, L. (2019). Data our bodies tell: Towards critical feminist action in fertility and period tracking applications. *Technical Communication Quarterly, 28*(4), 332–360. https://doi.org/10.1080/10572252.2019.1607907

Petersen, E. J., & Moeller, R. M. (2016). Using antenarrative to uncover systems of power in mid-20th century policies on marriage and maternity at IBM. *Journal of Technical Writing and Communication, 46*(3), 362–386. https://doi.org/10.1177/0047281616639473

Platt, D. (2016). *The joy of UX: User experience and interactive design for developers.* Addison-Wesley Professional.

Ranney, F. J. (2000). Beyond Foucault: Toward a user-centered approach to sexual harassment policy. *Technical Communication Quarterly, 9*(1), 9–28. https://doi.org/10.1080/10572250009364683

Rode, J. A. (2011). A theoretical agenda for feminist HCI. *Interacting with Computers, 23*(5), 393–400. https://doi.org/10.1016/j.intcom.2011.04.005

Royal, C. (2005). A meta-analysis of journal articles intersecting issues of internet and gender. *Journal of Technical Writing and Communication, 35*(4), 403–429. https://doi.org/10.2190/3rbm-xkeq-traf-e8gn

Saldaña, J. (2016). *The coding manual for qualitative researchers.* Sage.

Sauer, B. A. (1994). Sexual dynamics of the profession: Articulating the *ecriture masculine* of science and technology. *Technical Communication Quarterly, 3*(3), 309–323. https://doi.org/10.1080/10572259409364574

Smith, E. O., & Thompson, I. (2002). Feminist theory in technical communication: Making knowledge claims visible. *Journal of Business and Technical Communication, 16*(4), 441–477. https://doi.org/10.1177/105065102236526

Stenstrom, E., Stenstrom, P., Saad, G., & Cheikhrouhou, S. (2008). Online hunting and gathering: An evolutionary perspective on sex differences in website preferences and navigation. *IEEE Transactions on Professional Communication*, *51*(2), 155–168. https://doi.org/10.1109/tpc.2008.2000341

Stob, P. (2008). "Terministic screens," social constructionism, and the language of experience: Kenneth Burke's utilization of William James. *Philosophy & Rhetoric*, *41*(2), 130–152.

Tan, D. S., Czerwinski, M., & Robertson, G. (2003). Women go with the (optical) flow. In *Proceedings of the Conference on Human Factors in Computing Systems—CHI '03*. https://doi.org/10.1145/642611.642649

Tebeaux, E. (1990). Toward an understanding of gender differences in written business communications: A suggested perspective for future research. *Journal of Business and Technical Communication*, *4*(1), 25–43. https://doi.org/10.1177/105065199000400102

Thompson, I. (1999). Women and feminism in technical communication. *Journal of Business and Technical Communication*, *13*(2), 154–178. https://doi.org/10.1177/1050651999013002002

Thompson, I., & Smith, E. O. (2006). Women and feminism in technical communication—An update. *Journal of Technical Writing and Communication*, *36*(2), 183–199. https://doi.org/10.2190/4juc-8rac-73h6-n57u

Tong, A., & Klecun, E. (2004). Toward accommodating gender differences in multimedia communication. *IEEE Transactions on Professional Communication*, *47*(2), 118–129. https://doi.org/10.1109/tpc.2004.828221

van Oost, E. (2003). Materialized gender: How shavers configure the users' femininity and masculinity. In N. Oudshoorn & T. Pinch (Eds.), *How users matter: The co-construction of users and technology* (pp. 193–208). The MIT Press.

Wang, H. (2021). Chinese women's reproductive justice and social media. *Technical Communication Quarterly*, *30*(3), 285–297. https://doi.org/10.1080/10572252.2021.1930178

White, K., Rumsey, S. K., & Amidon, S. (2015). Are we "there" yet? The treatment of gender and feminism in technical, business, and workplace writing studies. *Journal of Technical Writing and Communication*, *46*(1), 27–58. https://doi.org/10.1177/0047281615600637

Young, I. M. (2005). *On female body experience: "Throwing like a girl" and other essays*. Oxford University Press.

Zdenek, S. (2007). "Just roll your mouse over me": Designing virtual women for customer service on the web. *Technical Communication Quarterly*, *16*(4), 397–430. https://doi.org/10.1080/10572250701380766

Contributors

Keshab Raj Acharya is a faculty member in the Department of Engineering Education at the University at Buffalo (SUNY), where he teaches courses related to technical and professional communication. His research interests include the rhetoric of health and medicine, usability studies, user localization, inter/cross cultural technical communication, international technical communication, and social justice. His publications have appeared in the journals *Technical Communication, Journal of Technical Writing and Communication, Communication Design Quarterly*, and *Present Tense*, among others.

Laura Becker is the ICARDA MEL Project Coordinator for the gender-transformative AI-Driven Climate-Smart Beekeeping (AID-CSB) for Women Project. Prior to her work with data and digital solutions to improve decision-making in the agriculture and nutrition sectors, she provided nutrition counseling to women and families in the WIC program.

Kristin C. Bennett (she/her) is Assistant Professor of Technical Communication at Sam Houston State University. She received her PhD in Writing, Rhetorics, and Literacies from Arizona State University. Her research examines the intersections between rhetoric, disability studies, and technical and professional communication across public and institutional discourse and design. Her previous work has appeared in *IEEE Transactions on Professional Communication, The Journal of Business and Technical Communication, Teaching English in the Two-Year College*, and *Composition Studies*.

Emily L. W. Bowers (she/her) is a graduate student at Michigan State University, pursuing an MA in Rhetoric and Writing. Her background is in digital and technical communication, and she has spent the past decade working as a communications professional, managing cross-disciplinary projects ranging from marketing to operations. She is currently part of a marketing and communication team for a construction management firm. Her research interests include experience architecture, accessibility, and corporate social responsibility.

Felicia Chong is a UX researcher with an academic background in rhetoric and technical communication. Her work has appeared in *IEEE Transactions in Professional Communication, Programmatic Perspectives, Technical Communication,* and *Technical Communication Quarterly.* She received the 2022 Nell Ann Pickett Award for her coauthored article, "Student Recruitment in Technical and Professional Communication Programs," in *Technical Communication Quarterly.*

Lin Dong is a researcher and lecturer of technical and professional communication at the University of International Business and Economics in Beijing, China, where she teaches business writing, technical writing, and user-experience design classes. Her research focuses on crisis communication in transcultural contexts, China's technical communication practices and education, and user-experience research for disadvantaged groups. She has published in *Communication Design Quarterly, IEEE ProComm Proceedings,* and *Technical Communication.*

Marci J. Gallagher is a graduate of Valdosta State University's MLIS program. Her current research explores information science, access, and usability. She has presented on communication accessibility at the STC SUMMIT and ACM SIGDOC. She is currently working on an accessibility workshop for STC. Her most recent publications are in the *2022 STC Summit Conference & Expo Proceedings* and in the *Journal of Plant Ecology.* She is a member of STC and the ALA.

Philip B. Gallagher is Assistant Professor of Technical Communication at Mercer University. His current research interests are accessibility and user experience in technical communication, user-centered design of visual information, and the influences of technology in writing ecology.

His recent publications appeared in *Computers & Composition, Computers & Composition Online,* and the *Journal of Interactive Technology and Pedagogy.* He is a member of STC, ATTW, and ACM.

Rob Grace is Assistant Professor at Texas Tech University, where he teaches courses in UX research, design, and writing. His research focuses on the design and use of information and communications technologies in emergency response and management and appears in journals such as *Safety Science* and the *International Journal of Disaster Risk Reduction.* In addition to his teaching and research, Rob serves as the current President of the International Association for Information Systems for Crisis Response and Management (ISCRAM).

Emma J. Harris (she/her) is an educator and scholar whose work is focused on advocacy and education for people who are impacted by carceral institutions, accessibility, disability justice, and community engagement. In all these areas, she is focused on pursuing social justice and improving people's everyday lives through writing and design. She is a recent graduate from Michigan State University's Critical Studies in Literacy and Pedagogy MA and has her BA in English with a concentration in Professional and Public Writing from Auburn University.

Mallory Henderson received her master's from the University of Central Florida (UCF) in English, with a track in Rhetoric and Composition. Her research interests are collectively centered in rhetoric of health and medicine, technical communication, and student advocacy, of which she presented at the 2021 RHM symposium and 2022 CCCC's conference. While at UCF, Mallory taught Composition II as a graduate teaching associate and now works as an Onsite Specialist for a medical technology company. Mallory received her bachelor's in English from the University of Lynchburg, where she was a collegiate athlete from 2015 to 2018, driving the inspiration behind this work.

Amy Hodges is Assistant Professor of English at The University of Texas at Arlington, specializing in technical writing and professional communication. Her research examines the language, writing, and communication strategies of multilingual engineers in transnational corporations, and she also researches what writing programs can do to create a more inclusive environment and prepare all writers for linguistically diverse workplaces. Her work has appeared in *IEEE Transactions on Professional Communication* and the *Writing Center Journal*.

Sarah Beth Hopton is Associate Professor in the Writing, Rhetoric & Technical Communication Program at Appalachian State University. Her research focuses on climate change, agriculture, and technical communication in global contexts. Her co-authored article with Dr. Rebecca Walton, "All Vietnamese Men are Brothers: Rhetorical Strategies and Community Engagement Practices used to Support Victims of Agent Orange," won the 2020 CCCC's best research award for qualitative or quantitative research in the field of technical communication. Also a creative nonfiction writer, her forthcoming memoir—*Blood, Bone, Breath, Earth,* a meditation on sustainable farming in the "age of loneliness"—was recently shortlisted for the Bridport Prize.

K. Alex Ilyasova is Associate Professor and Associate Dean in the College of Liberal Arts and Sciences. She has researched and published findings (chapters and manuscripts) in literacy and identity studies in composition programs, writing program administration and outcomes development and assessment, and empathy and emotions in understanding users' experiences, team building, dynamics, and professional burnout. Dr. Ilyasova has served as Director of the Technical Communication and Information Design Program and has worked with industry and community partners on establishing work-based experiences and internships for students. She teaches courses that include emotions in and diversity and inclusion topics in technical communication.

Carie S. Tucker King is Clinical Professor and Director of Rhetoric at The University of Texas at Dallas (UTD). Her background is in medical editing and clinical research, specializing in technical communication and ethics. Her classes include communication ethics, scientific publication, and intercultural communication, and she created the first usability course for the UX/game-design program at UTD. Carie researches pedagogy, usability, and ethics, with publications including her monograph, *The Rhetoric of Breast Cancer,* and articles in *International Journal of Intercultural Relations, Health Communication, Journal of Technical Writing and Communication, Technical Communication,* and *Programmatic Perspectives.*

Amber Lancaster is Associate Professor of Communication and Director of Professional Writing (PWR) at Oregon Tech. She has over twenty years of experience in UX and client-based research projects as academic and consultant. She teaches user research, usability testing, risk communication, and technical writing. Her research includes the intersections of user-centered design, ethics, and social issues as well as technology and writing pedagogy. She has published manuscripts and special issues for *Technical Communication, IEEE Transactions on Professional Communication, International Journal of Sociotechnology and Knowledge Development, Intercom,* and *Programmatic Perspectives.*

Jessica Nalani Lee is a full-time instructor of writing and English at Portland Community College's Rock Creek campus. She teaches technical writing and English literature courses. Dr. Lee's approach to teaching writing is informed, in part, by her research, which is continuously engaged in the rhetorical concept of agency as it functions to disrupt hierarchical power structures. She focuses primarily on medical rhetoric, specifically studying the ways in which knowledge is communicated in healthcare settings between providers and patients. Her commitment to honoring marginalized knowledges is also reflected in her work as a mentor to other faculty of color.

Jamie May is Principal Instructor in the Technical Communication and Information Design Program. She teaches technical communication, iterative design, and web and print design. Dr. May earned her BA and MA in English (professional writing) from Southeastern Louisiana University before her PhD in Technical Communication and Rhetoric at Texas Tech University. She has researched social justice and the technical communication service course and also UX and usability, emphasizing the use of pop culture to teach those concepts. Her article "YouTube Gamers and Think Aloud Protocols: Introducing Usability Testing" appeared in *IEEE Transactions on Professional Communication.*

Ruby Mendoza (they/them) is Assistant Professor of English at California State University, Sacramento. Ruby's research focuses on critical race theory, queer and transgender studies, cultural rhetorics, and technical and professional communication. Their work has been published in *Technical Communication and Social Justice, IEEE Transactions on Professional Communication Conference,* and *Literacy in Composition Studies.*

Brett Oppegaard is Professor in the School of Communication and Information at the University of Hawai'i at Manoa in Honolulu, HI. His scholarship focuses on media accessibility and locative media. His research has been published in *Technical Communication, IEEE Transactions on Professional Communication,* and *the Journal of Technical Writing and Communication.* He is available at brett.oppegaard@hawaii.edu.

Sushil K. Oswal is Professor of Human Centered Design and CREATE Faculty in the Center for Research and Education on Accessible Technology and Experiences at the University of Washington. The broad focus of his HCI research is on the employment of technology in the knowledge industry. His research has encompassed HCI design issues in distributed web environments, digital library databases, self-service kiosks, and learning management systems. His interdisciplinary, socially relevant research has appeared in the Association for Computing Machinery and technical communication journals and proceedings. He consults in accessible HCI, technology design, and digital accessibility of workspaces.

Timothy M. Ponce is Assistant Professor of Instruction in the Department of English at The University of Texas at Arlington, as well as the Coordinator of the Technical Writing and Professional Design Program. He specializes in programmatic administration, andragogy in technical writing and professional communication, and AI-aided writing instruction. He also coordinates the Department of English Internship Program. With the help of his team, he has grown both programs to be more community focused. His published work has appeared in *Pedagogy: Critical Approaches to Teaching Literature; Language, Composition, and Culture;* and *Hybrid Pedagogy.*

Johansen Quijano is Adjunct Assistant Professor of English at The University of Texas at Arlington and Associate Professor of English at Tarrant County College, specializing in digital rhetoric and multimedia narrative. His research focuses on games-as-text, writing and games, gaming culture, and social media discourse. He is also interested in education, the teaching and learning process, language acquisition, and the intersection between writing and technology. He has published and presented both nationally and internationally.

Michael K. Rabby serves as Scholarly Associate Professor at Washington State University Vancouver. His research explores the intersections of the human experience with technology in a variety of contexts, including online relationships and mobile app usage. He has published research in the *Journal of Social and Personal Relationships, Military Behavioral Health, Digital Journalism*, and *Communication Studies*. He can be reached at michael.rabby@wsu.edu.

Tammy Rice-Bailey is Associate Professor of Technical Communication and User Experience at the Milwaukee School of Engineering. Prior to her academic career, she managed corporate documentation and training for global software, financial, and retail corporations. She has been published in *Technical Communication, Technical Communication Quarterly, Communication Design Quarterly*, and *Technical Writing and Communication*. She is also the coauthor of the book *Interpersonal Skills for Group Collaboration*.

Max Rünzel is the CEO and Co-Founder of HiveTracks, a company that builds technology-powered products for beekeepers to connect people, nature, and data. Prior to joining HiveTracks, he worked for the Food and Agriculture Organization of the United Nations and CGIAR with a focus on digital extension services and smallholder producer empowerment.

Bethany Shaffer is a Lecturer in the Department of English at The University of Texas at Arlington. Her areas of focus are on technical communication, working-class literature, and women's literature.

Vince Sosko is an adjunct assistant professor of English at The University of Texas at Arlington (UTA), specializing in rhetoric and composition with a focus on pedagogy. He is also pursuing a PhD at UTA and is writing a dissertation on the influence of digital communication on the social, cultural, and epistemic outcomes of teaching composition. His research interests also include writing program administration, writing centers, genre theory, and gender studies.

Jason Tham is Associate Professor at Texas Tech University, where he co-directs the User Experience (UX) Research Lab, teaches design and digital rhetoric, and advises the student organizations Raider UI/UX and TTU STC student chapter. He is the author of *Design Thinking in Technical Communication* (Routledge, 2021). He is an associate editor of *IEEE Transactions on Professional Communication* and a review board member for *Computers and Composition, Kairos*, and *Xchanges*. He serves as the

Vice President of CPTSC and a member on the advisory board of the Association for Computing Machinery (ACM) Special Interest Group on Design of Communication (SIGDOC).

Elizabeth Topping is Assistant Professor at University of Pikeville in Pikeville, Kentucky. Her dissertation focused on the intersections of rhetoric, reproductive healthcare, public health, race, and class. Her current research includes the rhetoric of health and medicine, feminist research methodologies, maternal and public health, and public memory.

Hua Wang is Senior Lecturer in the College of Engineering at Cornell University. Her research focuses on technical and professional communication, rhetoric of health and medicine, cross-cultural communication, and engineering communication pedagogy. Her work has appeared in the *Journal of Technical Writing and Communication*, *Technical Communication Quarterly*, and the book *Strategic Interventions in Mental Health Rhetoric*.

James Wilkes is the founder of HiveTracks and Professor of Computer Science at Appalachian State University. For more than a decade, he has worked at the intersection of beekeeping and technology to improve beekeeping practices and honey bee research worldwide through the thoughtful development of digital platforms.

Index

Abrams campaign study, 357–380; comparison to UX methods, 358–359, 364, 365, 374–376; implications, 375–376; institutional action, 372–375; institutions' oppressive effects on TPC, 363–364; interpersonal action, 366–369; literature review, 359–361; structural action, 370–372

access-first design, 115, 122

accessibility: checklist approach, 59; deep, 79, 81; defined, 59–62, 78; failures to integrate into curricula, 49–50, 105–107; and social inclusion/exclusion, 275–276. *See also* audio description study; digital course design case study; introductory TPC course case study; visual communication course case study

Acharya, K. R., 167–191, 228, 302

actionability, in PEMAT, 235, 240–245, 247–248

AD (audio description). *See* audio description

adjunct faculty. *See* teaching faculty

advance directives case study, 193–221; ADs defined, 193–194; common recognized problems with ADs, 196–198; comprehension, 201, 205–206; federal and state requirements, 194–196, 198–199; formatting, 201, 204–205, 212; instruction types, 206–207, 211–212; key issues, 207–210, 212–215, 216; methods, 198–201; readability, 201–204, 210–211; recommendations and future directions, 215–217

advertisements, in DiDi Hitch app, 419–422

affective empathy, 139–140

Agboka, G. Y., 173, 228, 302

Agbozo, G. E., 302

agency: located, 348–349; of oppressed people, 365; of students, in TEACH framework, 53; of teaching faculty, 336, 343, 346–347; as thematic negotiation, 82, 83, 86, 87–88

AI (artificial intelligence), limiting bias in, 389–390

AID-CSB A&E (AI-Driven Climate-Smart Beekeeping for Women Advisory & Extension). *See* Beekeeper's Companion app case study

Albers, M. J., 57

algorithms, bias in, 389

alternative (alt) text, 114

443